Primer on Radiation Oncology Physics

T0303867

Primer on Radiation Oncology Physics

Video Tutorials with Textbook and Problems

Eric Ford

CRC Press
Taylor & Francis Group
Boca Raton London New York

CRC Press is an imprint of the
Taylor & Francis Group, an **informa** business

Visit the Companion Website: www.crcpress.com/cw/ford

First edition published 2020
by CRC Press
6000 Broken Sound Parkway NW, Suite 300, Boca Raton, FL 33487-2742
and by CRC Press
2 Park Square, Milton Park, Abingdon, Oxon, OX14 4RN

© 2020 Taylor & Francis Group, LLC

CRC Press is an imprint of Taylor & Francis Group, LLC

ISBN: 978-1-138-59438-8 (hbk)
ISBN: 978-1-138-59170-7 (pbk)
ISBN: 978-0-429-48888-7 (ebk)

In memory of Fran McCauley, my first physics teacher

Contents

Introduction

This book and video series provides a new way to learn radiation oncology physics. There are three components which are key to this: (1) Teaching Videos: each video is 10–15 minutes in length and corresponds to a specific section in the text. These videos, with embedded multiple-choice questions, are especially helpful for visual learners. (2) Textbook: the text follows a well-structured curriculum that builds on concepts. Chapter 7, for example, discusses how a charged particle loses energy when traveling through matter, and Chapter 9 builds on this concept to explain how photon beams are created in a linear accelerator. (3) Problem Sets: the textbook includes hundreds of problems (with answer explanation keys in the Educator Version); the videos include hundreds more problems. This provides an essential learning tool since physics is best learned by doing problems.

The material in this series has been extensively "field tested." That is, as an outgrowth of a class taught for over 15 years, it has been tried, tested, and refined with a large number of learners from a variety of backgrounds. Past students have included medical residents in radiation oncology, medical students, medical physics residents, graduate students, radiation therapists, dosimetrists, and others. The material has been well-received. Part of this is the material itself which is very rich and nuanced and can be learned repeatedly from different perspectives and at different levels.

Educators will find this series especially helpful. The text and videos can be used to create a "flipped classroom" where the traditional lecture is reviewed prior to class and then most of the in-class time is spent on interactive discussions and problem solving. This is becoming a mainstream method of teaching advanced topics. Data have shown that such active learning approaches result in a more thorough understanding and a better retention of knowledge. Active learning also reduces the achievement gaps between students. As attractive as active learning is, however, it has not been accessible for most educators in medical physics. This is because of the time and editing skill needed to develop high-quality content. This series fills that gap and makes it easier for medical physics educators to employ an active learning approach.

Each chapter includes Problem Sets (typically 10 problems each). A Solution Set is available in the Educator Version, which provides the Answer Key and also a detailed derivation of each answer with commentary. This will also help to promote active learning.

Physics is challenging to learn. There is no doubt about that. But it is also a rich and fascinating subject. My hope is that this book and video series will stimulate interest and provide the support to learn the material at a deep level. Ultimately, this should result in higher-quality patient care. And that, to me, is well worth the effort.

1

Basic Physics

1.1 Waves and Particles

1.1.1 Electromagnetic Waves

Electromagnetic waves come in many forms, e.g. radio waves, lights, and X-rays. In 1870, James Clerk Maxwell developed a formalism to describe these electromagnetic waves in which the changing magnetic field creates an electric field and vice versa to form a self-sustaining wave that propagates at a speed given the symbol c, the speed of light, which is $3\cdot10^8$ m/s in a vacuum. Three properties describe waves: the speed, c, wavelength, λ, and frequency, ν (sometimes written as f). See Figure 1.1.1. These are related by

$$c = \lambda \cdot \nu \tag{1.1}$$

The units are c (m/s), λ (m), and frequency, ν (1/s given the special unit Hertz, Hz). The type of wave is determined by the wavelength (or equivalently the frequency), see Figure 1.1.1. Note that optical light occupies a relatively narrow range of the spectrum from 400 to 700 nanometers (nm).

At very short wavelengths electromagnetic waves are X-rays. X-rays were first produced and characterized by Wilhelm Roentgen in 1895. They have a wavelength similar to the size of the atom itself. It is more common to describe these waves by their energy instead of their wavelength. The energy, E, is given by

$$E = h\nu \tag{1.2}$$

Here h is Planck's constant, a fundamental constant of nature whose value is $6.626\cdot10^{-34}$ m² kg/s. A common unit for energy useful for medical physics applications is the electron-volt, eV, which is the energy gained by one electron moving through a potential of one volt. X-rays and particles in medical physics applications often have energies in keV to MeV range.

1.1.2 Particles

The various particles of importance for medical physics applications are shown in Table 1.1 The electron was first discovered by J.J. Thompson in 1897. Somewhat later, in 1908, Robert A. Millikan and Harvey Fletcher performed a series of experiments which showed that the charge of the electron is quantized, i.e. is present discrete amounts. These experiments involved suspending oil drops in an electric field (see the video for further background). The charge of the electron, we now know, is $1.602\cdot10^{-19}$ Coulombs (C).

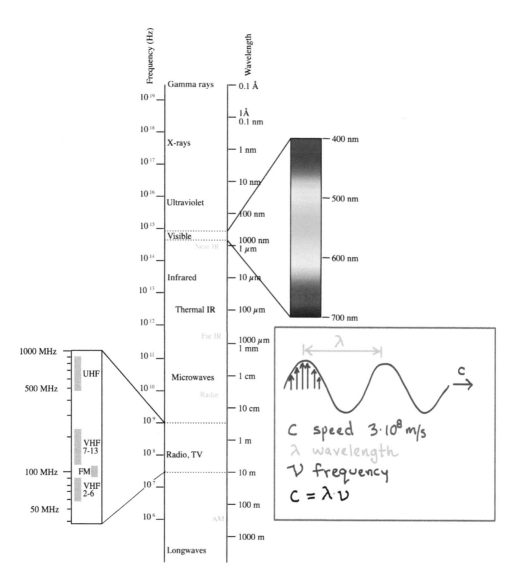

FIGURE 1.1.1
Properties of electromagnetic waves.

Another important property of particles is their electric charge. One way to measure this is to measure the charge-to-mass ratio. Then, if the mass is known, the charge is known. The charge-to-mass ratio can be found by measuring the movement of a particle through a magnetic field. Any charged particle moving in a magnetic field experiences a force perpendicular to its motion (Figure 1.1.2) given by $F = qvB$, where q is the charge, v is the speed, and B is the magnetic field. This force bends the particle into a curved trajectory. **The force on a charged particle moving in a magnetic field will have various practical applications in medical physics and forms a key concept.**

Table 1.1 shows the masses of the various particles. Note that because of Albert Einstein's energy–mass equivalence, $E = mc^2$, the rest mass of a particle can be written either as a

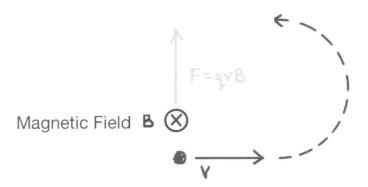

FIGURE 1.1.2
Force on a moving charged particle in a magnetic field, B.

TABLE 1.1

Fundamental Particles of Importance in the Radiation Therapy Context

Particle	Symbol	Charge	Rest Mass Energy $(E = mc^2)$
Electron	e⁻	–	0.511 MeV
Positron (antiparticle of the electron)	e⁺	+	0.511 MeV
Proton	p⁺	+	938.3 MeV
Neutron	n	None	939.9 MeV
Photon	γ	None	None

mass (e.g. kg) or as an energy (e.g. MeV). Note that the mass of a proton is nearly 2000 times that of an electron, a fact that will have many implications for radiation therapy physics.

1.1.3 Waves or Particles?

One of the key insights of early 20th-century physics is the fact that particles can behave as waves and, conversely, electromagnetic waves can behave as particles. In the 1920s French physicist Louis de Broglie suggested that particles have a wave nature and the wavelength, λ, is given by $\lambda = h/p$, where h is Planck's constant, and p is the momentum of the particle (i.e. the mass, m, times the speed, v). This wavelength, the de Broglie wavelength, can also be written as $\lambda = hc/E$, since we know that the energy and momentum are related by $E = pc$. This theoretical insight was confirmed in the mid-1920s by Clinton Davisson and Lester Germer in experiments at Bell Labs in which they observed diffraction patterns of electrons scattered from crystals, the same diffraction patterns that are seen from waves.

Electromagnetic waves can also behave as particles, so-called "photons," given the Greek symbol gamma, γ (Table 1.2). Photons have no mass and travel at the speed of light. The understanding of this dates back to 1905 when Albert Einstein proposed that waves can be thought of as packets of propagating particles (the photons). He put this theory forward in order to explain the photoelectric effect, in which electrons are ejected from a metal when the frequency (or energy) is above a certain threshold value. **The existence of photons is a key concept. Each photon has an associated energy given by Equation 1.2.**

1.2 Atomic Structure

1.2.1 Structure of the Atom and Coulomb's Law

The modern understanding of the structure of an atom is shown schematically in Figure 1.2.1, electrons orbiting around a nucleus. The nucleus consists of protons and neutrons. There are an equal number of protons (charge +1) and electrons (charge –1) in a neutral atom. The overall size of the atom is tens of nm (note that 10 nm = 1 Å, Angstrom). The nucleus, however, is much smaller, typically a few × 10^{-15} m. Thus, the atom is made up mostly of empty space.

It is useful to consider how we came to this understanding of the structure of the atom, since it also illustrates an important physical principle. In 1909 Earnest Rutherford, in collaboration with other physicists at the University of Manchester in the UK, designed an elegant experiment to decipher the structure of the atom using alpha particles (helium nucleus which consists of two protons and two neutrons). The alpha particles were scattered through a thin gold foil, and the foil was surrounded by a scintillating material. In this way, the path of the alpha particle could be visualized since it created a small flash of light when it hit the screen. Most of the scatter events were observed to occur along the beam path of the alpha particle. However, there were a small number of events in which the alpha particle was backscattered. That is, there was a head-on collision with something small and compact. For a further illustration of the experiment see the video.

Rutherford explained these results as an interaction of the alpha particle with the nucleus of gold atoms in the foil. As shown in Figure 1.2.2 the positively charged alpha

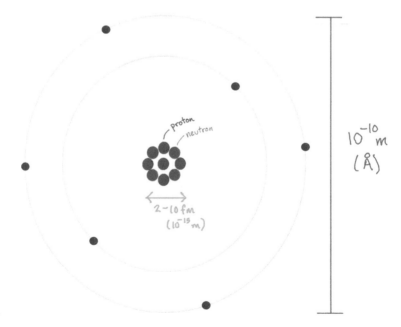

FIGURE 1.2.1
Structure of the atom.

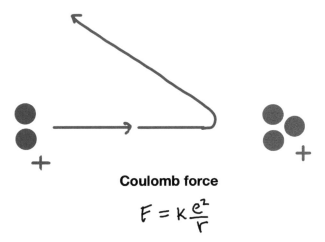

Coulomb force

$$F = k \frac{e^2}{r}$$

FIGURE 1.2.2
Repulsion of like charges and Coulomb's Law.

particle experiences a repulsive force when it approaches the nucleus (like charges repel each other). The force between two charged particles is described by Coulomb's Law

$$F = k \frac{q_1 q_2}{r^2} \qquad (1.3)$$

where k is Coulomb's constant (i.e. a number) and r is the distance between the two charge particles, and the numerator in Equation 1.3 is the charge of the two particles, q_1 and q_2. For the case of two electrons, $q_1 = e$ and $q_2 = e$. For an alpha particle and a gold nucleus, $q_1 = 4e$ and $q_2 = 79e$ since there are 79 protons in the gold nucleus.

Coulomb's Law (Equation 1.3) is an important key concept which will have many applications. Note that as the two particles get closer together (r becomes smaller) the force between them becomes larger.

1.2.2 Quantum (Bohr) Model of the Atom

The concepts outlined above mean that the structure of the atom should not be stable. The electron and proton in an atom would attract each other, and the electron would be pulled toward the nucleus. The closer the electron gets the stronger the attractive force should be so the electron should spiral into the nucleus. However, atoms are observed to exit, so an alternative model is needed.

This alternative model of the atom uses quantum mechanics, the theory that describes matter and energy on a very small scale. Originally proposed in 1913 by Niels Bohr, a Danish physicist, the quantum mechanical model of the atom states that electrons can only be found in discrete energy states. That is, the energies are quantized. Figure 1.2.3 shows the energy levels in the simplest atom with one proton in the nucleus, i.e. the hydrogen atom, H. **Quantum energy levels are a key concept.** Just as electrons in the atom are found in discrete quantum states the nucleus itself also has quantum energy levels, as we will see later.

In the atom, the energy levels are described by a principle quantum number, n, and the energy is proportional to $1/n^2$. The $n = 1$ quantum number is the ground state, that is, the

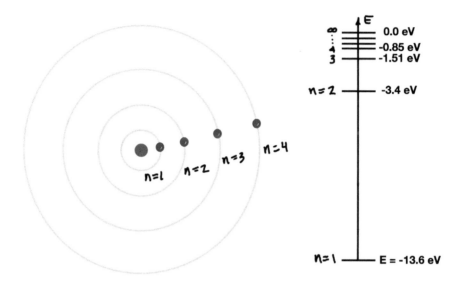

FIGURE 1.2.3
Quantum energy levels in the hydrogen atom.

lowest energy state. In the hydrogen atom the ground state has an energy of –13.6 eV. That is, an electron in this state would require 13.6 eV of energy to be removed from the atom. Electrons can transition between quantum levels. If an electron loses energy (e.g. transitioning from $n=2$ to $n=1$) then the extra energy is emitted as a photon. If an electron is to gain energy (e.g. $n=1$ to $n=2$) then a photon must be absorbed. These photons are also quantized in energy and have energies that are characteristic of the particular atom.

1.2.3 Quantities and Units

Some useful quantities are the following (in the International System of Units [SI]):

$h = 6.626 \cdot 10^{-34}$ m^2 kg/s

$eV = 1.602 \cdot 10^{-19}$ J

$c = 3 \cdot 10^8$ m/s

Coulomb's constant $k = 8.987 \cdot 10^9$ N m^2/C^2

Charge of electron $e = 1.602 \cdot 10^{-19}$ C

Common prefixes:

n nano	10^{-9}
μ micro	10^{-6}
m milli	10^{-3}
c centi	10^{-2}
k kilo	10^3
M mega	10^6
G giga	10^9

Further Reading

Kahn, F.M. and J.P. Gibbons. 2014. *Kahn's The Physics of Radiation Therapy*. 5th Edition. Chapter 1. Philadelphia, PA: Wolters Kluwer.
McDermott, P.N. and C.G. Orton. 2010. *The Physics of Radiation Therapy*. Chapter 1. Madison, WI: Medical Physics Publishing.

Chapter 1 Problem Sets

*Note: * indicates harder problems.*

1. What is the wavelength for the carrier signal from KUOW radio 94.9 FM?
   a. 3.16 m
   b. 10.5 m
   c. 105 m
   d. 316 m

2. Consider the n=3 to n=2 transition in a hydrogen atom. What is the energy difference (in eV)? (See Lecture Notes.) What wavelength of light does this correspond to? What color is this?
   a. 1.51 eV   A. 91.2 nm   1. Red
   b. 1.88 eV   B. 365 nm   2. Yellow
   c. 3.39 eV   C. 656 nm   3. Green
   d. 13.6 eV   D. 1028 nm   4. Blue

3. Consider the transitions to the n=1 state in a hydrogen atom. What is the highest energy photon that can emerge? What region of the electromagnetic spectrum is this?
   a. 1.51 eV   1. Radio wave
   b. 1.88 eV   2. Infrared
   c. 3.39 eV   3. Optical
   d. 13.6 eV   4. Ultraviolet

4. In the solar spectrum below, what is the origin of the dark bands (Figure PS1.1)?

FIGURE PS1.1
Solar spectrum.

5. How long would it take light to travel around the earth? Would the speed be different for blue vs. red light?

   a. 21 msec   1. Red is faster

   b. 133 msec   2. Blue is faster

   c. 21 sec   3. Both are the same speed

   d. 133 sec

6. What is the wavelength of the radiofrequency (RF) waves in an S-band linear acceleration (linac) operating at 3 GHz frequency?

   a. 1 mm

   b. 1 cm

   c. 10 cm

   d. 1 m

7. What is the wavelength of the RF waves in an X-band linac (frequency 11 GHz)?

   a. 27 mm

   b. 0.27 cm

   c. 2.7 cm

   d. 27 m

8. Consider two scenarios: (1) two electrons placed 10 nm apart, (2) an electron and a proton placed 10 nm apart. Which of the following statements describes the forces between the two particles?

   a. Larger force in 1

   b. Larger force in 2

   c. Same force

   d. Same force but opposite direction

9. * Calculate the acceleration that is experienced by an electron that is passing 10 nm away from the nucleus of a hydrogen atom. (Derive the formula for acceleration and calculate it in m/s^2.) How does this compare to an electron passing 10 nm from a lead nucleus?

   a. $1.33 \cdot 10^8$ m/s^2

   b. $2.54 \cdot 10^8$ m/s^2

   c. $1.33 \cdot 10^{18}$ m/s^2

   d. $2.54 \cdot 10^{18}$ m/s^2

10. * What is the rotation frequency of a proton in a cyclotron with a magnetic field of 1T? Derive the formula for the frequency in terms of charge, mass, and B. (Assume a low-energy proton which is non-relativistic.)

   a. 1.53 MHz

   b. 9.86 MHz

   c. 15.3 MHz

   d. 98.6 MHz

2

Nuclear Structure and Decay

2.1 Nuclear Structure and Energetics

2.1.1 Nuclear Structure and Nomenclature

In the previous chapter we reviewed the structure of the atom and the nucleus. The atom has a size of approximately 10^{-10} m (1 Å) and a very small nucleus, 2–10 fm (1 fm $= 10^{-15}$ m). It may be surprising that the atomic nucleus can be stable, given that the protons are in such close proximity to each other. (Remember that the repulsive force between two particles is described by Coulomb's Law and scales as $1/r^2$ where r is the distance between the particles. A small distance results in large forces.) There is, however, a balancing force, the strong force described by quantum mechanics that holds the nucleus together. It is beyond the scope of this series to talk about that force, but the reader may refer to Podgorsak (2016) for more information.

The standard nomenclature for elements is the following:

$$^A_Z Y$$

Z is the atomic number ($Z =$ number of protons), A is the atomic mass ($A =$ number of protons + neutrons), and Y is the element type. An example is $^4_2$He, i.e. a helium atom with two protons and two neutrons ($A = 2 + 2 = 4$). Because the Z and element name are redundant, the bottom number will sometimes be omitted, e.g. ^{4}He.

2.1.2 Chart of Isotopes and Binding Energies

A chart of isotopes is shown in Figure 2.1.1. This chart shows the number of neutrons in an element (y-axis) vs. the number of protons, or Z (x-axis). Recall that a given element can have different numbers of neutrons, and these are called "isotopes." An example is ^{4}He with two protons and two neutrons vs. ^{3}He with two protons and one neutron. More about isotopes can be found in Section 2.1.4.

There are several features to note from Figure 2.1.1:

1) Some isotopes are stable (boxes shown in black), and some are unstable and undergo decay (boxes in other colors).
2) Most elements are neutron rich (i.e. line above the black line where the number of neutrons equals the number of protons).

3) Higher Z elements are relatively more neutron rich (# neutron/# protons increased with Z).

4) An important extreme of this is hydrogen which has no neutrons (n/p = 0).

5) Nuclei with Z > 83 are unstable, i.e., have no stable form.

To understand why some nuclei are stable it may be helpful to review an example calculation of the energy of a carbon nucleus (Table 2.1). The stable atom of carbon (C) has six protons, six neutrons, and six electrons. The mass of these component parts can be tabulated as shown in Table 2.1 (using the atomic mass units [amu], which is a unit for mass).

The mass of all the components sums to 12.099006 amu. However, the total mass of the stable carbon atom is 12.00000 amu (that is actually the definition of an amu). In other words, when the components are bound in a carbon atom (protons, neutrons, and electrons

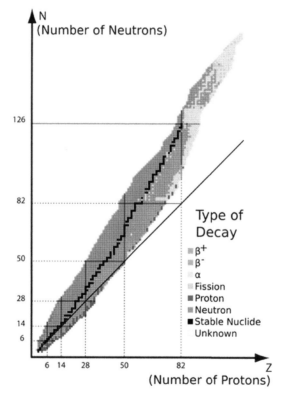

FIGURE 2.1.1
Chart of the isotopes. For a website with an interactive table see: www-nds.iaea.org/relnsd/vcharthtml/VChartHTML.html.

TABLE 2.1

Masses of the Component Parts of a Stable Carbon Atom

Component	Mass	Total Mass
Protons (6)	6 × 1.007277 amu	6.043662 amu
Neutrons (6)	6 × 1.008665 amu	6.051990 amu
Electrons (6)	6 × 0.000549 amu	0.003294 amu
Total		**12.099006 amu**

all together) they have a mass that is 0.099006 amu lower than when you simply add them all up individually. Usually, this is thought of not as a mass difference but as an energy difference (recall that mass and energy can be used interchangeably, $E = mc^2$). The energy difference for carbon is 0.099006 amu or 92.2 MeV.

What accounts for this smaller mass-energy of the stable carbon atom? It is the binding energy of the nucleus. The protons and neutrons are bound in a nucleus. In other words, it would take some energy to dissociate the nucleus. Therefore, the mass/energy of the stable nucleus is less than the mass/energy of its component parts.

2.1.3 Nuclear Decay and Energy Release

Given that there is some binding energy in a nucleus, it is perhaps not surprising that nuclei can decay. This decay happens when it is energetically favorable, i.e. when energy can be extracted in the decay. This is illustrated in Figure 2.1.2 which shows the binding energy for each element (plotted as a function of the number of nucleons, i.e. number of protons+neutrons).

As a nucleus decays from one isotope to another it moves up the curve of binding energy and energy is released. Consider the case of ^{235}uranium fission. Here a neutron combines with the ^{235}U to form ^{236}U, and this undergoes spontaneous decay as follows:

$$^{235}\text{U} + n \rightarrow\,^{236}\text{U} \rightarrow\,^{141}\text{Ba} +\,^{92}\text{Kr} + 3n + \text{energy}$$

In this fission process energy is produced.

Another way to extract energy is the fusion of low-Z nuclei which moves up the curve of binding energy from a light element to a heavier element. An example of this is the main reaction in the hot dense core of the sun:

$$^{2}\text{H} +\,^{2}\text{H} \rightarrow\,^{2}\text{He} + n + \text{energy (3.27 MeV)}$$

Here ^{2}H is the deuterium nucleus, i.e. hydrogen with the addition of one neutron.

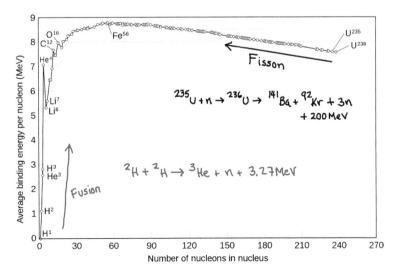

FIGURE 2.1.2
Curve of nuclear binding energy. (Public Domain, https://commons.wikimedia.org/w/index.php?curid=1540082.)

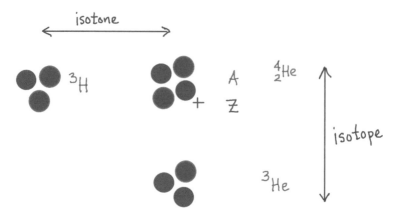

FIGURE 2.1.3
Example of nuclei with different numbers of protons and neutrons.

2.1.4 Nomenclature for Isotopes

Nuclei can have different numbers of protons and neutrons as illustrated in Figure 2.1.3. Two isotopes of helium are shown: $^4_2$He and $^3_2$He. These are isotopes of one another, i.e. they have the same number of protons ($Z = 2$), and are therefore the same element, but they differ in the numbers of neutrons. Shown: $^4_2$He and $^3_1$H are isotones of one another, i.e. they have the same number of neutrons.

2.2 Nuclear Decay Schemes

As noted above, some nuclei in Figure 2.1.1 are stable (black squares) but some are not stable and will decay. There are different decay modes depending on whether a nucleus is relatively neutron-rich (blue area in Figure 2.1.1) vs. neutron-poor (orange area). The following sections describe these various decays modes, the products that emerge after decay, and their energies.

2.2.1 Beta-Minus Decay

This section describes beta-minus decay which occurs in neutron-rich nuclei. Figure 2.2.1 shows an example beta-minus decay of $^{32}_{15}$P into $^{32}_{16}$S. A beta-minus particle is produced. Note that a "beta-minus" particle is the same as an electron. Also produced is an anti-neutrino ($\bar{v}$), which, for medical physics purposes, is not important. Energy is released as well, denoted with the symbol E. In this decay we started with 15 protons and ended with 16 protons. In other words, a proton was "produced" or, more accurately, a neutron was converted into a proton. Because the result was one extra proton, an electron must also be produced to have an equal electrical charge before vs. after the decay.

 In general, any beta-minus decay can be written as $^A_Z X \rightarrow\ ^A_{Z+1}Y + \bar{v}$, i.e. element "X" with Z protons decaying into element Y with $Z + 1$ protons. The decay scheme is also often shown graphically as in Figure 2.2.1 (bottom left, yellow) with the energy axis vertical and

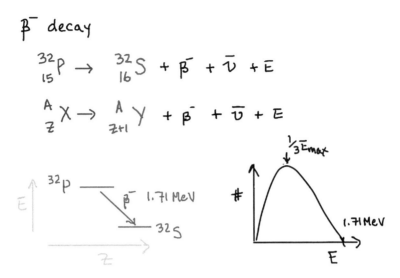

FIGURE 2.2.1
Example of beta-minus decay and associated energy spectrum of the emerging beta particles (electrons).

the number of protons increasing to the right. Often the axes in yellow are not included in such diagrams. Here the maximum energy of the beta-minus particle is labeled (1.71 MeV in the example decay shown here).

The electron that is emitted in beta-minus decay does not always emerge with the maximum possible energy. That is because the energy of the decay is shared between the beta-minus particle and the neutrino. There is a spectrum of energies of the emerging electron as shown in Figure 2.2.1 (bottom right). The mean energy of the emerging electrons is typically 1/3 of the maximum possible energy.

2.2.2 Beta-Plus Decay

Beta-plus decay is relevant for neutron-poor nuclei (orange squares in Figure 2.1.1). Figure 2.2.2 shows an example of the decay of ^{18}F, an isotope widely used in PET imaging. In beta-plus decay a proton is lost and a neutron is gained and a beta-plus particle is produced. Note that a "beta-plus" particle is the same as a positron which is the positively charged anti-particle of the electron. As with beta-minus decay, a neutrino is produced, and because the charged particle shares energy with this neutrino there is a spectrum of possible energies for the positron (Figure 2.2.2, bottom right).

2.2.3 Beta Decay: Production and Half-Life

Figure 2.2.3 provides a summary of beta-plus and beta-minus decay schemas. Isotopes that undergo beta-minus decay are relatively neutron-rich. They lose a neutron and gain a proton in the decay. The half-lives for such decays are relatively long. These isotopes are produced in a nuclear reactor (where a target is bombarded with neutrons to produce a neutron-rich isotope).

Isotopes that undergo beta-plus decay are relatively neutron-poor. They lose a proton and gain a neutron in the decay. The half-lives for such decays are relatively short. These isotopes are produced in a cyclotron (where a target is bombarded with a heavy charged particle such as a proton).

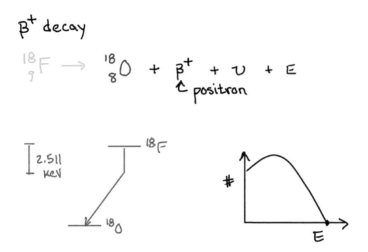

FIGURE 2.2.2
Example of beta-plus decay and associated energy spectrum of the emerging beta-plus particles (positrons).

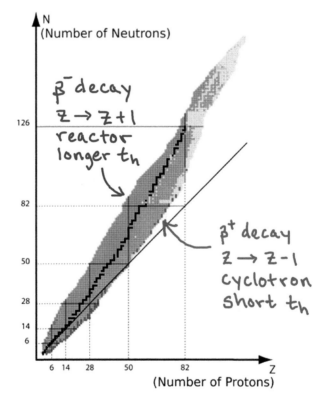

FIGURE 2.2.3
Overview of beta decay. Neutron-rich isotopes (blue) are made in nuclear reactors, undergo beta-minus decay, and have relatively long half-lives; while the proton-rich isotopes (orange) are made in cyclotrons, undergo beta-plus decay, and have relatively short half-lives.

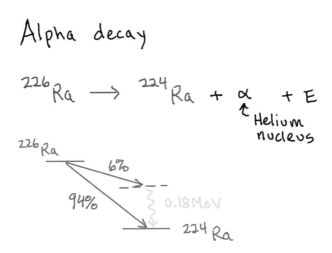

FIGURE 2.2.4
Example of alpha decay.

2.2.4 Alpha Decay

Some nuclei, particularly very heavy ones, undergo alpha decay (yellow boxes in Figures 2.1.1 and 2.2.3). A nucleus undergoing this decay produces an alpha particle, α, i.e. a particle with two protons and two neutrons which is the same as a helium nucleus, $_2^4$He. An example of such a decay is shown in Figure 2.2.4, i.e. the decay of ^{226}Ra (radium) into ^{222}Rn (radon). In 94% of such decays the result is a nucleus that is in the ground-state of ^{222}Rn, but 6% of the time the decay is into an excited state of ^{222}Rn at a higher energy. This excited ^{222}Rn nucleus then decays into the ground state of ^{222}Rn, releasing a photon of energy 0.18 MeV in this case. Historically, ^{226}Ra was often used in brachytherapy, and it was the 0.18 MeV photon that provided the therapeutic dose.

2.2.5 Other Decay Modes

Other possible nuclear decay modes are shown in Figure 2.2.5. In electron capture (EC) an inner-shell electron interacts with the nucleus and the result is that a proton is lost. An example of such a decay is ^{125}I (iodine) decaying into ^{125m}Te (tellurium) in a metastable state, i.e. a high energy state of the nucleus that lasts for a long time but eventually decays into a

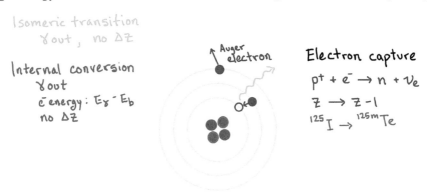

FIGURE 2.2.5
Other decay modes: electron capture, isomeric transition, and internal conversion.

ground state. After EC there is a vacancy in an inner shell electron and an outer shell electron can transition into this state, producing a characteristic photon. Auger electrons can also be emitted. For more on characteristic photons and Auger electrons see Section 5.1.2.

Another nuclear decay mode is isomeric transition in which a nucleus in an excited state transitions to the ground state and a photon is emitted. There is no change in Z.

The final decay mode is internal conversion in which an excited nucleus transitions to the ground state. However, instead of emitting a photon, an electron is ejected. The energy of the ejected electron is the energy of the transition (which would have been a photon) minus the binding energy of the electron

TABLE 2.2

Summary of Nuclear Decay Modes

Decay Mode	Z Change	Main Particles Emitted	Notes
Beta-minus	Z increases	Electron	Reactor generated, longer half-life
Beta-plus	Z decreases	Positron + γ (daughter)	Cyclotron generated, shorter half-life
Electron capture (EC)	Z decreases	Gamma ray, Auger electrons	
Alpha	Z -> Z − 4	Alpha particle (i.e. He nucleus) + γ (daughter)	
Isomeric transition (IT)	No change	Gamma ray	
Internal conversion (IC)	No change	Gamma ray	

"γ (daughter)" indicates that gamma rays are emitted by the decay of the daughter isotope into the ground state.

TABLE 2.3

A Selection of Isotopes of Medical Importance

Isotope	Half-Life	Decay Mode	Exposure Rate Constant Γ (R cm^2/h/mCi)	Notes
^{226}Ra	1600 years	α	8.25 R cm^2/h/mg	Product of the uranium decay series
^{137}Cs	30.2 years	β$^-$	3.28	1.18 MeV E$_{max}$ Photon also, 662 keV
^{192}Ir	73.8 days	β$^-$ (95.3%) or E.C. (4.7%)	4.69	0.67 MeV E$_{max}$ β$^-$ Numerous photons also, 295–468 keV
^{60}Co	5.27 years	β$^-$	13.0	1.48 MeV E$_{max}$ β$^-$ Photons at 1.17 MeV and 1.33 MeV (avg.: 1.25 MeV)
^{125}I	59.4 days	E.C.		Photons. Avg: 28.4 keV
^{103}Pd	17.0 days	E.C.		Photons. Avg: 20.7 keV
^{131}I	8.01 days	β$^-$		0.61 MeV E$_{max}$ Photons also, 80–723 keV
^{90}Mo	66 hours	β$^-$		
^{99m}Tc	6.0 hours	β$^-$		
^{90}Y	64.0 hours	β$^-$		2.28 MeV E$_{max}$
^{90}Sr	28.7 years	β$^-$		0.546 MeV E$_{max}$
^{223}Ra	11.4 days	α		5.78 MeV
^{177}Lu	6.71 days			0.5 E$_{max}$ Photons also, 113, 208 keV

A summary of decay modes is shown in Table 2.2. Table 2.3 lists common isotopes of medical importance, their decay modes, half-lives, and exposure constants (which will be disussed in subsequent chapters).

Further Reading

1. Kahn, F.M. and J.P. Gibbons. 2014. *Kahn's The Physics of Radiation Therapy*. 5th Edition. Chapters 1 and 2. Philadelphia, PA: Wolters Kluwer.
2. McDermott, P.N. and C.G. Orton. 2010. *The Physics of Radiation Therapy*. Chapter 3. Madison, WI: Medical Physics Publishing.
3. Podgorsak, E.B. 2016. *Radiation Physics for Medical Physicists*. 3rd Edition. Chapters 1 and 11. Switzerland: Springer.

Chapter 2 Problem Sets

*Note: * indicates harder problems.*
Useful links:

http://atom.kaeri.re.kr/nuchart/

www.ptable.com

For a website of isotopes see: http://www.nndc.bnl.gov/chart/

For Problems 1 to 3 refer to the decay chain of ^{226}Ra (Figure PS2.1).

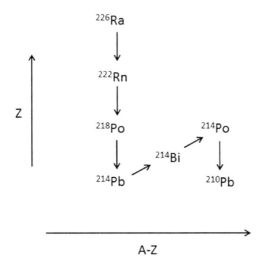

FIGURE PS2.1
Radium decay series.

1. ^{226}Ra -> ^{222}Rn is an example of what decay scheme?
   a. Beta-minus
   b. Beta-plus
   c. Alpha
   d. Electron capture

2. ^{214}Pb -> ^{214}Bi is an example of what decay scheme?
   a. Beta-minus
   b. Beta-plus
   c. Alpha
   d. Electron capture

3. Which types of decay are not represented above? (Explain why.)
   a. Beta-minus
   b. Beta-plus
   c. Alpha
   d. Nuclear spallation

4. ^{15}O -> ^{15}N is an example of what decay scheme?
   a. Beta-minus
   b. Beta-plus
   c. Alpha
   d. Electron capture
   How is this isotope used for medical purposes?

5. What reaction is used to create ^{15}O from ^{14}N?
   a. Beta-plus
   b. Beta-minus
   c. Bombardment with deuterons
   d. Fission product

6. The decay of ^{99}Mo -> ^{99m}Tc is which type of decay?
   a. Beta-minus
   b. Beta-plus
   c. Internal conversion
   d. Electron capture

7. The decay of ^{99}Tc -> ^{99}Ru is which type of decay? (Note: Tc $Z = 43$, Ru $Z = 44$.)
   a. Beta-minus
   b. Beta-plus
   c. Internal conversion
   d. Electron capture

8. Given the decay scheme of ^{137}Cs shown in Figure PS2.2, which particles are emitted (choose all that apply)?

   a. Photon
   b. Electron
   c. Positron
   d. Proton
   e. Alpha particle

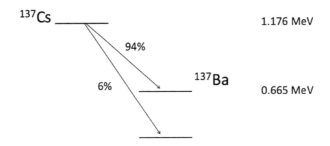

FIGURE PS2.2
^{137}Cs decay.

9. Describe how Auger electrons are produced.

10. *^{64}Cu can decay to either ^{64}Ni or ^{64}Zn. Match each of the spectra of particles shown below (Figure PS2.3 A & B) with the correct decay mode:

   a. Beta-plus
   b. Beta-minus

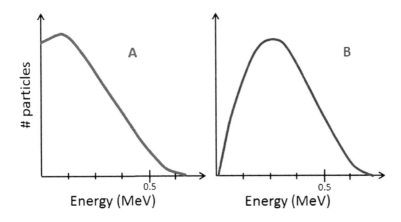

FIGURE PS2.3
Decay spectra of ^{64}Cu.

3

Mathematics of Nuclear Decay

3.1 Exponential Decay

The previous chapter discussed the physical processes of the nuclear decay of unstable isotopes including electron capture, beta-minus, beta-plus decay, and other decay modes. This chapter considers a mathematical description of that decay. To appreciate why this is important, consider the decay of ^{125}I decay into tellurium (Te) which happens by electron capture. This isotope is used in low-dose rate brachytherapy treatments of prostate and other applications. In order to calculate the dose in such an application, one needs to know the amount of radioactive ^{125}I that is present after some amount of time (more material results in a higher dose rate). This chapter presents a mathematical framework for understanding this.

3.1.1 Introduction to Exponential Decay

Here we consider the simple scenario of one isotope decaying into another. As shown in Figure 3.1.1 there is some number, N, of unstable nuclei in a sample. Consider some time interval which we write as Δt. During this time interval some of the nuclei decay. The number that decay we write as ΔN. Now the decay process has two properties. First, the number of decays is proportional to the total number of nuclei in the sample, N (more nuclei results in more decay events). Second, the number of decays is proportional to the time interval that we are considering, Δt (more time results in more decays). Therefore, we can write an equation for this as $\Delta N = -\lambda N \Delta t$. Here, λ is the proportionality constant representing the number of decays per second in a sample of size N, and the negative sign indicates that nuclei are lost, i.e. they decay into other nuclei. We can replace the symbols with calculus notation, Δt becomes dt and ΔN becomes dN (indicating small changes in each variable), and the equation becomes $dN = -\lambda N dt$ which we then rewrite as $\frac{dN}{dt} = -\lambda N$. There is a solution to this equation from calculus which is $N(t) = N_0 e^{-\lambda t}$, where we write N as $N(t)$, i.e. the number, N, at some time, t. This is the formula for exponential decay. The quantity, e, is just a number, though a special number to be sure. It is the natural logarithm of 1 or 2.71828 ... At a time t equals 0, the number of nuclei in the sample is N_0. The decay curve of this equation is shown graphically in Figure 3.1.1.

The exponential decay of activity is governed by the decay constant, λ. The larger the decay constant the faster the exponential decay. The decay constant is determined by the stability of the nucleus; less stable nuclei will have more decays per second or a larger decay constant.

It is worth noting that the exponential form of the decay curve is due to the independent and random process of nuclear decay. That is, any one decay event is completely unrelated to the decay events that happened before it or the ones that will occur after it. In such a

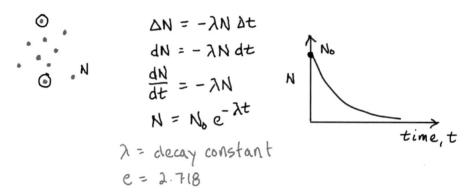

$$\Delta N = -\lambda N \, \Delta t$$

$$dN = -\lambda N \, dt$$

$$\frac{dN}{dt} = -\lambda N$$

$$N = N_0 \, e^{-\lambda t}$$

λ = decay constant

$e = 2.718$

FIGURE 3.1.1
Nuclear decay equation. There are N nuclei in a sample (red), some of which decay. Over time, t, the number decreases exponentially with a decay constant, λ.

situation we can write the simple proportionality equation, $\Delta N = -\lambda N \Delta t$, whose solution is the exponential form. The same mathematics will arise in other contexts so it is worth pausing to ponder this for a moment.

3.1.2 Activity and Units of Activity

It is difficult to measure or quantify the number of nuclei, N, in a sample. An easier quantity to work with is the activity of the sample, i.e. the number of decays per second, since this can be measured through the decay products that emerge (e.g. photons produced from an isomeric transition, etc.). Activity, A, is the number of decays per second which can be written as $A = \dfrac{dN}{dt}$. As with the equations above the solution for this is $A(t) = A_0 e^{-\lambda t}$, where $A(t)$ is the activity at some time t and the activity at time 0 is A_0 (Figure 3.1.2).

The SI units for activity are "decays per second" which has the special unit Becquerel, written with the symbol Bq. Another useful unit is the Curie, written with the symbol Ci. The conversion between the two units of activity is 1 Ci = $3.7 \cdot 10^{10}$ Bq or 1 mCi = $3.7 \cdot 10^7$ Bq = 37 MBq. This is useful to remember.

Activity, A

$$A = \frac{dN}{dt} = \lambda N$$

$$A = A_0 \, e^{-\lambda t}$$

Specific Activity

Activity / mass $= \lambda \frac{N_A}{A}$

Units

1 decay/s = Becqurel (Bq)

1 Ci = $3.7 \cdot 10^{10}$ Bq
  ↳ curie

1 mCi = 37 MBq

FIGURE 3.1.2
Activity. Activity is a measurable quantity and is defined as decays per second. Activity decays exponentially over time.

Another useful quantity is the specific activity, defined as activity per unit mass and given by $\lambda \frac{N_A}{A}$. Specific activity is a useful quantity because it indicates how "concentrated" a particular radioactive isotope is, i.e. for a given mass how much activity is produced by that isotope.

3.1.3 Half-Life

Conceptually, half-life indicates the time in which the activity of an isotope reaches half of its initial activity. Using the equations above this can be written as $A(t) = \frac{1}{2}A_0 = A_0 e^{-\lambda t}$.

This equation can be solved to find the time, t, as outlined in Figure 3.1.3. This time is the half-life which we write with the symbol, t_h. As outlined in Figure 3.1.3 the half-life is related to the decay constant, λ, by the equation:

$$t_h = \frac{0.693}{\lambda} \tag{3.1}$$

The decay constant (and half-life) are determined by the stability of the nucleus; less stable nuclei will have more decays per second or a larger decay constant, λ. This in turn results in a shorter half-life.

With these quantities then the equation for activity above can be rewritten as:

$$A(t) = A_0 e^{-\lambda t} = A_0 e^{-0.693 \cdot t / t_h} \tag{3.2}$$

The activity equation can be written in an alternative form, noting that for each half-life the activity is reduced by half. After n half-lives then the activity is given by:

$$A(\text{after } n \text{ half-lives}) = A_0 \left(\frac{1}{2}\right)^n \tag{3.3}$$

$$\underline{\text{Halflife}}$$

$$\text{Time when } A = \tfrac{1}{2} A_0$$

$$A_0 e^{-\lambda t} = \tfrac{1}{2} A_0$$

$$\ln(e^{-\lambda t}) = \ln(\tfrac{1}{2})$$

$$-\lambda t = -\ln 2$$

$$t_h = \frac{\ln 2}{\lambda} = \frac{0.693}{\lambda}$$

$$\text{After } n \text{ halflifes:}$$

$$A = A_0 \left(\tfrac{1}{2}\right)^n$$

FIGURE 3.1.3

Half-life. The exponential decay of the activity of a source is governed by the decay constant, λ, or, alternatively, the half-life, t_h. Isotopes that are less stable have larger decay constants and shorter half-lives.

Note that Equation 3.3 still describes exponential decay even though it has a different form than Equation 3.2. One can prove mathematically that these two equations are equivalent. Equations 3.2 and 3.3 will be used repeatedly in calculations in the problem sets below and in other places in the text. It is worthwhile to know and understand them well.

3.1.4 Mean-Life

The final quantity to consider is mean-life, sometimes referred to as average life. This construct is somewhat artificial and challenging to understand, but it will prove useful for dose calculations from permanent implants of radioactive sources.

First consider a radioactive source which starts out with N_0 nuclei. in the lifetime of this source there will be N_0 total decays, i.e. the source decays into other nuclei until there are no more of the original nuclei left. Now consider a "fake source" for conceptual purposes. This fake source does not undergo exponential decay but has constant activity which can be written as $A = \lambda N_0$. Over a total time, τ, the number of decays from this source will be $\lambda N_0 \tau$. If we set this value ($\lambda N_0 \tau$) to be equal to the total number of decays from a real source (N_0) we get the following equation: $\lambda N_0 \tau = N_0$. If we solve this equation for τ we find $\tau = 1/\lambda$, which can also be written as $\tau = t_h/0.693$, or:

$$\tau = 1.44\, t_h \qquad\qquad (3.4)$$

The mean-life, τ, then is the time over which the number of decays is equal to what it would have been for a source with a constant activity equal to the initial activity, A_0. This will prove useful for calculations of dose from permanent implants of radioactive sources (see Section 4.2.5).

3.2 Equilibrium of Isotopes

The sections above present the situation of one isotope decaying into its daughter product (another isotope). However, in some medical physics applications the situation is more complex. There may be one isotope which decays into another isotope, which in turn decays into another isotope, and so on. That is, there might be a chain of decays. The activity of these isotopes will come into equilibrium over time. This situation is considered in further detail in the video.

Further Reading

Kahn, F.M. and J.P. Gibbons. 2014. *Kahn's The Physics of Radiation Therapy*. 5th Edition. Chapter 2. Philadelphia, PA: Wolters Kluwer.

McDermott, P.N. and C.G. Orton. 2010. *The Physics of Radiation Therapy*. Chapter 3. Madison, WI: Medical Physics Publishing.

Podgorsak, E.B. 2016. *Radiation Physics for Medical Physicists*. 3rd Edition. Chapter 10. Switzerland: Springer.

Chapter 3 Problem Sets

*Note: * indicates harder problems.*
See Table 2.3 for values for useful values of half-lives.

1. Which source has the highest activity?
   a. 1 mCi ^{125}I
   b. 1 mCi ^{131}I
   c. 1 mCi ^{192}Ir
   d. All are the same

2. What is the activity (in Bq) of a 10 mCi injection of ^{18}F-FDG for PET scanning?
   a. 2.70×10^{-13}
   b. 2.70×10^{-10}
   c. 3.70×10^{8}
   d. 3.70×10^{11}

3. A ^{137}Cs seed is purchased with an activity of 38 mCi. How much activity remains after ten years?
   a. 12.6 mCi
   b. 27.3 mCi
   c. 30.2 mCi
   d. 37.1 mCi

4. What is the reference dose rate of a ^{60}Co teletherapy unit ten years after installation if the initial dose rate was 2 Gy/min?
   a. 0.30 Gy/min
   b. 0.54 Gy/min
   c. 1.75 Gy/min
   d. 3.80 Gy/min

5. By regulation, radioisotopes can be disposed of with no special precaution after ten half-lives. What is the activity of a 0.5 mCi ^{125}I seed after ten half-lives?
   a. 0.02 μCi
   b. 0.49 μCi
   c. 0.02 mCi
   d. 0.49 mCi

6. According to NRC NUREG-1556 regulations, patients with a ^{125}I implant may be released from the hospital if the exposure rate measured at 1 m is less than 1 mR/hr. How long would a patient technically have to be held if the measured rate post-implant is 1.1 mR/hr?

    a. 2.3 min
    b. 5.9 days
    c. 8.2 days
    d. 54.0 days

7. How many total disintegrations happen in the course of a ^{223}RaCl$_2$ (Xofigo) treatment if the injection is 130 μCi, assuming that no ^{223}RaCl$_2$ is cleared from the body?

    a. 7692
    b. 8.43×10^7
    c. 6.84×10^{12}
    d. 4.74×10^{18}

8. * What is the activity of ^{90}Sr in equilibrium with its daughter isotope ^{90}Yr if the initial activity of ^{90}Sr is 10 mCi? (See video for formalism.)

    a. 1 mCi
    b. 5 mCi
    c. 10 mCi
    d. 15 mCi

9. * Prove that these formulae for activity are equivalent: $A = A_0 \exp(-0.696 \, t/t_h)$ and $A = A_0 (1/2)^n$. Where n is the number of half-lives.

10. Describe how ^{137}Cs can be used to determine the age of old wines. (Hint: The ^{137}Cs is a by-product of the nuclear fission of ^{235}U.)

4

Brachytherapy I

4.1 Brachytherapy Sources and Isotopes

The previous chapters have focused on a few important basic physical concepts, the structure of atoms, nuclei and the decay of unstable isotopes. This chapter describes a therapeutic application of these concepts, namely brachytherapy, which is treatment delivered with radioactive sources implanted in tissue. This chapter serves as an introduction to this topic with more detail to follow in subsequent chapters, in particular Chapter 26 which will present practical examples of brachytherapy procedures (e.g. prostate implants and high-dose rate brachytherapy for gynecological cancers).

4.1.1 Common Isotopes: LDR/HDR Brachytherapy

Brachytherapy can be delivered in two different ways: low-dose rate (LDR) brachytherapy at < 2 Gy/hour or high-dose rate brachytherapy (HDR) at > 12 Gy/hour. These definitions and criteria are spelled out in ICRU Report No. 38 (Wyckoff et al. 1985).

The most common method of delivering brachytherapy uses radioactive isotopes which decay and emit particles. The most commonly used brachytherapy sources emit photons, and it is these photons that deliver the dose. The isotopes used in LDR have relatively short half-lives and emit photons at relatively low energies, while isotopes for HDR have somewhat longer half-lives and emit relatively high-energy photons. LDR implants are typically permanent (i.e. sources are implanted in the patient and left in) while HDR implants are temporary (i.e. catheters are implanted in the patient and sources are temporarily inserted in these catheters to deliver dose).

Example isotopes are shown in Figure 4.1.1. Two example isotopes used in LDR treatments are ^{125}I and ^{103}Pd which both decay by electron capture and emit low-energy photons. The half-life of ^{103}Pd is slightly shorter than that of ^{125}I and the photon energies are slightly lower. Both of these sources are used in brachytherapy treatment of prostate cancer (Chapter 26). In HDR a commonly used isotope is ^{192}Ir which decays by beta-minus decay to an excited state of ^{192}Ro which then decays down into the ground state, emitting photons of various energies. The average energy of photons emitted is 380 keV.

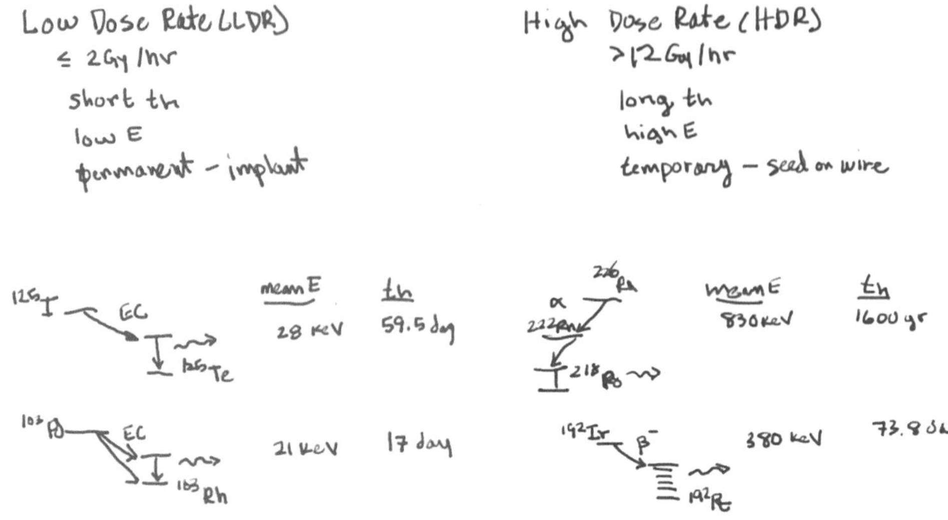

Low Dose Rate (LDR)
  ≤ 2Gy/hr
    short th
    low E
    permanent - implant

High Dose Rate (HDR)
  >12 Gy/hr
    long th
    high E
    temporary - seed on wire

^{125}I ⟶ EC ⟶ I ⟶ ^{125}Te

mean E   th
28 keV   59.5 day

^{226}Ra ⟶ α ⟶ ^{222}Rn ⟶ ^{218}Po

mean E   th
830 keV   1600 yr

^{103}Pd ⟶ EC ⟶ I ⟶ ^{103}Rh

21 keV   17 day

^{192}Ir ⟶ β⁻ ⟶ ^{192}Pt

380 keV   73.8 day

FIGURE 4.1.1
LDR and HDR brachytherapy and example isotopes for each.

4.1.2 LDR Source Design

Many brachytherapy applications rely on sources encapsulated in a "seed," a small metal-lic delivery container about the size of a rice grain (Figure 4.1.2). This source design serves several purposes. It contains the radioactive isotope, it filters out unwanted particles (e.g. the electrons created in ^{192}Ir from beta-minus decay which are not useful therapeutically),

FIGURE 4.1.2
Brachytherapy seeds. (From Wikimedia Commons Open Source. https://commons.wikimedia.org/wiki/File: Radioactive_Seeds_(7845754328).jpg.)

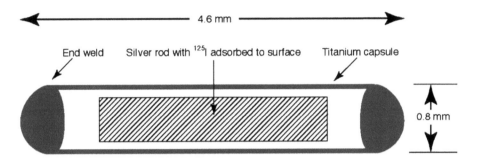

FIGURE 4.1.3
Example LDR brachytherapy source. The ¹²⁵I Amersham Model 6711 seed. (From AAPM TG43 update, Rivard et al. 2004.)

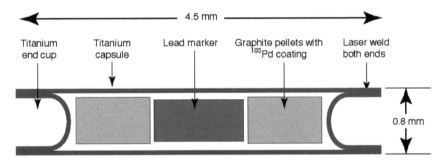

FIGURE 4.1.4
Example LDR brachytherapy source. The ¹⁰³Pd Theragenics Model 200 seed. (From AAPM TG43 update, Rivard et al. 2004.)

and the seed serves as a vehicle for radiographic markers which can be visualized on X-ray or CT scans.

Figure 4.1.3 shows an example of one such source, the ¹²⁵I seed sold as Model 6711 seed by Amersham Inc. The active isotope is deposited onto the surface of a silver rod and is encapsulated in a 50 μm-thick titanium sheath which serves to protect the source and to absorb Auger electrons created in the beta-minus decay of the isotope. The silver rod serves as a radiographic marker. Note the small dimensions of the source which is designed to fit in a 17 or 18 gauge needle.

Figure 4.1.4 shows another example of an LDR source, the Model 200 ¹⁰³Pd seed from Theragenics Inc. Here the isotope is coated on a graphite pellet on one end. A small lead marker is included for radiographic visualization.

4.1.3 HDR Source Design

Figure 4.1.5 shows an example of an HDR source, the ¹⁹²Ir from Nucletron Inc. Again this source is small in size, designed to fit in a catheter, and is enclosed in a stainless steel capsule. Here, however, the source capsule is laser welded onto a stainless-steel cable. The cable allows the source to be temporarily inserted into the catheter and then removed with an automatic device called the remote afterloader (see Chapter 26). The reason for this remote afterloading technology is that this is a very high-activity course (installed at approximately 10 Ci) in order to deliver high dose rates. The procedure, therefore, cannot be done manually by staff but must be performed in a shielded room with remote control of the source.

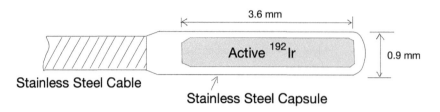

FIGURE 4.1.5
Example HDR brachytherapy source. The ^{192}Ir source from Nucletron Inc. (Adapted from G. Douysset et al. 2008. Comparison of air kerma standards of LNE–LNHB and NPL for ^{192}Ir HDR brachytherapy sources: EUROMET Project No 814. *Phys Med Biol* 53:N85–N97 doi:10.1088/0031-9155/53/6/N02.)

4.1.4 Other Forms of Brachytherapy

Not all brachytherapy is delivered with seeds. Other forms include the following:

- "Electronic brachytherapy" (eBT). There are various ways to deliver this including a miniaturized X-ray tube inserted into a catheter (e.g. Axxent®, Xoft/iCad Inc.) or electron targets enclosed in a spherical applicator (Intrabeam®, Carl Zeiss Meditec AG). More information is available in the forthcoming AAPM Task Group Report 182.

- Radionuclide therapy. Sometimes classified as brachytherapy, this treatment is accomplished by injecting unsealed sources into the patient. Different forms include ^{131}I for thyroid cancers, ^{223}Ra-chloride for bone metastases, and ^{90}Yr-coated microspheres for liver. More information is found in Chapter 26.

4.2 Brachytherapy Exposure and Dose

The key to using brachytherapy sources in clinical applications, of course, is to understand what dose they produce. Chapter 3 presented the concept of the activity of a source, i.e. the number of decays per second, but this does not tell us anything directly about dose in tissue. To get from activity to dose we must move through an intermediate step of exposure. That is, activity (A) -> exposure (X) -> dose (D). If one knows the activity of a source, one can calculate the dose. This section serves as an introduction to this topic.

4.2.1 Exposure

The concept of exposure is important not only for brachytherapy but for other applications in medical physics as well. To understand exposure, consider the experiment shown in Figure 4.2.1. Here some source of photons (green) is used to irradiate an air-filled chamber (black circle). These photons might be from a brachytherapy source or some other source. When photons interact with the air in the chamber they ionize it, i.e. cause electrons to be released from the molecules in the air. This process will be described in more detail in Chapter 5. In this chamber we have placed an electrode at high voltage (shown in red). This creates an electric field in the chamber. The electrons are accelerated by this electric field toward the electrode and are collected by it. The end result is an electrical charge (symbol "Q") collected on the electrode. The SI units for charge are Coulombs, C.

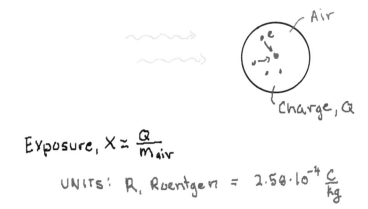

Exposure, $X \approx \dfrac{Q}{m_{air}}$

UNITS: R, Roentgen $= 2.58 \cdot 10^{-4} \dfrac{C}{kg}$

FIGURE 4.2.1
Definition of exposure. In this example experiment photons (green) ionize air molecules in an ionization chamber (black circle). This results in some charge, Q, in the chamber. Exposure is defined as charge per unit mass of air.

Exposure is defined as the charge per unit mass of air created in the chamber and is written with the symbol, X:

$$\text{Exposure, } X = \frac{Q}{m_{\text{air}}} \tag{4.1}$$

So if we know the mass of air in the chamber (or can derive it through some calibration) and we measure the charge we would know the exposure created by the incident photons. The units of exposure are Roentgens, R, defined as $2.58 \cdot 10^{-4}$ C/kg. This somewhat strange number is explained by the fact that in the old unit system (i.e. non-SI, cgs units) 1 R was defined as 1 esu/cm^3.

4.2.2 Exposure Rate from Brachytherapy Sources

We now consider exposure from brachytherapy sources with the goal of going from activity to exposure to dose.

Activity (A) -> exposure (X) -> dose rate (D)

For brachytherapy applications we typically consider not exposure but exposure *rate*, and not dose but dose *rate*. This is because brachytherapy sources are continuously emitting and are characterized by their activity which is a rate (number of decays per second). Exposure rate is written as $\dot{X} = \dfrac{dX}{dt}$, i.e. the change in exposure (dX) per unit time (dt). Note that the dot over the quantity indicates change per unit time (e.g. $\dot{D}$ is dose rate).

The exposure rate is proportional to the activity divided by the square of the distance, r, from that source: $\dot{X} = \Gamma \dfrac{A}{r^2}$. The constant of proportionality, Γ, is called the exposure rate constant, and it is different for different sources because the energy of photons emitted from different sources is different (see Figure 4.1.1). **This is a key concept, that different brachytherapy sources of the same activity produce different exposure rates.** Sources which emit high-energy photons (e.g. [192]Ir) produce much more ionization in air than sources that emit low-energy photons (e.g. [103]Pd). The units of the exposure rate constant

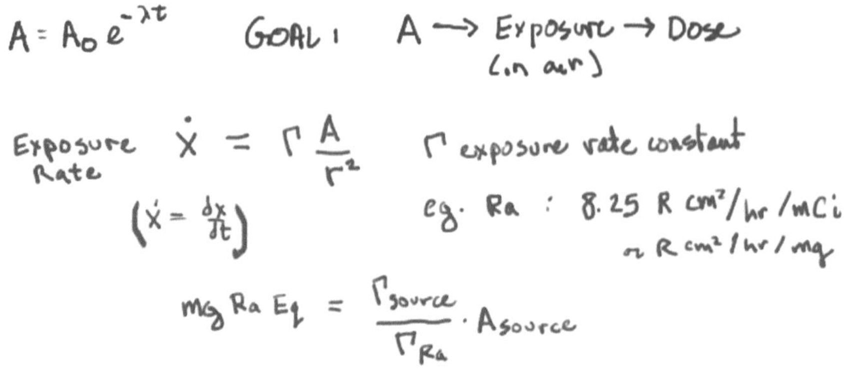

$$A = A_0 e^{-\lambda t} \qquad \text{GOAL}_1 \quad A \longrightarrow \text{Exposure} \longrightarrow \text{Dose}$$
$$\text{(in air)}$$

$$\text{Exposure Rate} \qquad \dot{X} = \Gamma \frac{A}{r^2} \qquad \Gamma \text{ exposure rate constant}$$

$$\left(\dot{X} = \frac{dx}{dt} \right) \qquad \text{eg. Ra} : 8.25 \ R \ cm^2 / hr / mCi$$
$$\sim R \ cm^2 / hr / mg$$

$$mg \ Ra \ Eq = \frac{\Gamma_{source}}{\Gamma_{Ra}} \cdot A_{source}$$

FIGURE 4.2.2
Exposure from brachytherapy sources.

are R cm^2/hr/mCi. Some example values are shown in Table 2.3. One historical unit for activity is milligram-radium-equivalent (mg-Ra-Eq). That is the mass of radium that would give the same exposure as the source under question. To find the activity of the source in mg-Ra-Eq we simply calculate $\frac{\Gamma_{source}}{\Gamma_{Ra}} A_{source}$. Exposure rate concepts are important to understand but are no longer directly used for brachytherapy dose calculations, since they have been replaced by a new formalism shown in the next section.

4.2.3 Inverse Square Falloff

In the equations above it will be noted that the exposure depends on 1 over the distance from the source squared, $1/r^2$, i.e. the inverse square of the distance. **Inverse square falloff is a key concept that underlies much of radiation therapy physics.** It is essential to understand it well.

The concept of inverse square is illustrated in Figure 4.2.3. Here a source is emitting photons (green) equally in all directions. There is some intensity of photons emerging

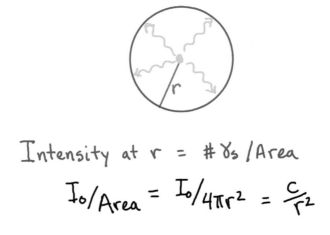

$$\text{Intensity at } r = \#\gamma s / \text{Area}$$
$$I_0 / \text{Area} = I_0 / 4\pi r^2 = \frac{c}{r^2}$$

FIGURE 4.2.3
Inverse square falloff. The area of the shell at a distance *r* away from the source is $4\pi r^2$. Therefore, the intensity per unit area is proportional to $1/r^2$.

from the source (i.e. the number of photons per second). We give this the symbol I_0. Now we want to find the intensity of photons per unit area at a distance r from the source. The intensity per unit area (or "fluence") will determine the exposure and other quantities. To find this let's draw a shell around the source at a distance r. Recall from basic geometry that the area of this shell is $4\pi r^2$. Therefore, the intensity per unit area on this shell is given by $\frac{I_0}{4\pi r^2}$. This can be written as $\frac{C}{r^2}$ where C is just some constant, some number. This then is the essence of inverse square falloff, the intensity per unit area falls off as $1/r^2$.

4.2.4 TG43 Formalism and Air Kerma Strength

In Section 4.2.2 we considered exposure from a brachytherapy source in air, but that is not useful clinically. Of more interest is the dose in tissue. We need a formalism to connect these two quantities together. The modern formalism for dose calculation is provided by AAPM Task Group #43 (Nath et al. 1995) which has since been updated several times to include new sources and slight modifications (e.g. Rivard et al. 2004).

TG43 formalism is based on kerma which stands for "kinetic energy released in matter." In air, the kerma is simply the exposure (charge/mass) times the energy needed to ionize air molecules which is given the symbol (W/e). The air kerma rate then is $\dot{K} = \dot{X}\left(\frac{W}{e}\right)$. TG43 defines a special quantity air kerma strength, S_k, which is simply the air kerma rate at some distance l, taken to be 1 m. We then have $S_k = \dot{K}l^2 = \dot{X}\left(\frac{W}{e}\right)l^2$. In dry air one can calculate the numerical quantities: $\left(\frac{W}{e}\right)l^2 = 0.876 \text{ cGy} \cdot \text{cm}^2/\text{R}$ and since $\dot{X}$ has units of R/hr we have $S_k = 0.876 \cdot \dot{X} \frac{\text{cGy} \cdot \text{cm}^2}{\text{hr}}$. There is a special unit for air kerma strength, "u", i.e $u = \frac{\text{cGy} \cdot \text{cm}^2}{\text{hr}}$.

4.2.5 TG43 Dose Calculation Formalism

Air kerma strength, S_k, can be used to calculate dose rate. Under the TG43 formalism, dose rate is given by $\dot{D} = S_k \cdot \Lambda \cdot G \cdot g \cdot F$. Here Λ is the dose rate constant which is a quantity that is specific to each brachytherapy source model. It is measured and calculated by manufacturers and research labs, and consensus values are published by the AAPM. The units of the dose rate constant are cGy/hr/u, and it can be seen that $S_k \cdot \Lambda$ has units of cGy/hr.

The factors G, g, and F characterize the geometry of the source and the dose distribution in tissue (Figure 4.2.4). Details can be found in the video, but the basic principles are as follows:

- G, geometry function, accounts for inverse square falloff with distance from the source, r. G can be either a point source (in which case $G = 1/r^2$) or a line source, in which case $G = \beta/Lr\sin\theta$, where β is the angle that the source subtended by the source as see from the calculation point and θ is the angle from a line perpendicular to the seed axis (see video).

- g is the radial dose function which accounts for scatter and attenuation in tissue. For a high-energy source this falls off relatively slowly with r. For lower energy sources the falloff is more rapid.

- F is an anisotropy function which accounts for the fact that photons are absorbed by the shielding capsule of the source. For points along the axis of the seed therefore the dose will be lower. F is a function of both r and θ (the angle from the perpendicular). This quantity can be averaged over θ to yield an anisotropy factor $\phi(r)$.

TG43 Brachytherapy Dose Calculations

$$\text{Dose rate} \quad \dot{D} = S_k \cdot \Lambda \cdot G \cdot g \cdot F$$

air kerma strength · dose rate constant · geometry · radial dose function · anistropy

$$\text{Total Dose} = \dot{D}_0/\lambda = \dot{D}_0 \cdot 1.44\, t_h$$
(permanent implant)

$$\text{Dose after a time, } t \quad D(t) = \frac{\dot{D}_0}{\lambda}(1 - e^{-\lambda t})$$

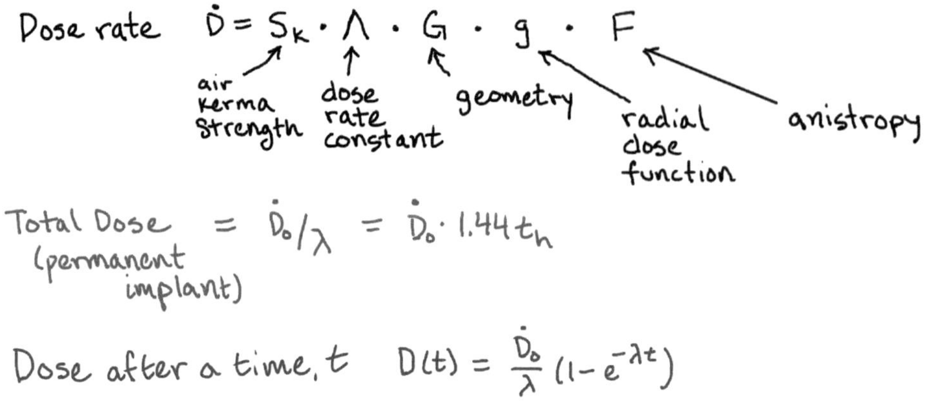

FIGURE 4.2.4
Brachytherapy dose calculation formalism from AAPM Task Group #43 (Nath et al. 1995).

For permanent implants it is useful to calculate the total dose delivered over the lifetime of the implant. This is given by $D_{\text{total}} = \dot{D}_0 \cdot 1.44 t_h$ where $\dot{D}_0$ is the initial dose rate. It is easy to show that this equation is true by calculating the integral of the dose rate $\dot{D}_0 e^{-\lambda t}$ over time from 0 to infinity. Note that $1.44\, t_h$ is the mean-life discussion in Section 3.1.4.

Further Reading

Dieterich, S., E. Ford, D. Pavord and J. Zeng. 2016. *Practical Radiation Oncology Physics*. Chapter 8. Philadelphia, PA: Elsevier.

Kahn, F.M. and J.P. Gibbons. 2014. *Kahn's The Physics of Radiation Therapy*. Edition 5. Chapter 15. Philadelphia, PA: Wolters Kluwer.

McDermott, P.N. and C.G. Orton. 2010. *The Physics of Radiation Therapy*. Chapter 16. Madison, WI: Medical Physics Publishing.

Nath, R., et al. 1995. Dosimetry of interstitial brachytherapy sources: Recommendations of the AAPM Radiation Therapy Committee Task Group No. 43. *Med Phys* 22(2):209–234.

Rivard, M.J., et al. 2004. Update of AAPM Task Group No. 43 Report: A revised AAPM protocol for brachytherapy dose calculations. *Med Phys* 31(3):633–674.

Wyckoff, H.O., et al. 1985. *ICRU Report No. 38, Dose and Volume Specification for Reporting Intracavitary Therapy in Gynecology*. Bethesda, MD: ICRU.

Chapter 4 Problem Sets

*Note: * indicates harder problems.*
For useful information on half-lives and exposure rate constants see Table 2.3. Problems 12 and on may require reference to TG43 update 2004 Report (Rivard et al. 2004). This has tables for Λ (Table I), g (Table II), and F (Tables IV–X) for the various seeds. The formula for G(r,θ) can be found (Equation 4) and a drawing for the geometry (Figure 1).

1. What is the exposure rate in air at 1 m for a ^{137}Cs source with an activity of 40 mCi?
   a. 13.1 mR/hr
   b. 33.0 mR/hr
   c. 1.31 R/hr
   d. 3.30 R/hr

2. What is the activity of a ^{192}Ir source if the measured exposure rate in air at 1 m is 5 R/hr?
   a. 0.61 mCi
   b. 1.07 mCi
   c. 6.1 Ci
   d. 10.7 Ci

3. What is the exposure rate in air at 2 m from a ^{192}Ir source assuming the exposure rate at 1 m is 5 R/hr?
   a. 0.25 R/hr
   b. 0.50 R/hr
   c. 1.25 R/hr
   d. 2.50 R/hr

4. A gynecological implant calls for a ^{137}Cs source with activity of 15 milligrams radium equivalent (mg-Ra-Eq). What would be the activity of this source in mCi?
   a. 1.80 mCi
   b. 5.96 mCi
   c. 26.4 mCi
   d. 37.7 mCi

5. What is the ratio of exposure rate in air from a 10 Ci ^{60}Co source to a 10 Ci ^{192}Ir source?
   a. 0.36
   b. 0.57
   c. 1.75
   d. 2.77

6. What physical factors account for the difference in exposure rates from a ^{60}Co vs. ^{192}Ir. (Hint: Reference to decay scheme and what emerges from the decay.)
   a. More decays per second for ^{60}Co
   b. Higher energy photons emitted in ^{60}Co
   c. Number of atoms in the ^{60}Co sample
   d. Lower linear energy transfer (LET) of photons from ^{60}Co

7. What is the initial dose rate for a ^{125}I permanent prostate implant if the prescribed total dose is 125 Gy?

   a. 6.1 cGy/hr
   b. 8.8 cGy/hr
   c. 21.2 cGy/hr
   d. 146.1 cGy/hr

8. What is the dose rate of a ^{103}Pd implant after one month (30 days) if the initial dose rate was 21.0 cGy/hr?

   a. 3.59 cGy/hr
   b. 6.18 cGy/hr
   c. 11.8 cGy/hr
   d. 14.8 cGy/hr

9. What is the air kerma strength of a source with an exposure rate of 5 R/hr at 1 meter?

   a. 0.0044 cGy cm^2/hr
   b. 0.17 cGy cm^2/hr
   c. 4.37 cGy cm^2/hr
   d. 17.5 cGy cm^2/hr

10. * Prove that the total dose in a permanent implant is given by $\dot{D_0}\cdot 1.44t_h$.

11. * What dose is delivered in 30 days from a ^{103}Pd prostate implant if the total prescription dose is 125 Gy?

    a. 0.25 Gy
    b. 0.34 Gy
    c. 70.8 Gy
    d. 88.3 Gy

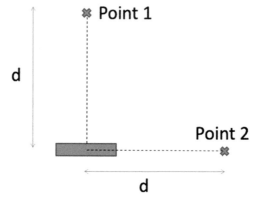

FIGURE PS4.1
Seed geometry.

12. What is the initial dose rate at point 1 shown in Figure PS4.1 from a ^{125}I Model 6711 seed with an air kerma strength of 1 U if the distance d is 1 cm? Use the line source approximation with an active source length L=3.0 mm.

    a. 0.20 cGy/hr

    b. 0.97 cGy/hr

    c. 1.32 cGy/hr

    d. 4.62 cGy/hr

13. What is the initial dose rate at point 2 shown in Figure PS4.1 from a ^{125}I Model 6711 seed with an air kerma strength of 1 U if the distance d is 1 cm? Use the line source approximation with an active source length L=3.0 mm. What accounts for the difference in the dose rate in this problem vs. Problem 12?

    a. 0.37 cGy/hr

    b. 0.85 cGy/hr

    c. 1.32 cGy/hr

    d. 2.29 cGy/hr

14. What is the initial dose rate at point 1 shown in Figure PS4.1 from a ^{125}I Model 6711 seed with an air kerma strength of 1 U if the distance d is 2 cm? Use the line source approximation with an active source length L=3.0 mm.

    a. 0.15 cGy/hr

    b. 0.20 cGy/hr

    c. 0.84 cGy/hr

    d. 0.91 cGy/hr

15. What is the initial dose rate at point 2 shown in Figure PS4.1 from a ^{125}I Model 6711 seed with an air kerma strength of 1 U if the distance d is 2 cm? Use the line source approximation with an active source length L=3.0 mm. Compare the answers to Problems 12 through 15 and explain what accounts for the differences.

    a. 0.088 cGy/hr

    b. 0.30 cGy/hr

    c. 0.97 cGy/hr

    d. 1.55 cGy/hr

16. What is the total dose delivered to point 1 in Problem 12 over the lifetime of a permanent implant?

    a. 0.57 cGy

    b. 58 cGy

    c. 1.9 Gy

    d. 19.9 Gy

17. How will the dose rate change for Problem 12 if the seed is replaced with ^{103}Pd Theragenics Model 200 seed of equal air kerma strength?

   a. Increase

   b. Decrease

   c. Stays the same

   d. Depends on anisotropy factor

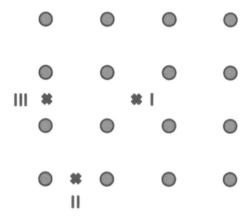

FIGURE PS4.2
Brachytherapy implant.

18. For the implant shown in Figure PS4.2 rank the dose to point I (increasing order) if seeds have the same $S_K \Lambda$ in each case. (Note: seeds are shown in blue in Figure PS4.2. The plane shown is perpendicular to the long axis of the seeds.)

   a. ^{192}Ir source (GammaMed Inc.)

   b. ^{125}I Model 6711

   c. ^{103}Pd Theragenics Model 200

19. For the implant shown in Figure PS4.2 rank the dose to points I, II, and III (increasing order).

20. * Plot the scattering and attenuation factor (g) as a function of distance r along the transverse axis ($\theta = 90$) for a ^{125}I Model 6711 seed and for a ^{103}Pd Theragenics Model 200 seed.

21. * Plot the effective inverse square law correction, $G_L(r, \theta=90)/G_L(r_0=1, \theta_0=90)$ as a function of r for (1) a line source approximation with source lengths of 3 and 5 mm, and (2) a point source.

5

Photon Interactions with Matter

Introduction: Photons and Particles in Radiation Therapy

The interaction of photons and particles with matter is fundamental to radiation therapy physics, since photons and particles are used to treat and image patients. Before considering this topic in detail it is useful to consider the broader context of photons and particles in radiation therapy. Figure 5.1.1 provides an overview of the treatment approaches for various targets (shown in red) when using external beam radiation therapy. Superficial targets are often treated directly with electron beams (see Chapter 15). Targets that are located deeper in the patient may be treated with high-energy photons (denoted with the symbol γ) or protons (see Chapter 24). There are also other more exotic modalities such as heavy ions or fast neutrons which are not shown.

If we zoom in to see how a photon beam interacts in tissue (gray circle in Figure 5.1.1) there are several processes at work. The photon may scatter off an electron and, in the process, give energy to that electron, and itself change energy and direction. It may also be absorbed. Not all photons are absorbed in the patient, however. Some are transmitted, and these are useful for imaging.

5.1 Low-Energy Photons

There are four main processes by which photons interact with matter as outlined in Table 5.1: coherent scattering, photoelectric effect, Compton scattering, and pair production. The energy of the incoming photon largely determines which process will dominate. Though there is not a sharp cutoff, it is useful to consider two energy regimes, "low-energy" (below approximately 100 keV) and "high-energy" (above approximately 100 keV). Recall that the energy of a photon can be expressed in electronvolts, eV (or keV or MeV), and this is related to wavelength. If these concepts are unfamiliar it may be useful to review Chapter 1.

5.1.1 Coherent Scattering

At the lowest energies coherent scattering is the main interaction process. In this interaction no energy is gained or lost in matter so it is not relevant for medical applications. However, it is included here for completeness.

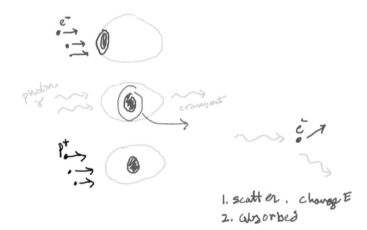

FIGURE 5.1.1
Radiation therapy treatment approaches for various targets (shown in red). Superficial targets are often treated with electron beams. Targets deeper in the patient may be treated with high-energy photons (γ) or protons.

TABLE 5.1

Photon Interaction Processes Shown in Order of Increasing Probability with Energy

Process	Description
Coherent scattering	Scatters from a charged particle (the electron). No energy gained or lost. Not relevant for medical applications.
Photoelectric effect ($\lesssim$ 100 keV)	Interacts with electron in an atom. This may excite the electron into a higher energy level or ionize the electron.
Compton scattering	Scatters from electron in an atom. Electron gains some energy from the photon and is ejected. Photon loses some energy and changes direction.
Pair production	Undergoes transformation in the presence of the nucleus to convert into an electron/positron pair.

In the coherent scattering process the incident photon interacts with an electron (Figure 5.1.2). The photon causes the electron to oscillate up and down (recall that photons can be thought of as electromagnetic waves and it is the oscillating electric field that produces a force on the electron). An accelerating charged particle always produces an electromagnetic wave. This is something we will consider in more detail in a later chapter. Therefore, as the electron oscillates, a second wave (or photon) is produced. This wave emerges in some other direction. The wavelength (or energy) of the emerging wave is equal to that of the incoming wave. That is, no energy is gained or lost in this interaction. There are two flavors of coherent scattering: Thompson scattering, in which the photon scatters from a free electron, and Rayleigh scattering, in which it scatters from an electron bound in an atom. It is the wavelength-dependent process of Rayleigh scattering from molecules in the atmosphere that makes the sky appear blue on the earth.

5.1.2 Photoelectric Effect Process

In the photoelectric effect a photon interacts with an electron in an atom (Figure 5.1.3). If the energy of the incoming photon is high enough to overcome the binding energy of the electron, the electron may be ejected from the atom. That is, the atom is ionized. This ejected electron is called a photoelectron.

Thompson scattering

Rayleigh scattering

FIGURE 5.1.2
Coherent scattering, a process which dominates at low energies. The wavelength (or energy) of the emerging photon is equal to the incident photon. Since no energy is gained or lost, this interaction is not relevant for medical applications.

The energy of the photoelectron is related to the energy of the incident photon. If the incident photon has an energy E_γ then the emerging photoelectron will have an energy $E_\gamma - E_b$, where E_b is the binding energy of the electron in the atom (Figure 5.1.3A). That is, some of the energy in the incident photon is used to liberate the electron from the atom and the rest goes into the photoelectron. Recall that electrons occupy specific, discrete quantum energy levels in the atom. The binding energy, E_b, therefore, is an energy specific to the atom and the shell that the electron is in.

After the photoelectron is ejected there will be an empty quantum level that the electron formerly occupied. Electrons from a higher energy state can therefore transition down to this energy state and release energy in the process. This energy is released in the form of a photon. This photon is called a "characteristic X-ray" (Figure 5.1.3B). That is, it is characteristic of the particular atom that it comes from. The photon has a very specific energy equal to the difference in energy of the higher vs. lower energy quantum states.

There is another type of process that can happen in the photoelectric interaction which results in the ejection of electrons at specific energies, so-called Auger electrons. This can be thought of in the following way. The characteristic X-ray, instead of leaving the atom, can interact with an outer-shell electron. If the energy of the X-ray is larger than the binding energy of the outer-shell electron, then the electron can be ejected from the atom. The electron thus ejected is called an Auger electron and emerges with very specific energy

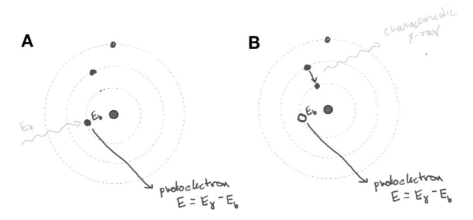

FIGURE 5.1.3
Photoelectric effect. (A) A photon incident on an atom ejects an electron from the atom. (B) An electron in an outer shell quantum level transitions to an inner shell and releases energy in the form of a photon, a characteristic X-ray.

equal to the difference between the energy of the characteristic X-ray and the binding energy of the outer-shell electron. The problems provide a few examples of the energetics of this interaction.

To summarize then, the following particles may be produced in a photoelectric interaction:

- Photoelectrons
- Auger electrons
- Characteristic X-rays
- Ionized atoms

5.1.3 Photoelectric Effect: Interaction Probabilities

The previous section presented what happens during the process of photoelectric interaction. Now we consider how often such interactions happen. As we will see, this depends sensitively on the energy of the incident photon and the atomic number of the atoms that it is interacting with.

Figure 5.1.4 plots a quantity that is proportional to the probability of interaction vs. the energy of the incident photon. The quantity shown is the mass absorption coefficient in units of cm²/g, a quantity which will be considered in more detail in Section 6.1.1. The photoelectric effect is the most dominant type of photon interaction below approximately 100 keV (diagnostic energy range). In this energy range the mass absorption coefficient decreases as energy increases. The scaling is approximately $1/E^3$, where E is the energy.

The photoelectric interaction probability depends sensitively on the material type. Figure 5.1.4 shows the mass absorption coefficient for interactions of the photons with muscle (red). For lead, the plot looks quite different. First, overall the mass absorption coefficient is much larger for lead than muscle. This is because of the higher atomic number of

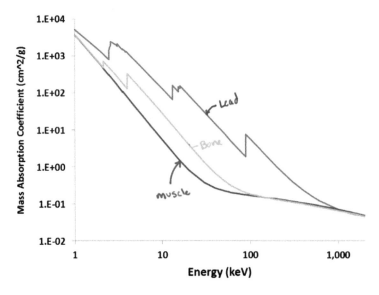

FIGURE 5.1.4
Mass absorption coefficient vs. energy for three different materials: muscle, lead, and bone.

lead ($Z = 82$) vs. muscle. Note that the mass absorption coefficient scales as approximately Z^3. Also notable in the curve for lead is the presence of various "edges," i.e. discontinuities where the mass absorption coefficient suddenly changes. These are due to the various principle quantum numbers of electrons in the atom ($n = 1, 2, 3...$) which in X-ray spectroscopy are labeled as K (the highest binding energy), L, M, and so on. To understand this, consider the K edge. If an incident photon has an energy less than the K edge energy it will not be able to ionize an electron in the K shell. However, as soon as the energy of the photon exceeds this threshold the electron can be liberated. The mass absorption coefficient, therefore, suddenly increases since more photons can be absorbed. This accounts for the sudden discontinuities in the mass absorption coefficient. Aside from the shell transitions the mass absorption coefficient for lead also decreases like $1/E^3$. Note that there are no K or L shells visible on the curves for muscle because it is composed mostly of carbon, oxygen, and hydrogen which are low-Z materials and have binding energies which are very small.

In the plot for lead in Figure 5.1.4 it is clear that the L and M edges are not single jumps but have a structure to them. This is due to the orbital subshells that are present within the principle quantum number (recall from chemistry the atomic structure $1s^2 2s^2 2p^2$, etc.). Each of these subshells has slightly different energies. So, for example, there are L_1, L_2, and L_3 shells each at a slightly different energy. Data on X-ray absorption coefficients and energy levels can be found on a variety of websites (see Section 5.3).

Figure 5.1.4 also shows the mass absorption coefficient bone (green curve). The curve for bone is higher than muscle because bone has a higher effective Z, Z_{eff}, i.e. the effective atomic number found by averaging together all the elements in the material. For bone $Z_{eff} \approx 12.3$, while for muscle $Z_{eff} \approx 7.6$ and for lead $Z = 82$. The mass absorption coefficient curve for bone also shows K and L shell edges at approximately 4 keV and 0.3 keV. This is due to the calcium ($Z = 20$) in the bone. Aside from the shell transitions the mass absorption coefficient for bone, like all materials, decreases like $1/E^3$ in the low-energy regime where the photoelectric process dominates. **Overall, photoelectric absorption is much more likely at low energy and high Z and the mass absorption coefficient scales roughly as $\sim Z^3/E^3$.**

The properties and dependencies of the photoelectric effect have many implications which will be seen throughout this text. One example is contrast in diagnostic imaging (Chapter 19). Diagnostic X-ray imaging operates at low energies (less than approximately 100 keV), and in this regime high-Z materials attenuate more (the $\sim Z^3/E^3$ scaling). Therefore, there will be a much larger difference in the attenuation in muscle vs. bone, and this will be accentuated as the energy is lowered. That is, the contrast between bone vs. muscle is highest at low energies. This is just one of many example applications of the basic principles outlined in this section.

5.2 Interaction of High-Energy Photons

5.2.1 Compton Scattering

As the energies of photons start to exceed approximately 100 keV the photoelectric effect becomes less important and photons tend to interact more through the Compton scattering process. This effect was discovered in 1923 by Arthur Compton (1892–1962), an American physicist who was awarded the Nobel Prize in 1927 for this discovery.

In Compton scattering an incoming photon interacts with an electron in the atom and scatters off at some angle, ϕ (Figure 5.2.1). The electron is also scattered off at some angle, θ, and has energy. Equations can be written for the energies and angles involved. Though these equations

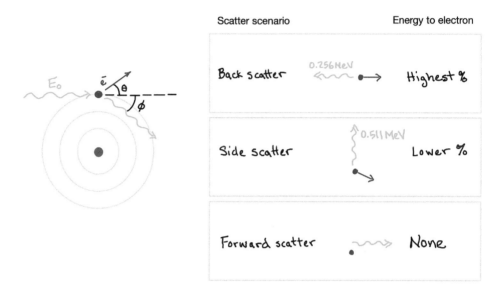

FIGURE 5.2.1

Compton scattering. (Left) An incident photon of energy E_0 scatters with an electron in an atom. (Right) In the high-energy regime relevant to radiation therapy ($E_0 \gg 0.511$ MeV) the equations predict for the scattering scenarios shown.

are beyond the scope of this text, they are not difficult to derive and rely on two fundamental principles of physics, namely conservation of energy (energy after the interaction =energy before) and conservation of momentum (momentum after the interaction =momentum before). For our purposes the most important dependency is the energies: the initial energy of the incoming photon, E_0, and the energies of the photon and electron that emerge.

The equations that describe Compton scattering become simplified if we consider a high-energy regime, i.e. when the energy of the incident photon is much higher than the rest mass of the electron (i.e. $E_0 \gg m_e c^2$, note that this is a good approximation for therapy energies since E_0 is usually larger than 0.511 MeV). In this regime the scattering scenario is greatly simplified as shown graphically in Figure 5.2.1. It is illustrative to consider scattering at three canonical angles and the energies involved:

- Backscatter. The incident photon scatters directly backward, and the electron goes directly forward. This is a "head-on collision." This is the case where the maximum amount of energy is given to the electron. In this case it can be shown from the equations that the energy of the scattered photon is 0.256 MeV (i.e. half of the rest mass of the electron) and the energy of the scattered electron is $E_0 - 0.256$.

- Side scatter. The incident photon scatters at 90 degrees. In this case the energy of the scattered photon is 0.511 MeV (i.e. the rest mass of the electron), and the energy of the scattered electron is $E_0 - 0.511$.

- Forward scatter. In this case the incident photon does not give any energy to the electron but scatters in the forward direction. The energy of the scattered photon is E_0 and the energy of the electron is 0.

In a real therapy beam there is not, of course, just one isolated Compton scattering event but rather a cascade of multiple scattering events (Figure 5.2.2). The first Compton

$$range\,(cm) \simeq \frac{E\,(MeV)}{2}$$

FIGURE 5.2.2
Multiple scattering events. Electrons produced by Compton scattering travel some distance in material and can produce other photons of their own.

scattering event produces an electron and photon. The photon may travel some distance and then undergo other Compton scattering events. The electron may wander some distance in the material, undergoing Coulomb-mediated scattering events as it does. It may also produce photons through the process of Bremsstrahlung radiation which will be described further in Section 7.1. Eventually all of the energy from the incident photon will have been absorbed in the material. **There are two aspects of this process which are key concepts:**

1) **It is the electrons that deposits the dose in the material through charged particle interactions. Photons themselves do not directly deposit dose.**

2) **The dose is not deposited directly at the site of the first interaction but rather is spread out over some relatively large distance. For a sense of scale, note that the range of an electron in cm is roughly equal to half of its energy in MeV. Range (cm) ≈ E(MeV)/2.**

5.2.2 Compton Scattering: Interaction Probabilities

The previous section presented what happens in the process of Compton scattering. Now we consider *how often* such interactions happen. The dependencies are quite different than for the photoelectric effect.

The Compton scattering probability depends somewhat on the energy of the incident photon and, more importantly, depends on the number of electrons per unit gram in the material (more electrons results in more scattering events). Materials with higher hydrogen content have more electrons per unit gram. This is because the hydrogen atom has one electron per nucleon (i.e. one electron and one proton) whereas most other elements have one electron per two nucleons (i.e. an electron, a proton, and a neutron). Also recall that higher-Z elements have relatively fewer protons since these elements tend to be more neutron-rich (Chapter 2). This means they also have fewer electrons. **The overall result is that there are more electrons per gram in low-Z, hydrogen-rich materials such as plastics, water, or tissues compared to high-Z materials like bone or metals. This is a key concept.** It means that the Compton interaction has a higher probability in low-Z materials.

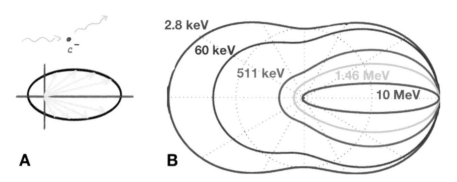

FIGURE 5.2.3
Angular dependence of Compton scattering. (A) Polar plots show the probability of scattering. Length of the arrow (yellow) is proportional to the probability of scattering in that direction. (B) Polar plots for Compton scattering show a strong energy dependence. (Adapted from https://commons.wikimedia.org/wiki/File:Klein-Nishina_distribution.png under common use.)

5.2.3 Compton Scattering: Directionality Dependence

Now we consider the question of what scattering angle is most likely in a Compton scattering event. The answer to this question, as we will see, depends sensitively on the energy of the incident photon. The theoretical treatment of this is given by the Klein–Nishina formula, which is beyond the scope of this text, but the key features can be understood graphically.

Figure 5.2.3 shows polar plots for scattering probability. To read these plots imagine drawing an arrow from the origin to the line on the plot at the angle of interest as shown in Figure 5.2.3A. The length of the arrow corresponds to the probability of scattering in this direction. Figure 5.2.3B shows the polar plots for Compton scattering.

Figure 5.2.3B shows that the scattering depends sensitively on energy. At low energies (e.g. 1 keV) the scattering is relatively symmetric. That is, it is as likely for a photon to scatter forward as to scatter backward. However, at higher energies this probability becomes more asymmetric. At the highest energies shown (e.g. 10 MeV) it is much more likely for a photon to be Compton scattered in the forward direction rather than the backward direction. **This is another key concept, i.e. at high energies Compton scattering is more forward directed.**

5.2.4 Pair Production

The final photon–matter interaction that we consider is pair production which happens at the highest energies. This is a quantum mechanical interaction of the photon in the field of the nucleus (Figure 5.2.4). The photon is converted to a pair of particles: one electron (e^-) and one positron (e^+). Recall that the positron is the positively charged antiparticle of the electron. Because of conservation of energy, the total energy of the pair that is created is equal to the energy before (i.e. the energy of the incident photon). This means that the energy of the incident photon must be at least twice the rest mass energy of the electron ($E_0 > 1.022$ MeV). If it is not then there will not be enough energy to create the electron/positron pair. The total kinetic energy of the pair that is created is $E_0 - 1.022$ MeV. The probability of pair production increases rapidly with energy above 1.02 MeV and is also larger for higher Z nuclei.

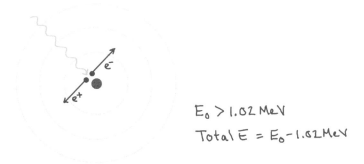

$$E_0 > 1.02 \, MeV$$
$$Total \, E = E_0 - 1.02 MeV$$

FIGURE 5.2.4
Pair production process. A photon interacts in the field of the nucleus and produces an electron/positron pair.

5.2.5 Interaction Cross-Sections: Putting It All Together

Figure 5.2.5 plots the mass attenuation coefficient for photons interacting in water through all of the above three processes. At low energies the photoelectric interaction dominates, but above about 30 keV the Compton interaction begins to be important. Above 100 keV almost all the interaction is from Compton scattering. Finally, at the highest energies pair production plays a role as well.

There is also a strong Z-dependence to these interactions, and this can be appreciated in Figure 5.2.6 which plots the mass absorption coefficient for titanium along with water. The photoelectric interaction is larger for titanium than water, especially above the K edge of titanium. Compton scattering, however, is approximately equal in titanium vs. water. It is slightly larger in water due to the larger number of electrons per gram in this low-Z, hydrogen-rich material. Finally, pair production again plays a role at the highest energies and is larger in titanium ($Z = 22$) than water ($Z_{eff} = 7.2$) due to the Z dependence.

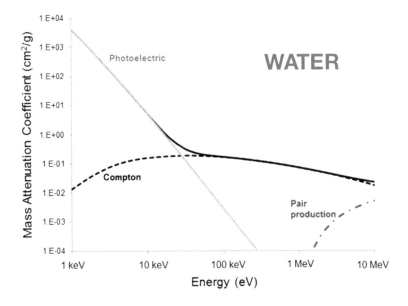

FIGURE 5.2.5
Mass attenuation coefficient for photons interacting in water.

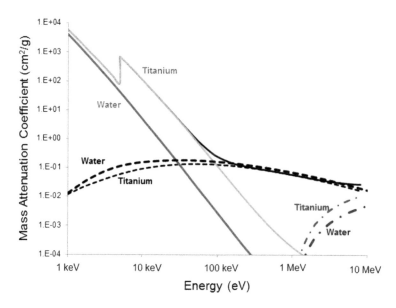

FIGURE 5.2.6
Mass absorption coefficient for photons interacting in titanium and water. Interactions are photoelectric (blue), Compton scattering (dashed black), or pair production (dashed red).

5.2.6 Photonuclear Reaction

One final interaction of note is photonuclear reaction (or photodisintegration). The interaction of most importance for therapy physics is a high-energy photon (>10 MeV) interacting with a nucleus to eject one of the neutrons. This can happen in high-Z material in and around the head of a linear accelerator. This creates an extra neutron component which may need to be considered in shielding for high-energy linear accelerators (Section 25.3).

5.3 Useful Reference Information: Photon Data

- www.nist.gov/pml/data/xraycoef/

  The website from the National Institute of Standards and Technology (NIST) from the US Department of Commerce contains very useful materials data and mass attenuation coefficients for different materials. Tables 3 and 4 give mass attenuation coefficients and Tables 1 and 2 provide data on the density of materials.

- http://xdb.lbl.gov/xdb.pdf

  The X-ray data booklet from Lawrence Berkeley Labs provides information on shell structures. See page 91.

- https://physics.nist.gov/PhysRefData/XrayTrans/Html/search.html

  This NIST website also provides a database of X-ray transitions.

Further Reading

Kahn, F.M. and J.P. Gibbons. 2014. *Kahn's The Physics of Radiation Therapy*. 5th Edition. Chapter 5. Philadelphia, PA: Wolters Kluwer.

McDermott, P.N. and C.G. Orton. 2010. *The Physics of Radiation Therapy*. Chapter 6. Madison, WI: Medical Physics Publishing.

Metcalfe, P., T. Kron and P. Hoban. 2007. *The Physics of Radiotherapy X-rays and Electrons*. Chapter 2. Madison, WI: Medical Physics Publishing.

Podgorsak, E.B. 2016. *Radiation Physics for Medical Physicists*. 3rd Edition. Chapter 7. Switzerland: Springer.

Chapter 5 Problem Sets

*Note: * indicates harder problems.*
See Section 5.3 for references to data of mass attenuation coefficients.

1. Which of the following is the highest energy of characteristic X-rays emitted after a 30 keV X-ray interacts with iodine? (Consider only the K, L1-3, and M1 shells. For energies consult Table 3 in www.nist.gov/pml/data/xraycoef/.)
   a. 1.07 keV
   b. 4.12 keV
   c. 5.19 keV
   d. 30 keV

2. What is the highest energy of Auger electron emitted after a 30 keV X-ray interacts with iodine? (Consider only the K, L1-3, and M1 shells.)
   a. 1.07 keV
   b. 3.05 keV
   c. 28.9 keV
   d. 30.0 keV

3. What is the ratio of mass attenuation coefficients of iodine vs. water at 30 keV?
   a. 1.5
   b. 3.2
   c. 9.7
   d. 22.8

4. What is the ratio of mass attenuation coefficients of iodine vs. water at 50 keV?
   a. 0.85
   b. 1.0
   c. 20.9
   d. 54.3

5. Which material has the largest mass attenuation coefficient at 2 MeV?
   a. Water
   b. Muscle
   c. Bone
   d. Titanium
   e. They are the same

6. At which incident photon energy and material is Compton scattering most probable?
   a. 2 MeV, muscle
   b. 20 keV, muscle
   c. 2 MeV, titanium
   d. 20 keV, titanium
   e. a and c are the same
   f. b and d are the same

7. In which material will pair production be the largest at 2 MeV?
   a. Water
   b. Muscle
   c. Bone
   d. Lead
   e. They are approximately the same

8. Approximately how far does an electron travel in tissue after undergoing a Compton scattering interaction that leaves it with a kinetic energy of 1.5 MeV?
   a. 0.8 mm
   b. 1.5 mm
   c. 0.8 cm
   d. 1.5 cm

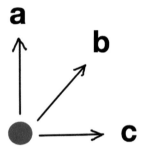

FIGURE PS5.1
Electron scattering.

9. * In Figure PS5.1, an electron (blue) is scattered in one of three directions (a, b, or c) after interacting with a 2 MeV photon which is incoming from the left. Draw the direction of the emerging photon corresponding to a, b, and c. Match the energy of the scattered electron with the direction (a, b, c) choosing from the following list of energies: 0, 0.47, 1.77, 2 MeV.

    a. _____

    b. _____

    c. _____

10. * For the incident spectrum shown in Figure PS5.3, sketch the spectrum of photons scattered in the beam direction (i.e. scatter angle 0) and the spectrum of photons scattered perpendicular to the beam direction (i.e. scatter angle 90). Which spectrum is softer?

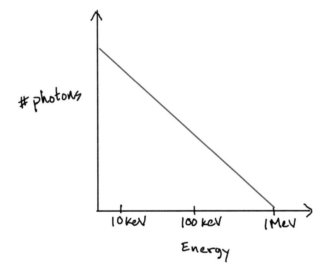

FIGURE PS5.3
Photon spectrum.

6

Photon Beams, Dose, and Kerma

6.1 Beam Attenuation and Spectra

The previous chapter presented a microscopic view of the interactions of photons with matter. Now we move to a more macroscopic level and consider how these interactions relate to measurable quantities in clinical applications. This will apply to photon beams used for both imaging and therapeutic applications.

6.1.1 Photon Beams: Exponential Attenuation

In Chapter 5 we used a quantity called the "mass attenuation coefficient" (e.g. Figure 5.1.4). This quantity was related to the probability of interaction of the photon. Now we consider in greater detail the meaning of the mass attenuation coefficient and how it relates to the properties of a photon beam.

Consider a situation in which there are a total number of N photons incident on a slab of some material of thickness Δx (Figure 6.1.1). The number of photons lost, ΔN, is proportional to the thickness of the material, Δx, and the total number of photons incident, N. That is, for more photons in more photons will be lost or, alternatively, the fractional number of photons lost $\Delta N/N$ is constant. The proportionality constant is called μ, the linear attenuation coefficient, and note that there is a minus sign in the equation denoting that photons are lost from the beam. This equation can be written in calculus notation as $dN/dx = -\mu N$. The solution for this equation is $N = N_0 e^{-\mu x}$. Note that instead of considering the number of photons, N, we can consider the number of photons per second, $\dot{N}$, which is a more natural quantity to consider since it is closer to something that can be measured. This is referred to as the intensity, I, the number of photons per second. Therefore, we have:

$$I = I_0 e^{-\mu x} \tag{6.1}$$

Here I is the intensity transmitted through some thickness x, if I_0 is the incident intensity. Equation 6.1 is known as Beer's Law (or the Lambert–Beer Law) and describes exponential attenuation. Note that Equation 6.1 is analogous to Equation 3.2 for exponential decay of the activity of a radioactive source, i.e. $A = A_0 \cdot e^{-\lambda t}$. Here instead of λ we have μ and instead of time, t, we have distance, x. The mathematical derivations, however, are identical.

Note that the units of μ are 1/cm, i.e. the fractional number of photons removed per unit cm. The units work out so that the quantity in the exponent, $\mu \cdot x$, is unitless as it should be, since a unit like $e^{(cm)}$ would be nonsensical. A more common unit than linear attenuation coefficient, μ, is a quantity called the mass attenuation coefficient (μ/ρ), which is simply the linear attenuation coefficient divided by the density of the material, ρ. The mass

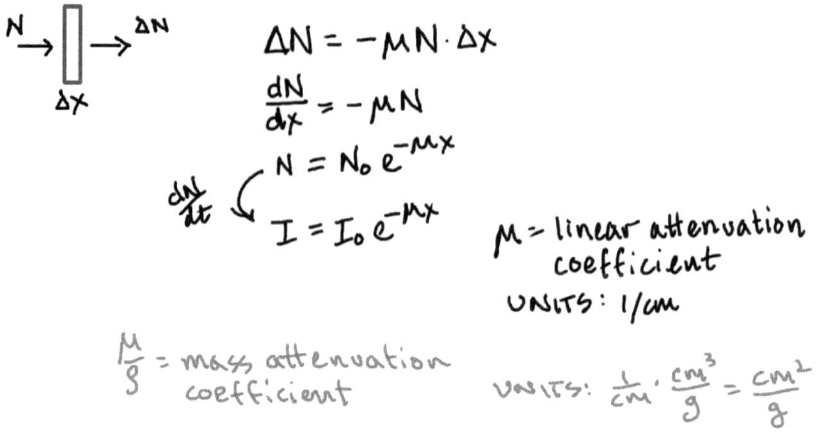

$$N \rightarrow \boxed{} \rightarrow \Delta N$$
$$\Delta x$$

$$\Delta N = -\mu N \cdot \Delta x$$

$$\frac{dN}{dx} = -\mu N$$

$$N = N_0\, e^{-\mu x}$$

$$\frac{dN}{dt} \Bigg\{ \quad I = I_0\, e^{-\mu x}$$

μ = linear attenuation coefficient

UNITS: $1/cm$

$\frac{\mu}{\rho}$ = mass attenuation coefficient

UNITS: $\frac{1}{cm} \cdot \frac{cm^3}{g} = \frac{cm^2}{g}$

FIGURE 6.1.1

Attenuation in an idealized beam. The intensity of photons, I (photons per second) transmitted through a thickness of material, x.

attenuation coefficient is an intrinsic quantity, depending only on the material type and not on how much of the material is present, so it is a more fundamental quantity. Equation 6.1 can then be rewritten using $(\mu/\rho) \cdot \rho$ in place of μ.

$$I = I_0 e^{-\left(\frac{\mu}{\rho}\right) \cdot \rho \cdot x} \tag{6.2}$$

The mass attenuation coefficient of various materials is tabulated and the density is known, so Equation 6.2 can often be calculated. I is the intensity transmitted. To find the intensity absorbed we would calculate $I_0 - I$ which is $I_0\,(1 - e^{-(\mu/\rho)\cdot\rho\cdot x})$.

The above derivation and equations are for an idealized beam. In reality, a purely exponential attenuation is almost never observed in a photon beam because there are other physical processes at work. First is the scatter of photons within the beam. This contributes an intensity that is not well-described by Equation 6.2. Second are the effects of the energy spectrum of the beam. The beam is usually not typically monoenergetic (one energy), so there is not one single value of (μ/ρ) for the beam (not that μ/ρ depends on the energy). These complexities are important and will be considered in more detail in future chapters. The key point is that exponential attenuation is a useful construct for learning but realistic photon beams have a more complex behavior.

6.1.2 Half-Value Layer (HVL) and Tenth-Value Layer (TVL)

The half-value layer (HVL) is the thickness of material after which the beam intensity is reduced to half of its initial value. Each material will have a particular HVL for a particular energy of photon. The HVL is related to the ability of that material to attenuate photons (μ/ρ) and its density, ρ. For the idealized beam considered above the HVL can be found by setting the final intensity, I, in Equation 6.1 equal to half of the initial intensity $(\frac{1}{2}I_0 = I_0 e^{-\mu x})$ and then solving for x. This yields $x = 1\,\mathrm{HVL} = \dfrac{\ln 2}{\mu}$ or:

$$\mathrm{HVL} = \frac{0.693}{\mu} \tag{6.3}$$

Note that the linear attenuation coefficient, μ, depends on the energy of the photon, as described in Section 5.2.5, so the HVL also depends on the energy of the photon. At the energies where the photoelectric effect dominates ($\lesssim$100 keV) μ increases as energy decreases so HVL decreases as energy decreases. HVL also, of course, decreases as the density of the material increases.

The equation for transmitted intensity can be written in terms of the number of half-value layers as:

$$I = I_0 \left(\frac{1}{2}\right)^{\# \text{HVL}} \tag{6.4}$$

Equation 6.4 and Equation 6.1 may look different in form but are equivalent. This can be proven mathematically.

Similar to half-value layer is the concept of tenth-value layer (TVL), the thickness at which the intensity is reduced to one-tenth of its initial value. Equations 6.3 and 6.4 become TVL $= \ln(10)/\mu$ and $I = I_0(1/10)^{\#\text{TVL}}$.

Note that Equation 6.3 is analogous to Equation 3.1 for half-life, t_h, in the decay of the activity of a radioactive source, i.e. $t_h = 0.693/\lambda$. Here instead of λ we have μ. Also Equation 6.4 is analogous to Equation 3.3, i.e. the activity of a radioactive source after n half-lives which is $A_0 \cdot (1/2)^n$. Though the physical processes are very different in radioactive decay vs. photon beam attenuation, the mathematical derivations are identical.

6.1.3 Z-Dependence in the Compton Regime

The photoelectric effect dominates at low energies below approximately 100 keV, and here μ decreases as energy increases and increases as atomic number, Z, increases (recall the Z^3/E^3 dependence of the photoelectric effect in Section 5.1.3). At higher energies, where Compton scattering dominates, there is a weaker dependence on atomic number, Z, but this dependence is important to understand.

Recall that Compton scattering occurs from the electrons in the material. Therefore, materials with more electrons per unit gram produce more Compton scattering. The number of electrons per gram decreases with increasing atomic number, Z, as shown in Table 6.1. There are two effects at work here. First, materials with a high hydrogen content have more electrons per gram because the hydrogen atom has one electron per nucleon whereas most other elements have one electron per two nucleons (i.e. proton and neutron).

TABLE 6.1

Materials, Atomic Number or Effective Atomic Number, and Number of Electrons per Gram

Material	Z or Z_{eff}	# Electrons/gm ($\times 10^{23}$)
Hydrogen	1	5.97
Fat	6.46	3.34
Water	7.51	3.34
Muscle	7.64	3.31
Air	7.78	3.01
Bone	12.3	3.19
Lead	82	2.38

Adapted from Table 5-3, H. Johns and J. Cunningham, *Physics of Radiology*, 1983.

Second, higher-Z elements tend to be more neutron-rich (Chapter 2) which means they have relatively fewer protons and relatively fewer electrons. The overall result is that there is relatively less Compton scattering in high-Z materials like bone or metals. **This is a key concept, and it means that in the Compton regime the mass attenuation coefficient decreases slightly with increasing Z.**

6.1.4 Beam Hardening and Attenuation

Here we consider a photon beam passing through a material. Figure 6.1.2 shows an energy spectrum from an example photon beam from a diagnostic X-ray tube (see Section 8.1 for more detail). The spectrum of the beam incident on the material is shown in red. The low-energy photons in this beam experience more attenuation since the mass attenuation coefficient is larger at low energies. Figure 6.1.2 also shows a plot of the mass attenuation coefficient of lead. As the beams passes through some t of material there are fewer low-energy photons present because of the larger attenuation coefficient. Photons in the

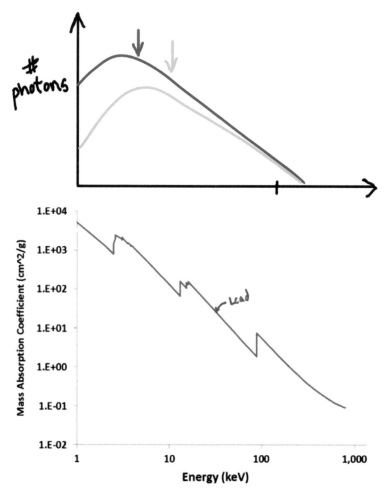

FIGURE 6.1.2
Energy spectrum of a photon beam from a diagnostic X-ray tube. The spectrum incident on the material (red) and after passing through the material (green).

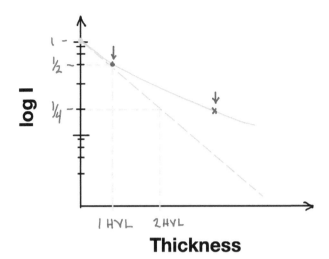

Thickness

FIGURE 6.1.3
Intensity of a low-energy photon beam as it penetrates through material. Without beam hardening the intensity would follow the dashed line. However, as the beam penetrates through material it gets harder and there is less attenuation (the HVL increases).

beam at higher energies, however, are relatively less attenuated. Therefore, the spectrum changes as the beam passes through the material. The spectrum after passing through some distance of material is shown in green in Figure 6.1.2 (top). Note that the average energy of photons in the beam shifts to higher energy (arrows) as the beam passes through more material. **This is referred to as beam hardening and is a key concept.** The more material the beam traverses the harder it becomes.

6.1.5 Beam Hardening: Effect on HVL

Beam hardening results in a change in the HVL as the beam penetrates through material. This can be visualized as shown in Figure 6.1.3. If the beam were monoenergetic or were not to harden then the intensity curve would follow the dashed line in Figure 6.1.3. Note that on a log-linear plot Equation 6.1 describes a line. After a thickness of one HVL the beam intensity decreases to half of its initial intensity.

As the spectrum of the beam hardens the average energy shifts to a higher energy. This means that the average mass attenuation coefficient is reduced, since the mass attenuation coefficient is smaller at higher energy. Alternatively, this means that the HVL is larger, since HVL and μ are inversely related as can be appreciated from Equation 6.3. The result is that the HVL becomes progressively larger as the beam penetrates through material and the intensity vs. thickness curve decreases less rapidly.

6.2 Dose and Kerma

Dose is a crucial concept in radiation therapy and medical physics, and a closely related quantity is kerma, "kinetic energy released per unit mass." This section explores these two quantities and their relationship.

6.2.1 Dose and Kerma in a Photon Beam

The units of both dose and kerma are the same, energy per unit mass. The SI units are J/kg which has the special name Gray, symbol Gy. One also often sees cGy in the radiation therapy context since this is a meaningful amount of dose.

Kerma is the kinetic energy released when a photon enters material. Recall that at therapy energies this usually occurs through the Compton scattering process (Section 5.2.1) where some of the energy of the photon is given to an electron in the form of kinetic energy. The photon can undergo further Compton scattering, and the electron can make photons of its own that then Compton scatter. All this happens in a cascading process, and energy is released in the form of moving charged particles, i.e. electrons (Figure 5.2.2). This is kerma.

To be exact, here we are considering the collisional kerma of the electron moving through the medium. That is, the energy that is deposited through charged-particle interactions as they move through the medium. There is another energy release mechanism called radiative energy loss in which the electron can produce a photon. This process, however, does not deliver dose since the photon itself does not deposit dose. Therefore, here we consider collisional kerma and not radiative kerma. More background about these energy loss processes will be presented in Chapter 7.

Now we consider the quantity dose. The dose is the energy absorbed at a particular location. It is similar to kerma, but different because the location where the energy is released (kerma) may not be the location where the energy is absorbed (dose). This is because the high-energy electrons wander some macroscopically large distance as they deposit dose (recall the cartoon in Figure 5.2.2).

Figure 6.2.1 shows one manifestation of this in terms of the plot of percent depth dose (PDD). Here kerma (red line) is seen to be steadily decreasing with depth. This is because photons are absorbed and removed as the beam penetrates through material. Gradually fewer photons with depth means there is gradually less kerma with increasing depth. The curve for dose, however, does not exactly track the kerma curve. That is because kerma released at some depth, say 10 cm, is not absorbed exactly at that depth. Rather the electrons travel downstream and deposit their dose more distally in the beam direction. Therefore, the dose and kerma curves are offset, and the amount of offset depends on the

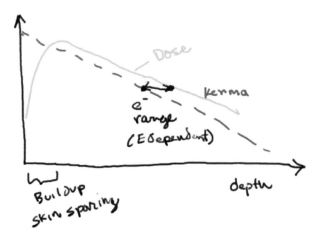

FIGURE 6.2.1
Dose and kerma as a function of depth in a high-energy photon beam.

range of the electrons in tissue. Higher energy beams result in higher energy electrons which have a longer range and therefore the offset is more.

6.2.2 Electronic Equilibrium and Buildup

Figure 6.2.1 shows an important effect, namely the build up of dose at superficial depths. This is important because it results in a skin-sparing effect which is one of the key characteristic advantages of therapy with a high-energy beam.

This effect can be understood from the cartoon Figure 6.2.2. Here three photons are incident on the surface of the patient. They undergo Compton scattering near the surface and produce high-energy electrons (dashed line). These electrons wander through the medium, depositing dose as they go. At some point, all the energy of the electrons has been dissipated and they stop.

Counting the number of electrons that cross each layer in depth (dotted lines in Figure 6.2.2), we observe that the first layer has three electrons, the second layer has six, the third layer six, and so on. If we take it that the number of electrons is proportional to the dose deposited at that layer, which is approximately true, then the dose starts out low near the surface and increases with depth. **This is called dose buildup and is a key concept.** Beyond a certain depth the number of electrons leaving a particular layer is balanced by the number of electrons entering that layer. **This is called electronic equilibrium and is another key concept.**

Now we can appreciate the dose buildup effect in Figure 6.2.1. At or near the surface the dose is low because the photons themselves do not deposit dose. They have to first produce electrons which deposit dose. As one goes deeper in the medium, the dose builds up until there is an equilibrium. At depths beyond this, the dose begins to decrease due to the attenuation of photons. The point where equilibrium is first reached and the dose is maximum is called the depth of maximum dose or d_{max}. (Note that d_{max} is a depth in cm and not a dose. It is the depth where the dose reaches a maximum.) This is a key quantity and something that we will revisit in Section 10.1 when considering the properties of megavoltage photon beams in more detail.

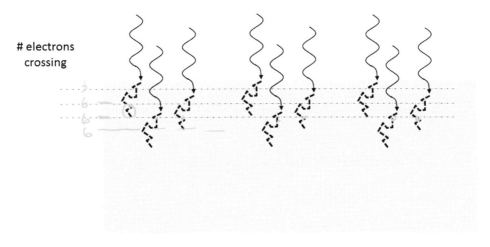

FIGURE 6.2.2
Dose buildup. Photons entering a medium undergo Compton scattering to produce high-energy electrons (dashed lines). These electrons wander through the medium and deposit dose.

Note that the d_{max} depth will depend on the energy of the beam. This is because higher energy beams produce higher energy electrons which travel farther in the medium. It therefore requires a longer distance for equilibrium to be reached.

Further Reading

Kahn, F.M. and J.P. Gibbons. 2014. *Kahn's The Physics of Radiation Therapy*. 5th Edition. Chapters 5 and 8. Philadelphia, PA: Wolters Kluwer.

McDermott, P.N. and C.G. Orton. 2010. *The Physics of Radiation Therapy*. Chapters 6 and 7. Madison, WI: Medical Physics Publishing.

Metcalfe, P., T. Kron and P. Hoban. 2007. *The Physics of Radiotherapy X-rays and Electrons*. Chapters 2 and 4. Madison, WI: Medical Physics Publishing.

Podgorsak, E.B. 2016. *Radiation Physics for Medical Physicists*. 3rd Edition. Chapters 7 and 8. Switzerland: Springer.

Chapter 6 Problem Sets

*Note: * indicates harder problems.*
See Section 5.3 for references to data of mass attenuation coefficients.

1. How does the linear attenuation coefficient depend on the material density, ρ?
   a. Decreases linearly with density (ρ)
   b. Decreases linearly as ρ^3
   c. Increases linearly with density (ρ)
   d. Increases linearly as ρ^3
   e. Does not depend on density (ρ)

2. What is the percent transmission of a 30 keV photon through 2 cm of cortical bone?
   a. 0.6%
   b. 6.0%
   c. 86.6%
   d. 92.8%

3. What is the percent transmission of a 3 MeV photon through 2 cm of cortical bone?
   a. 6.0%
   b. 23.9%
   c. 86.6%
   d. 92.8%

4. Rank the following materials in order of percent attenuation per cm for a 30 keV beam: water, cortical bone, breast tissue.

5. Rank the following materials in order of percent attenuation per cm for a 3 MeV beam: water, cortical bone, breast tissue. Explain the difference in relative ordering in this problem vs. Problem 4.

6. What is the half-value layer in aluminum for a monoenergetic X-ray beam at 20 keV?
   a. 0.13 mm
   b. 0.75 mm
   c. 1.1 mm
   d. 2.8 cm

7. What is the half-value layer in aluminum for a monoenergetic X-ray beam at 40 keV?
   a. 1.1 mm
   b. 4.5 mm
   c. 2.8 cm
   d. 4.0 cm

8. What is the HVL in tungsten for a monoenergetic photon beam at 1.25 MeV?
   a. 0.25 cm
   b. 0.64 cm
   c. 0.85 cm
   d. 1.23 cm

9. How many HVLs of tungsten are required to attenuate a high-energy photon beam down to 3% of its initial intensity?
   a. 1 HVL
   b. 3 HVL
   c. 5 HVL
   d. 7 HVL

10. At a depth of 5 cm in tissue which beam exhibits the largest difference in dose and kerma?
    a. Orthovoltage beam at 300 kVp
    b. ^{60}Co beam (average photon energy 1.25 MeV)
    c. Therapy beam from a linear accelerator at 18 MV
    d. All are approximately the same

11. In what combination of photons beams and tissues is the depth of maximum dose the largest?
    a. 6 MV therapy beam in lung
    b. 6 MV therapy beam in muscle
    c. 18 MV therapy beam in lung
    d. 18 MV therapy beam in muscle

7

Particle Interactions with Matter

One remarkable example of charged particle interaction is the aurora borealis, the beautiful display of lights seen at high northern latitudes on earth. In the aurora borealis charged particles from the sun (electrons and protons) interact with the upper atmosphere of the earth, exciting the oxygen and nitrogen molecules there. As these excited atoms relax, they give up energy by releasing photons that are at specific energies (colors) related to the quantum levels in that atom. These lights are strongest at northern latitudes due to funneling of the charged particles along the magnetic field lines of the earth. For an example of these remarkable northern lights see the video.

In the context of medical physics there are two basic ways that a charged particle can lose energy: (1) radiative loss and (2) collisional loss. Section 7.1 provides an overview of charged particles and specifics on the radiative loss mechanism. Section 7.2 focuses on collisional loss. Finally, Section 7.3 considers energy loss from neutrons. The neutron is not a charged particle. It loses energy through other processes. This is important to understand in the radiation therapy context.

7.1 Radiative Energy Loss

7.1.1 Introduction: Charged Particle Interactions

Consider an electron interacting with an atom as shown in Figure 7.1.1. The negatively charged electron experiences a repulsive Coulomb force with the electrons in the atom, the magnitude of this force being given by $F = k\dfrac{ee}{r^2}$, where k is a constant, e is the charge of the electron, and r is the distance between the two electrons. This interaction will impart energy to the electrons in the atom, and energy will be lost from the incident electron. This process is called collisional energy loss.

The other process is radiative energy loss. As shown in Figure 7.1.1, an electron incident toward an atom might normally move in straight line as it passes. However, as the electron gets close to the nucleus it experiences the force between two charged particles. Here the force is attractive because the electron has a negative charge and the nucleus has a positive charge. As a result, the trajectory of the electron is bent as it passes the nucleus. This bending is a form of acceleration, and any charged particle undergoing acceleration produces radiation (this will be seen again in future chapters). This process is called radiative energy loss. The photons produced through this process are called "bremsstrahlung" photons (a German word meaning "braking radiation"). Note that we have focused here on an electron interacting with matter but the same forces and interactions apply to heavier charged particles such as the proton or the nucleus of an atom.

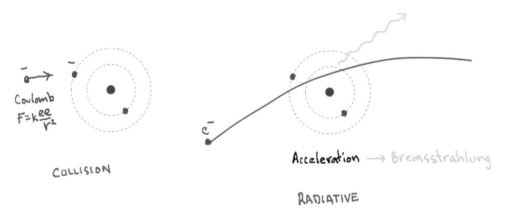

FIGURE 7.1.1
The two energy loss mechanisms for charged particles.

7.1.2 Stopping Power

There needs to be some system for quantifying the energy lost through the processes described above. Consider the cartoon shown in Figure 7.1.2. Here an electron is wandering through a medium. As it does so it loses energy through either collisional or radiative processes. The amount of energy loss we write as ΔE, in some length of travel Δx, or in differential form, $-\dfrac{dE}{dx}$. Note that the minus sign is used as a reminder that this is energy loss. $\dfrac{dE}{dx}$ has units of energy loss per unit length or MeV/cm or, as will prove useful later, keV/μm. An intrinsic quantity is more informative in many cases, so we divide the above quantity by the density, ρ, of the medium, $-\dfrac{1}{\rho}\dfrac{dE}{dx}$. This removes the density dependence and makes it easier to directly compare the properties of various materials. This quantity is called the stopping power, S.

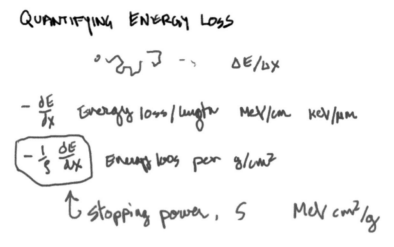

FIGURE 7.1.2
Stopping power. A quantity which describes energy loss.

$$S = -\frac{1}{\rho}\frac{dE}{dx}\left[\text{MeV}\cdot\text{cm}^2\Big/\text{g}\right] \tag{7.1}$$

Note that stopping power has units of MeV·cm²/g which can be seen from the units in the equation, i.e. $\dfrac{\text{MeV}}{\dfrac{\text{g}}{\text{cm}^3}\cdot\text{cm}}$.

We can now write an expression for the stopping power in terms the fundamental processes that are occurring, i.e. the stopping power from collisional energy loss, S_{coll}, and the stopping power from radiative energy loss, S_{rad}. The total stopping power, S_{tot}, is simply $S_{\text{tot}} = S_{\text{coll}} + S_{\text{rad}}$. That is, the total stopping power is the sum of the contribution from collisional and radiative energy losses. Note that sometimes S_{coll} is referred to as the "electronic stopping power" since it involves the electrons in the atom, and S_{rad} is referred to as the "nuclear stopping power" since it involves an interaction with the nucleus.

Collisional stopping power is closely related to a quantity called linear energy transfer (LET) which is considered in more detail in Section 7.3.6. LET is essentially a description of energy deposited locally and is useful because it is related to biological effects operating at the level of the cell.

7.1.3 Radiative Stopping Power, S_{rad}: Mass and Z Dependence

Here we will develop a formalism to describe the radiative energy loss, S_{rad}, that comes from the bremsstrahlung process. This will use a semi-classical approach relying on a collection of fundamental physics equations. First, for an accelerated charged particle, the power released is proportional to acceleration squared, $P \propto a^2$. Second, charged particles experience a force described by Coulomb's Law, $F = k\dfrac{Ze^2}{r^2}$. Here Z is the atomic mass of the nucleus and so the nucleus has a charge $Z\cdot e$. Finally, we have Newton's Law, $F = ma$, i.e. force equals mass times acceleration. Since the power (energy per unit time) is proportional to acceleration squared, the radiative stopping power will also be proportional to acceleration squared, $S \propto a^2$. Combining the above equations we find:

$$S_{\text{rad}} \propto \left(\frac{Z}{m}\right)^2 \tag{7.2}$$

That is, the radiative stopping power is proportional to the square of the atomic number, Z, of the medium and to the square of the mass, m, of the incident particle.

Equation 7.2 has several implications. First, for proton beams the radiative stopping power will be relatively small. This is because the mass, m, of the proton is relatively large. Second, for electron beams the radiative stopping power is large. This is because the mass, m, of the electron is relatively small. Third, the radiative stopping power will be higher in high-Z materials. These dependencies make intuitive sense, if one considers that acceleration from the attractive force from the nucleus will be larger if the particle is a light particle (like the electron) or if the nucleus has a larger charge (high Z).

Figure 7.1.3 shows these dependencies quantitatively for the radiative stopping power of an electron interacting with two different materials, water or lead. The stopping power in lead is much higher than in water (note that the graph uses a logarithmic scale). This is the reflection of the Z dependence described above. Also it is apparent that S_{rad} increases

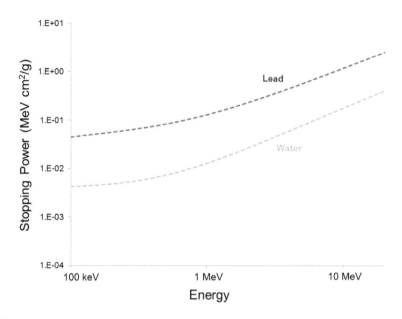

FIGURE 7.1.3
Stopping power from the radiative energy loss process for electrons interacting in two different materials.

as the energy of the electron increases. This makes intuitive sense as well, since the power released is proportional to the acceleration and higher energy electrons will have higher acceleration.

7.2 Collisional Energy Loss

7.2.1 Collisional Stopping Power, S_{coll}, of Electrons

We now turn our focus to collisional energy loss, S_{coll}, and consider specifically the case of electrons interacting with a medium (Figure 7.1.1). There are several properties to note in this context. First, the electron typically undergoes many charged particle interactions with electrons in various atoms as it wanders through the medium, depositing small amounts of energy each time. However, it is possible that a single collision of the electron can produce an enormous energy loss. This can be up to 50% of the energy of the electron. Second, the collisional stopping power depends on the number of electrons per unit gram in the material. This should make intuitive sense, since fewer electrons in the medium results in fewer interactions. Now, the number of electrons per unit gram decreases in higher atomic number materials. Therefore, there is a slight Z dependence of the collisional stopping power. Collisional stopping power is somewhat lower in higher-Z materials. This is very similar to the Z dependence that is observed in Compton scattering which also depends on the number of electrons per unit gram (see Section 5.2.2 for a description of this).

These dependencies can be seen in Figure 7.2.1 which provides a helpful summary of the effects. For collisional stopping power, it can be appreciated that there is not a strong dependence on energy. High-Z materials such as lead have a slightly smaller collisional stopping power than low-Z material like water. Also it can be appreciated

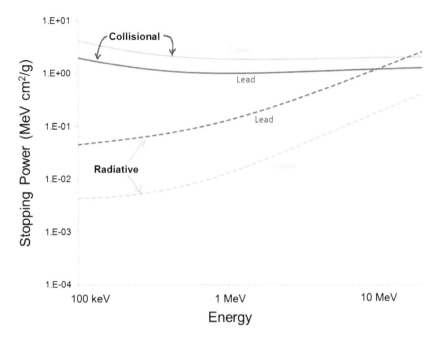

FIGURE 7.2.1
Radiative and collisional stopping power for electrons interacting in two different materials.

from Figure 7.2.1 that for electrons the collisional stopping power is almost always higher than the radiative stopping power particularly at lower energies. The exception is high-energy electrons in high-Z materials which can have a relatively high radiative stopping power. This will have many implications for the production and properties of X-ray beams.

7.2.2 Energy Loss of Protons

The above discussion and examples focused on electrons interacting with the material. Now we turn our attention to protons. The stopping power for protons is given by the Bethe–Bloch formula. A full description of this is beyond the scope of this text, but the essential elements are the following:

$$S \propto \frac{z^2 Z}{v^2 A} B_{col} \qquad (7.3)$$

Here z is the atomic number of the incident particle ($z=1$ for the proton), Z is the atomic number of the medium, v is the velocity of the incident particle, A is the atomic mass of the medium, and B_{col} is a term that characterizes the atomic stopping power and depends on the speed of the particle and the ionization potential of the medium. Almost all of this stopping power is in collisional losses. Radiative losses for heavy charged particles are very small.

The dependencies of Equation 7.3 are important to understand. The stopping power depends on the atomic number of the medium and also, importantly, on the speed of the particle, but it does not depend on the mass of the particle. Also the collisional stopping power for heavy charge particles like the proton is much higher than for electrons.

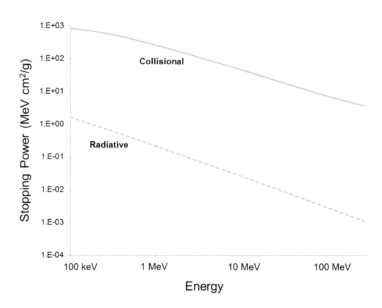

FIGURE 7.2.2
Stopping power for protons in water.

Figure 7.2.2 shows a plot of stopping power for the proton in water. Note that the stopping power increases as the energy of the particle decreases. That comes from the $1/v^2$ dependency seen in Equation 7.3, i.e. as the speed (or energy) decreases the loss of energy is higher. This will have important practical implications which we will see later.

7.2.3 Path of Charged Particles

The path of charged particles in matter depends most sensitively on their mass. As Figure 7.2.3 demonstrates, the electron, which has a very small mass, does not travel in a straight line. Rather it wanders through the material undergoing many charged particle

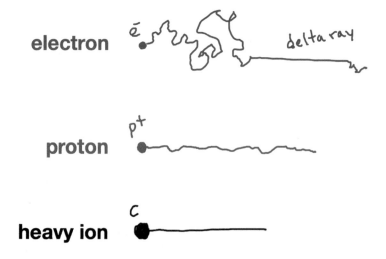

FIGURE 7.2.3
Pathway of charged particles in matter. The mass of the particle is the largest determinant, with more massive particles traveling in straighter lines.

collisions that change its direction. It can also undergo "hard" collisions in which a substantial amount of energy is transferred to the electron. These are sometimes called delta rays, or high-energy electrons, which can travel relatively large distances. For heavy particles like the proton, the pathway is quite different. Larger mass particles are accelerated less by charged particle interactions and so the pathway through the medium is straighter. More massive particles, like the carbon nucleus, travel in straighter pathways.

7.3 Neutron Energy Loss and LET

7.3.1 Neutrons

Neutrons, being uncharged particles, do not undergo interactions mediated by the Coulomb force. Instead they deposit energy via capture and collisions. For a full description of this topic see the video, but a few key points are the following:

- There are two energy regimes: (1) "thermal" neutrons (energies $\lesssim 0.025$ eV), where the interactions are a capture mostly by H or N. (2) Fast neutrons (energies $\gtrsim 100$ eV), where the interactions are "billiard ball" collisions with the nucleus. After such a collision, the nucleus travels off and deposits dose through charged particle interactions.
- Neutrons are produced in photon therapy beams with energies >10 MV. These neutrons from interactions of photons with a high-Z material in the linear accelerator. Neutrons are also produced in proton therapy beams.
- A few beamlines are available in the world for neutron therapy. Boron neutron capture therapy (BNCT) is one modality for enhancing the effect, due to the high interaction cross-section of boron. The boron nucleus captures the neutron and a high-LET alpha particle is created which deposits dose.

7.3.2 Linear Energy Transfer (LET) and Relative Biological Effectiveness (RBE)

LET is a quantity that describes the energy deposited locally near the point of interaction of a particle. As such it is useful because it can be tied to biological effects operating at the level of the cell. LET is equal to the collisional stopping power, S_{coll}.

Further details about the underlying physics of LET can be found in the video, but a few key concepts are the following:

- LET depends strongly on the type of radiation (e.g. X-ray vs. light charged particle vs. heavier charged particle).
- LET is relatively low for X-rays, gamma rays, and electrons (<few keV/μm).
- LET is much higher for more massive charged particles (up to >100 keV/μm). This can be seen in Equation 7.3, which shows a z^2 dependence for collisional stopping power. Note that neutrons are a high-LET modality because the products they produce (protons and various nuclei) are high-LET.

LET has a strong impact on the biological effectiveness of radiation modalities. The metric used to quantify this is relative biological effectiveness (RBE). RBE is defined as the ratio of doses of two types of radiation that achieve the same biological effect. As LET increases RBE also increases up to approximately 100 keV/μm where the RBE can be 3 or more. For further details see the video.

7.4 Useful Reference Information: Charged Particle Data

- www.nist.gov/pml/data/star/
  The website from the National Institute of Standards and Technology (NIST) from the US Department of Commerce contains useful information on the stopping powers of electrons and protons in various materials.
- http://physics.nist.gov/PhysRefData/Star/Text/ESTAR.html
  "ESTAR": Stopping powers for electrons.
- http://physics.nist.gov/PhysRefData/Star/Text/PSTAR.html
  "PSTAR": Stopping powers for protons.

Further Reading

McDermott, P.N. and C.G. Orton. 2010. *The Physics of Radiation Therapy*. Chapter 6. Madison, WI: Medical Physics Publishing.
Podgorsak, E.B. 2016. *Radiation Physics for Medical Physicists*. 3rd Edition. Chapter 6. Switzerland: Springer.

Chapter 7 Problem Sets

*Note: * indicates harder problems.*
See Section 7.4 for references to data of stopping powers for electrons and protons in various materials.

1. What is the ratio of collisional stopping power in air (dry, near sea level) to water for an electron with energy 6 MeV? (Hint: Use data from the ESTAR website in Section 7.4.)
   a. 0.10
   b. 0.98
   c. 1.02
   d. 9.90

2. Match the radiative stopping powers in tungsten with the corresponding electron energy, selecting from the following list: 0.08162, 0.6523, 0.1159, 1.132, 1.759 MeV cm^2/g.
   1 MeV   _____
   6 MeV   _____
   10 MeV   _____
   15 MeV   _____

3. What is the ratio of radiative stopping power of a 6 MeV electron in lead to that in water? Discuss the clinical implications.

    a. 0.14

    b. 0.60

    c. 1.67

    d. 7.09

4. What is the ratio of collisional stopping power of a 6 MeV electron in lead to that in water?

    a. 0.14

    b. 0.60

    c. 1.67

    d. 7.09

5. What is the ratio of collisional stopping power in tungsten for a 40 keV electron vs. a 6 MeV electron?

    a. 0.39

    b. 1.08

    c. 3.17

    d. 4.84

6. What is the ratio of collisional stopping power in water for a 225 MeV proton vs. a 6 MeV electron?

    a. 0.0138

    b. 0.459

    c. 2.18

    d. 42.3

7. What is the ratio of radiative stopping powers in water for a 225 MeV proton vs. a 6 MeV electron?

    a. 0.0138

    b. 0.459

    c. 2.18

    d. 42.3

8. Which material would provide the most effective shielding for thermal neutrons?

    a. Lead

    b. Tungsten

    c. Glass

    d. Water

9. * For an ionization chamber in a water tank, which filling gas will provide the highest charge reading for a 6 MeV electron beam?

   a. Air

   b. Carbon dioxide

   c. Methane

   d. Approximately the same

10. * A 1 MeV electron beam interacts with a $1 \times 1 \times 1$ cm block of material. Which type of material will produce the highest exposure at 2 m from the block?

   a. Muscle

   b. Water

   c. Aluminum

   d. Tungsten

8

X-Ray Tubes and Linear Accelerators

This chapter describes the fundamental physical processes and basic engineering components by which X-ray and electron beams are produced. These systems can be categorized by energy since the technologies that are employed differ greatly depending on the energy. At low energies, tubes are used to produce X-rays. These are used for diagnostic imaging and other purposes (Section 8.1). At higher energies, linear accelerators ("linacs") are used to produce both electron beams and high-energy photon beams used for therapeutic applications (Section 8.2). This chapter will consider both systems.

8.1 X-Ray Tubes

8.1.1 Electron Acceleration and Energy

To understand the operation of tubes and linacs it is important to first understand how electrons move in an electric field. Figure 8.1.1 shows two metal plates ("electrodes") with a voltage applied to them. One electrode is held at a positive voltage (the "anode"), and one electrode is at a negative voltage (the "cathode"). The voltage between the two plates creates an electric field. An electron placed between the two plates will experience a force from this electric field and will be accelerated toward the positive electrode, i.e. the anode. As it does so it will gain some energy. The energy gained by one electron traveling through a voltage of 1 volt (V) is 1 "electron volt" (eV). **The electron volt (eV) as a unit of energy is a key concept.** As a unit of energy, it can also be expressed in joules (J), $1 \text{ eV} = 1.602210^{-19}$ J.

8.1.2 X-Ray Tubes: Physical Processes

X-ray tubes are used to create X-rays with energies in the tens to hundreds of keV range. The most common X-ray tubes are those used in diagnostic imaging which produce X-rays up to approximately 200 eV. Less common are "orthovoltage" tubes which produce higher energy X-rays up to 500 keV and can be used for radiation therapy of superficial lesions.

The fundamental process by which X-rays are produced in a tube is shown in Figure 8.1.2. Electrons are "boiled" off a cathode (through a process called thermionic emission). The high voltage between the anode and the cathode creates an electric field which accelerates the electron toward the anode. It gains energy and collides with the metallic anode. This produces X-ray photons through the bremsstrahlung process (Section 7.1).

8.1.3 Anode Design and Materials

It is important to understand the design of the tube and the anode in these devices in some detail. As shown in Figure 8.1.3A the anode and cathode assembly is enclosed in a

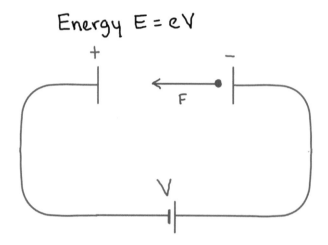

FIGURE 8.1.1
Two metal plates (red) have a voltage (V) applied to them. An electron (blue) in this region experiences a force and moves to the positive electrode (anode). The energy gained by one electron traveling through 1 volt is 1 "electron volt" (eV).

X-ray Tube

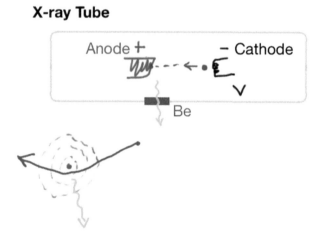

FIGURE 8.1.2
X-ray tube. X-rays are produced when the electron (blue) interacts with the anode (red) and produces bremsstrahlung photons through radiative loss.

vacuum chamber typically made from glass. This serves the purpose of eliminating air which would cause sparking due to the high voltages used and would cause the electrons to scatter. Below the anode is a thin window made of beryllium through which the photons emerge, and below that is a collimator (not shown).

The design of the anode is shown in Figure 8.1.3B. There are several aspects to note. First, the anode face is at an angle relative to the electron beam. Bremsstrahlung photons are produced in the anode in all directions, but the only photons to escape are those emerging in the direction of the window, roughly perpendicular to the electron beam. Second, electrons impact the anode in a roughly circular area, the "focal spot" (Figure 8.1.3B, brown). The size of this spot is an important parameter which influences the resolution of imaging that is achievable. The size of the focal spot as viewed from below is smaller than the

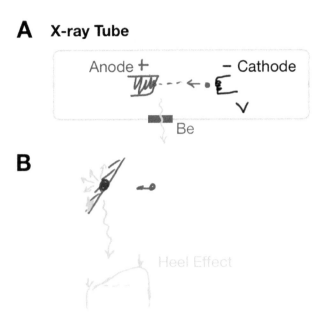

FIGURE 8.1.3
X-ray tube and anode design. (A) Anode and cathode are sealed in a glass vacuum tube (yellow). (B) Electrons impact the anode in a focal spot and emerge in a perpendicular direction.

actual size because of the angle of the anode. Thus the angle of the anode determines the apparent focal spot size. This is sometimes referred to as the "line focus principle." Third, there is differential attenuation of photons by the anode itself. That is, photons produced on the more distal side (left in the figure) have to travel through more anode material before they emerge, resulting in attenuation. Thus there is a lower flux of photons on the left side relative to the right side. This is called the "heel effect" and also depends on the anode angle.

The choice of material for the anode is also important. A key design goal is to achieve the highest efficiency production of X-rays possible. Recall that the bremsstrahlung photons are produced through a radiative loss process (Section 7.1). Recall also that the energy loss through radiative processes is much higher in high-Z materials vs. low-Z materials (Figure 7.2.1). This means that a high-Z material should be used for the anode. The other design consideration is heating in the anode. If the heating is too high the anode may melt. Heating in the anode is caused by the collisional losses of the electrons. Note from Figure 7.2.1 that the collisional stopping power can be quite large especially when the energy of the electrons is low. The first way to address this is by using a metal for the anode which has a high melting-point. The anode in most diagnostic tubes is made from tungsten (Z = 74). It has the highest melting point of any element (3422°C), making it a good choice. The second way to mitigate the tube-heating effect is to rotate the anode. In this way the electron spot does not always impact in the same region on the anode but is distributed over a larger area.

8.1.4 X-Ray Tube Spectra

When considering the spectrum of the photons emerging from an X-ray tube (i.e. number of photons vs. energy) there is a strong dependence on the voltage between the cathode and the anode. Higher voltages result in more acceleration of the electrons which produces higher energy photons. Figure 8.1.4B (green) shows the basic bremsstrahlung spectrum (number vs. energy in keV). The spectrum has more photons at low energies, and the number decreases up to some maximum energy, a characteristic of a bremsstrahlung spectrum from a thick target.

The maximum photon energy is determined by the voltage between the cathode and the anode. If the voltage were, for example, 100 kilovolts (100 kV) then the maximum energy of any electron striking the anode would be 100 keV so the maximum photon energy would also be 100 keV.

The voltage of the tube deserves special consideration. It is not simply a constant voltage supplied by a battery or the like. Rather voltage is provided by an AC source, i.e. alternating-current, which is then converted by a rectifier circuit into a roughly constant voltage. However, even with a high-voltage rectifier there is a small oscillation of the voltage over time (Figure 8.1.4C). In X-ray tubes, therefore, we refer to the "kVp," that is the voltage (in kV) at the peak of the oscillation. The kVp of a tube determines the maximum photon energy.

A second parameter of importance is the electron current in the tube. The current (number of electrons per second) determines the number of photons that are produced. The total number of photons increases linearly with the current. Recall that the unit of current is Ampere (A). In X-ray tubes that range of currents is in the milli-Amp range, so we refer to the "mA" setting of the tube.

This forms a key concept, namely that the mA controls the total number of photons and the kVp controls the maximum energy of photons from an X-ray tube. The mA and kVp can be varied to control the photon spectrum and, thereby, ultimately influence the imaging performance of a system. This is referred to as changing the "technique" in imaging.

The low-energy photons from the bremsstrahlung spectrum are not useful for most imaging applications since they do not penetrate through the patient. Recall that photons in this energy range interact through the photoelectric process, and the mass attenuation

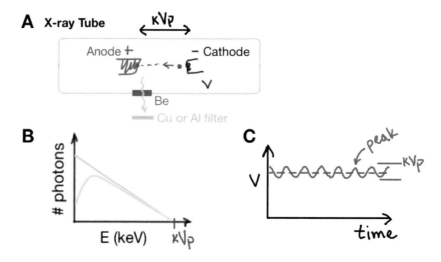

FIGURE 8.1.4
X-ray tube and spectrum. (A) Tube with filter. (B) Unfiltered (green) vs. filtered spectra (yellow). (C) Voltage supply determines kVp.

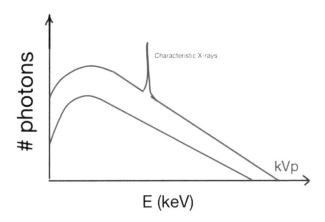

FIGURE 8.1.5
X-ray tube spectra varying mA and kVp.

coefficient is very large at low energies (Section 5.1.3). Furthermore, they deposit dose in the patient especially superficially. It is therefore useful to reduce this low-energy component of the spectrum. To accomplish this, thin metal plates are placed just after the exit window of the tube (Figure 8.1.4A). These plates are most often copper ($Z = 29$) or aluminum ($Z = 13$) which "filter" the beam, that is they reduce the contribution of low-energy photons to the spectrum. The higher energy photons are relatively less attenuated by these materials (recall the approximately $1/E^3$ dependence of the mass attenuation coefficient, Section 5.1.3). Filtration results in a spectrum that is "harder," i.e. more biased toward higher energies. Thicker and higher-Z filters result in harder beams. **Beam hardening is a key concept**, and Section 6.1.4 considered this in more detail. Recall that harder beams will have a larger HVL (Section 6.1.5). The spectra of X-ray tubes are often characterized by their HVL since this is a quantity which can be measured, though such measurements can be somewhat complex.

Putting all these concepts together, Figure 8.1.5 shows spectra at two different kVp and mA settings. The red spectrum has a higher kVp and, therefore, the maximum photon energy is higher. The average energy of photons is also higher, i.e. the higher kVp spectrum is harder. A useful rule of thumb is that the energy at which the spectrum peaks in most tubes is approximately $E_{max}/3$, where E_{max} is the maximum energy. A tube operating at 120 kVp, therefore, would have a peak photon energy of 40 keV. The red spectrum in Figure 8.1.5 also has a higher mA resulting in a higher overall number of photons.

Figure 8.1.5 also shows an "emission line." These are characteristic X-rays produced by K- or L-shell transitions in the anode (for further discussion of characteristic X-rays see Section 5.1.2). These characteristic X-rays are an incidental feature of the X-ray production process in tubes, but they are also used in mammography imaging. These tubes use molybdenum anodes with appropriate filtration to produce strong K- and L-shell lines at low energy.

8.2 Beam Production in Linear Accelerators

The previous section considered kV beams which are used for imaging. Here we consider megavoltage (MV) beams which are used for therapy. As we will see in this section, the technology for producing MV beams is quite complex, so it is worthwhile first to motivate their need from the clinical point of view.

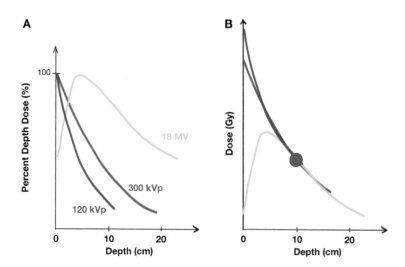

FIGURE 8.2.1
Depth doses from kV vs. MV beams. (A) Percent depth doses in water. (B) Dose scaled in order to deliver the same dose to a tumor at a depth of 10 cm.

8.2.1 Rationale for Megavoltage (MV) Beams

Figure 8.2.1A shows depth-dose data for various beams. Two curves are shown for X-ray tubes, one from a diagnostic X-ray tube operating at 120 kVp (blue) and one from an ortho-voltage tube operating at 300 kVp (red). Also shown is a megavoltage beam (green). Note that the lower the energy of the beam the more rapidly the dose falls off with depth.

To appreciate how this might translate into a clinical delivery consider Figure 8.2.1B in which each beam is scaled in order to deliver the same dose (in Gy) to the tumor at a depth of 10 cm. In order to achieve a therapeutic dose in the tumor with a kV beam (blue or red), a very high radiation dose is required in the superficial region. By contrast the superficial dose with the MV beam is much lower. Also, importantly, the dose to the very superficial-most layer, the skin, is also much lower with a megavoltage beam. **The reduced dose in the superficial depths with an MV beam is called "skin-sparing" and is a key concept.** Skin-sparing is the key reason that MV therapy beams were developed starting in the 1950s.

8.2.2 Acceleration with RF Waves

One might think that a method for creating a megavoltage beam is to simply increase the voltage in an X-ray tube up to the MV range. It turns out, however, that this is not feasible. The chief reason is that it is not possible to maintain a very high voltage gradient in the tube without sparking. Some other method is needed. This is where the linear accelerator (or "linac") plays a role.

The linac relies on a series of metal cavities through which electrons are accelerated. Figure 8.2.2 shows an example with four cavities (labeled C1, C2, C3, and C4) each of which is held at some voltage supplied by the device labeled "G." Each cavity has a voltage oppo-site to the adjacent one. This creates an electric field in the gap between cavities. Just as in the X-ray tube, an electron in this field will experience a force and be accelerated. To understand this, consider the bunch of electrons labeled "A" in the figure. This bunch of

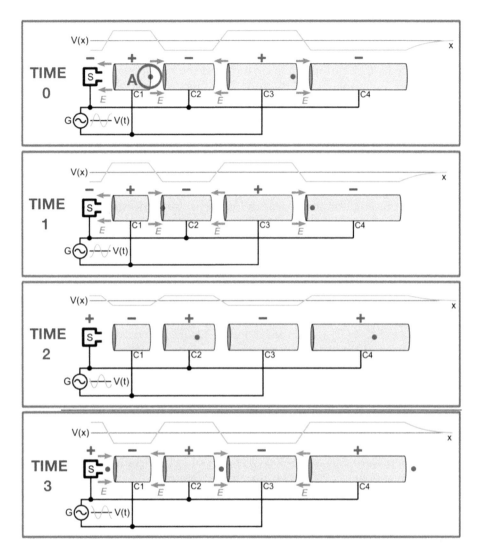

FIGURE 8.2.2
Schematic of acceleration of charged particles in a linac.

electrons is injected into the tube from the device labeled "S" (the source). Let's follow them through the tube at various time points:

- Time point 0: The electron bunch is accelerated from cavity C1 to C2 because there is an electric field present between the two. (Note that the voltage convention is different than in Figure 8.1.1 because these figures were actually generated for protons which have an opposite charge).

- Time point 1: The electron bunch continues to travel. Once the electrons enter the cavity marked "C2" there is no acceleration because there is no voltage gradient within the cavity C2 so the electrons just continue to drift down the tube.

- Time point 2: The voltage sense has changed. C2 is now at a positive voltage, and C3 is at a negative voltage. Still the electron bunch continues to travel. There is no voltage gradient within the cavity C2.

- Time point 3: The electron bunch is now between C2 and C3 and is accelerated by the electric field between these two cavities, gaining energy.

This process repeats through multiple cavities and the result is a very high-energy electron bunch emerging at the end of the tube. For a movie illustration of this process, see Video 8-2.

The key to the acceleration process described above is that an oscillating voltage is applied to the accelerating structures. The oscillation frequency is such that the electron reaches the gap between cavities just as the voltage gradient is at its maximum.

8.2.3 Linac Waveguides

The schematic representation in Figure 8.2.2 is useful for understanding the acceleration process, but medical linacs do not employ separate accelerating cavities; rather there is one single metallic structure called a waveguide (Figure 8.2.3). The electron bunches travel through the center of this waveguide, and the quasi-toroidal cavities serve the function of the separate cavities described above.

Because the electrons are high-energy they are moving fast, very near the speed of light ($v \lesssim c$). The oscillation frequency, therefore, also has to be very high in order to synchronize the voltage changes with the arrival of the electrons. The frequencies relevant for medical linacs are the S band (2–4 GHz) and the X band (8–12 GHz). The cavities are designed such that one wavelength of the RF spans exactly one cavity length as shown in Figure 8.2.2. Note that the wavelength of the electromagnetic waves in an X-band linac is shorter than S-band so the accelerator can be made more compactly. However, the manufacturing tolerances are more difficult to achieve so typically the S band is employed unless there is an applications need for the structure to be compact.

The electromagnetic wave in a linac is in the radiofrequency range (i.e. 30 kHz to 300 GHz). They are therefore often referred to as "RF" waves. Specifically they are a type of RF wave called a microwave (300 MHz–300 GHz). An everyday example of this is the microwave oven which operates at 2.45 GHz, also in the S band.

There are two main types of waveguide in a linac, the standing waveguide (shown in Figure 8.2.3) and the traveling waveguide (not shown). The standing waveguide uses resonant cavities where the RF oscillates as a standing wave. It can be made more compact especially with the side-coupling cavities shown in Figure 8.2.3 in which one node of the wave (i.e. zero voltage crossing) can be located.

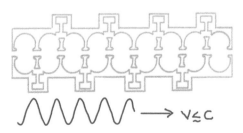

FIGURE 8.2.3
Linac waveguide. A radiofrequency (RF) wave is applied to the waveguide.

8.2.4 Microwave System

In linacs there is a subsystem which produces microwaves and introduces them into the waveguide. The first part of this is the microwave generator or amplifier. Two technologies are available for this, either a klystron or a magnetron. Depending on the linac design, one or the other is used. The klystron is an amplifier which takes an RF input which has a power in watts (W) and amplifies it to a power of megawatts (MW). It is the larger of the two systems, is bulky and expensive, sits in an oil bath, and is stationary (can't rotate with the gantry). The magnetron is the smaller of the two, and it generates microwaves.

Figure 8.2.4 illustrates the operation of the magnetron. This device consists of an anode which is a metal block with resonant cavities in it and a central cathode. The cathode is heated and electrons "boil" off it and are accelerated toward the anode which is at a voltage that is 4 kV higher than the cathode. At the same time, there is a magnetic field pointing in a direction out of the page in the figure. The trajectory of the electrons is bent by the magnetic field (recall that charged particles moving in a magnetic field travel in curved trajectories; Section 1.1.2). Therefore, instead of hitting the anode directly, the electrons instead circle around the cathode. Recall that accelerating charge particles create electromagnetic waves. The electron accelerating around the cathode therefore produces microwaves which are picked up by the output antenna shown on the top of the figure.

The system for introducing the microwave into the waveguide is shown in Figure 8.2.5 (red section). Here we see that the microwaves from the klystron or magnetron are not introduced directly into the waveguide, but rather go first through a circulator, travel around the left arm of the circulator and enter the waveguide. RF waves are also reflected out of the waveguide back into the circulator, and these are useful for tuning the system. Some of these waves go into the automatic frequency control (AFC) system, and this provides information for tuning the frequency of the microwaves. The rest of unused RF power is dumped in a load. This schematic is for a standing waveguide. In a traveling waveguide the microwave system is somewhat different. One major engineering challenge is the RF power requirements. The wall-to-RF efficiency in current linacs is low (~50%), and the wall-to-beam efficiency is even lower (10%).

One detail to note about the linac is that, like the X-ray tube, most components are maintained at very low vacuum pressures. This is to prevent sparking or arcing associated with

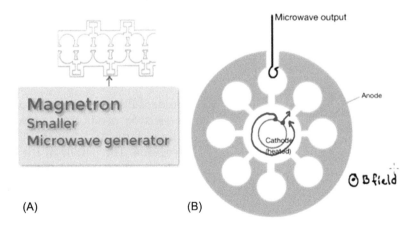

(A)   (B)

FIGURE 8.2.4

Microwave generation systems. (A) Microwaves are produced or amplified by either a klystron or magnetron depending on the system. (B) Magnetron generates microwaves through the circular motion of electrons.

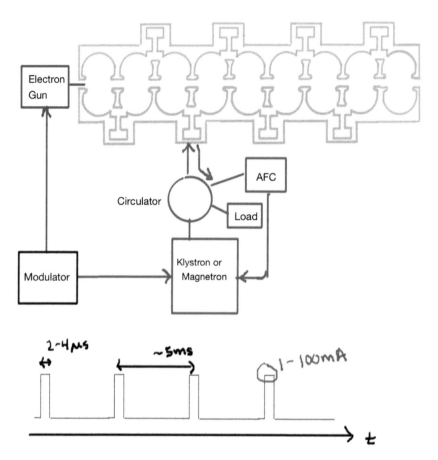

FIGURE 8.2.5
RF components of the linac system.

high voltages. (If there is no gas in the RF system there are no atoms to ionize and carry the current of a spark.) However, the microwave generation system is an exception to this. It is, instead, filled with sulfur hexafluoride (SF_6) gas at high pressure. This is less costly than a low-pressure vacuum and easier to maintain, though it should be noted the SF_6 is toxic. There is a glass window between the microwave system and waveguide to separate the SF_6 region from the vacuum region, and there is a ceramic window to separates out the klystron/magnetron.

The result of the above system is that the waveguide is filled with microwaves that will accelerate electrons through the waveguide. The electron needs to be introduced into the waveguide in some way, and this starts with an electron gun (Figure 8.2.5 left). The gun introduces bunches of electrons into the waveguide. The details of the gun design are beyond the scope of this text, but it is essentially a cathode from which electrons are emitted.

Electrons do not come through the accelerator in a steady stream, but rather they are grouped in pulses. An important component of the linac, therefore, is the pulse formation network (PFN). The pulse formation network starts with the modulator which has a fast switch to turn components on and off. In early linac systems this switch was a thyratron (a gas chamber) but in more recent designs solid-state components are used such as an IBGT or FET. The modulator introduces pulses into the klystron/magnetron and also, at the same time, into the gun. In this way the electron system and the RF system are switched on and off in synchrony with each other.

An example pulse train is shown in Figure 8.2.5. Each pulse is very short in duration (approximately 2–4 µs) while the time between pulses is much longer (e.g. 5 ms). The pulse repetition rate can be adjusted. The output of the machine (dose rate) is determined by the pulse repetition rate and also the current in the peak of the pulse. The current of electrons within the pulse itself is high, either 1 mA for electron mode or up to 100 mA for photon mode. The power within the pulse, therefore, is extremely high. However, because the pulse is only on for approximately 0.1% of the time, the average power is relatively lower.

8.2.5 Bending Magnets and Targets

The result of all the systems above is a high-energy electron beam emerging from the waveguide. There may be a problem with this geometry, however, since the electron beam may not be directed toward the patient. A solution to this is to employ a bending magnet which redirects the energy beam downward toward the patient (Figure 8.2.6). Recall that a charged particle moving in a magnetic field experiences a force perpendicular to its direction of motion. In this case the magnetic field points out of the page, and the electron experiences a perpendicular force which directs it in a circle. This electron beam can be used to treat the patient directly (electron treatment mode), or it can be converted into photons in a tungsten (W) target through the bremsstrahlung process (photon treatment mode).

Figure 8.2.7 shows a further consideration in the bending magnet design and the impact of the spectrum of electrons. First note that not all electrons have exactly the same energy but there is some spread in energy, as shown by the spectrum. The different energy

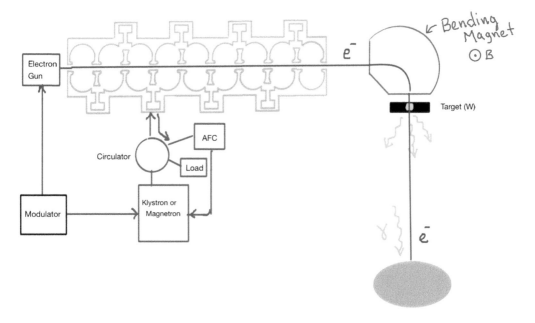

FIGURE 8.2.6
Bending magnet and target. The bending magnet directs the beam toward a target which produces photons which are directed toward the patient (orange).

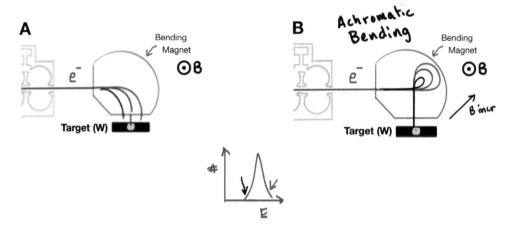

FIGURE 8.2.7
Achromatic bending magnets. (A) In the 90-degree bending magnet shown the electrons strike the target at different positions depending on their energy. (B) An achromatic bending magnet in which all electrons strike the target at the same position.

electrons will have different bending radii. Recall that the bending radius is $r = \dfrac{\sqrt{2mE}}{eB}$ for a non-relativistic particle, which is found by setting the centripetal force, mv^2/r, equal to the force of the magnetic field, evB. Therefore, higher energy electrons (red) will strike the target at a more distal location than low-energy electrons (black) as shown in Figure 8.2.7A. This will create a large focal spot on the target which will have a negative impact on the penumbra (Section 9.1.6).

An alternative design is the 270-degree magnet (Figure 8.2.7B). As before, the higher energy electrons (red) travel in a larger circle. However, the magnetic field also increases going out radially from the center. Therefore, as the higher energy electrons travel through a larger orbit they then experience a larger force due to the increased magnetic field and are bent back toward the same location on the target. This is called an "achromatic" bending magnet, achromatic because the bending does not depend on the color, or equivalently, the energy.

Figure 8.2.8 shows an alternative bending magnet design, the slalom magnet. It relies on three magnet sectors with bending angles of 45-45-112.5. As with the 270-degree design, higher energy electrons are bent through larger angles because of the gradients and strike the target at a single spot. An advantage of this magnet design is that the vertical height is smaller which means the gantry head can be made more compact.

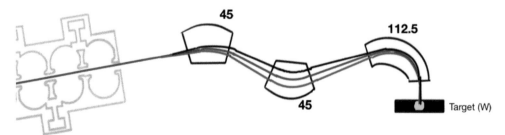

FIGURE 8.2.8
Slalom bending magnet.

8.2.6 X-Ray Production in Linac Targets

In a linac, electrons are converted to photons in the target through the bremsstrahlung process as described in Section 7.2.1. The target is typically made from tungsten (W). The radiative and collisional stopping power for this material are shown in Figure 8.2.9. At the energies where linacs operate, say 6 MeV (red line in the figure), the radiative stopping power in tungsten is quite large. This means that photons are produced quite efficiently. Also recall that at these high energies the electrons are more forward directed. So these targets are "transmission" targets, i.e. the photons emerge in the same direction as the pathway of the electrons. This is different than a lower energy X-ray tube where photons emerge at 90 degrees to the electron beam path (Figure 8.1.2).

Figure 8.2.10 shows an image of a linac target. The silver-colored circle is the tungsten target. It is embedded in a copper structure which is cooled by the two intake/outlet tubes that allow for a flow of water. This cooling is necessary to dissipate the heat generated in the target. Note that the collisional energy losses (heat) generated are also quite large as can be appreciated in Figure 8.2.6.

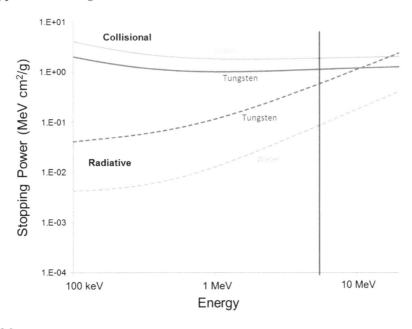

FIGURE 8.2.9
Radiative and collisional stopping power for electrons interacting in tungsten and, for comparison, water.

FIGURE 8.2.10
Linac target. The tungsten target (arrow) is embedded in a copper structure with a water-cooling system (tubes).

8.2.7 Linac Beam Energy

The energy of the electrons or photons emerging from the linac is determined not by the flux (number of electrons per second), but rather by the energy of the electron as it emerges from the waveguide. This can be controlled by switching on or off sections of the waveguide using a device called an "E-switch" or energy switch. Various designs are available. In this way, the linac can operate in different energy modes. Some linacs only have one energy mode, but others allow for the selection of various modes. For photon beams, the common nomenclature is to label these by the maximum energy of the photons that emerge. So, "6 MV" mode corresponds to a beam that has a maximum photon energy of 6 MeV. Similarly, "18 MV" mode corresponds to a maximum photon energy of 18 MeV. Most modern S-band linacs operate in this range of 6 to 18 MV.

Subsequent chapters will describe further the design of linacs and their collimation systems (Chapter 9), the properties of megavoltage photons beams (Chapter 10), and the properties of megavoltage electron beams (Chapter 15).

Further Reading

X-Ray Tubes

Kahn, F.M. and J.P. Gibbons. 2014. *Kahn's The Physics of Radiation Therapy*. 5th Edition. Chapter 3. Philadelphia, PA: Wolters Kluwer.
McDermott, P.N. and C.G. Orton. 2010. *The Physics of Radiation Therapy*. Chapter 4. Madison, WI: Medical Physics Publishing.
Podgorsak, E.B. 2016. *Radiation Physics for Medical Physicists*. 3rd Edition. Chapter 14. Switzerland: Springer.

Linacs

Greene, D. and P.C. Williams. 1997. *Linear Accelerators for Radiation Therapy*. New York: Francis and Taylor.
Kahn, F.M. and J.P. Gibbons. 2014. *Kahn's The Physics of Radiation Therapy*. 5th Edition. Chapter 4. Philadelphia, PA: Wolters Kluwer.
McDermott, P.N. and C.G. Orton. 2010. *The Physics of Radiation Therapy*. Chapter 9. Madison, WI: Medical Physics Publishing.
Metcalfe, P., T. Kron and P. Hoban. 2007. *The Physics of Radiotherapy X-rays and Electrons*. Chapter 1. Madison, WI: Medical Physics Publishing.
Podgorsak, E.B. 2016. *Radiation Physics for Medical Physicists*. 3rd Edition. Chapters 13 and 14. Switzerland: Springer.

Chapter 8 Problem Sets

*Note: * indicates harder problems.*

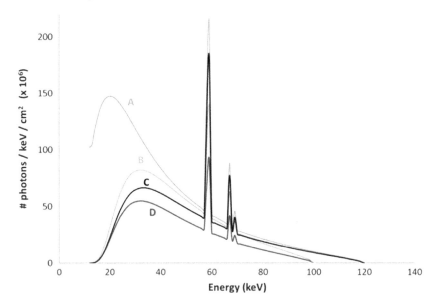

FIGURE PS8.1
X-ray spectra. (Calculated with SpekCalc v. 1.1, G.G. Poludniowski, F. DeBois, G. Landry, and F. Verhaegen. C.f. Poludniowski, G.G. and Evans, P.M., *Med Phys* 34(6), 2164–2174, 2007.)

1. Match the spectra shown in Figure PS8.1 with the following choices of kVp, mA, and filtration.
   a. 100 kVp, 10 mA, 2 mm Al filter
   b. 100 kVp, 15 mA, 2 mm Al filter
   c. 120 kVp, 10 mA, no filter
   d. 120 kVp, 10 mA, 2 mm Al filter

2. What is responsible for the line emission at 58 keV?
   a. Bremsstrahlung in the cathode
   b. Bremsstrahlung in the anode
   c. Characteristic X-rays in the aluminum filter
   d. Characteristic X-rays in the tungsten anode

3. Which spectrum in Figure PS8.1 results in the best visualization between muscle and iodine?
   a. A
   b. B
   c. C
   d. D

4. What are the disadvantages of spectrum A compared to spectra B or D? (Select all that apply.)
   a. Larger spot size on the anode
   b. Larger heating in the anode
   c. Larger output
   d. Larger superficial (skin) dose to patient

5. What is the impact of using a 0.4-mm spot size in an X-ray tube anode vs. a 3-mm spot?
   a. Improved image resolution
   b. Improved contrast
   c. Higher output
   d. Higher maximum X-ray energy

6. Discuss the implications for Problem 5 in the Problem Sets for Chapter 7 for X-ray tubes and for linacs, i.e. "What is the ratio of collisional stopping power in tungsten for a 40 keV electron vs. a 6 MeV electron?"

7. How does the copper housing of the linac target affect the spectrum of emerging photons?
   a. Hardens the beam
   b. Softens the beam
   c. Reduces overall output
   d. Increases overall output

8. What is the effect of not synchronizing the pulses of electrons from the gun with the RF waves in the waveguide?
   a. Lower energy electron emerging from the waveguide
   b. Higher energy electron emerging from the waveguide
   c. Lower average beam current
   d. Higher average beam current

9. * What would be the effect of increasing the overall magnetic field in a 270°-bending magnet?
   a. Output decreases
   b. Target spot location moves away from the gun direction
   c. Beam energy increases
   d. Spot size decreases

10. * Calculate the magnetic field required for a 6 MeV electron to travel in a circle of radius 10 cm.

9

Medical Linear Accelerators

9.1 Linac Collimation System

The previous chapter discussed the production of megavoltage electron beams in a linac up to the point in the system where the beam hits the target and photons are produced (see Figure 8.2.6). This chapter describes the system components after the target as well as the overall mechanical design of various commercial medical linacs. Though somewhat detailed, these engineering aspects are important for later understanding of the beam properties.

9.1.1 C-Arm Linac Geometry

One type of medical linac is the "C-arm" linac, an example of which is shown in Figure 9.1.1. The electron beam moves in the waveguide (blue in Figure 9.1.1) and reaches the target (green) which then produces a wide beam directed down toward the patient. The entire unit can rotate ±180 degrees around the center of the circle. For an animation of this see the video. The rotating assembly is called the "gantry."

As the linac gantry rotates around a patient, the center of the beam is always directed to a virtual point in space called the isocenter. In Figure 9.1.2 the gantry will rotate in and out of the page. One can define the distance between the source (i.e. the target where the photons are produced) and the isocenter. This is the so-called "source-to-axis distance" or SAD. The SAD in most modern C-arm linacs is 100 cm. One often hears of an "SAD setup" for a patient, meaning that the intent is to set up to a known isocenter point in the patient and treat around that point. One can also define another quantity called the "source-to-surface distance" or SSD. Depending on the location of the isocenter point in the patient and the thickness of the patient the SSD can vary. It is also possible to have an "SSD setup" for a patient, meaning that the SSD to the patient surface is set at 100 cm for one or more beams, although this is a less common setup technique. **These geometrical aspects of a linac are key concepts, especially the concept of an isocenter and the associated SAD and SSD.**

9.1.2 Components of the Linac Head

This section describes the components of the linac "head," i.e. the collimation system of a C-arm linac that is just distal to the target along the beam path. Here we focus particularly on the linac operating in photon mode. The first component is the primary collimator, which acts to restrict the beam to a circular region (Figure 9.1.3). It is conical in shape, designed to match the divergence of the beam, i.e. the edges are along the beam path of

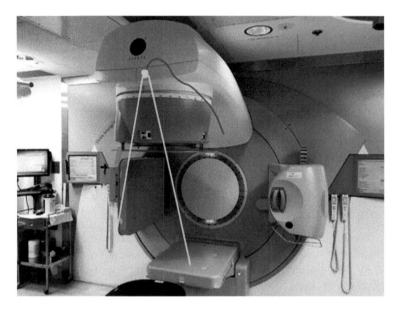

FIGURE 9.1.1
C-arm linac. This particular system employs a traveling waveguide and a slalom magnet design to direct the electron beam (blue) to the target (green). The gantry can rotate around ±180 degrees.

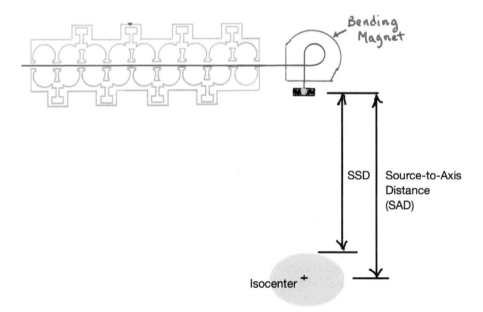

FIGURE 9.1.2
Key geometrical aspects of the linac.

photons emerging from the source. If there were no other devices in the beam path then the flux photons (and correspondingly the dose) would be much higher at the center of the beam than at the edges (see green profile). This is because the photons produced in the target are more forward directed. There are more photons produced in forward directions toward the center of the beam than at angles away from the center. It is, however, often more desirable to

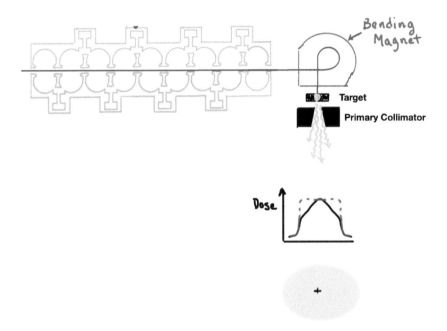

FIGURE 9.1.3
Unfiltered linac beam. With no flattening filter, the dose at the center of the beam is much higher (green). A flat beam (grey) is often desired, so a flattening filter is employed (Figure 9.1.4).

have a beam profile that is more uniform across the beam (grey). Therefore, a flattening filter is used. This is the first component in the beam path (Figure 9.1.4).

The flattening filter is roughly cone-shaped and attenuates the beam more along the central axis and less along the edges in order to produce a flat beam profile. The shape of the filter will depend on the spectrum of the beam. Therefore, different shaped filters are used for different energy beams. It is also important that the filter is aligned with the central axis of the beam. Otherwise an asymmetry will be introduced on the beam (more attenuation on one side than the other). One disadvantage of the flattening filter is that the dose rate is reduced since the photon flux is attenuated. Therefore, modern linac systems offer an optional flattening filter free beam (FFF beam) which produces a higher dose rate at the expense of no longer providing a flat beam.

The next component along the beam path after the flattening filter is the monitor chamber. This is an ionization chamber which monitors the dose of the beam passing through. **A key concept here is the idea of monitor units or "MU." An MU is essentially one unit or one "tick" in the chamber that represents a certain amount of dose delivered. The number of MU registered in the MU chamber can be calibrated to dose; typically is set so that 1 cGy is equal to 1 MU under some standard calibration conditions.** The MU chamber is divided into two sections, one on the top (MU chamber 1) and one on the bottom (MU chamber 2). MU chamber 1 is designed to shut the linac off after some dose. If MU chamber 1 was to fail or be miscalibrated for some reason then MU chamber 2 would shut the beam off. This serves as a backup safety system. The monitor chambers are also divided into various sectors. As seen from the beams-eye view, they occupy different regions of the beam and the signals in the different chambers therefore provide information about the shape of the beam. This is used as feedback to the linac to magnetically steer the electron beam into a location on the target that provides the desired beam profile.

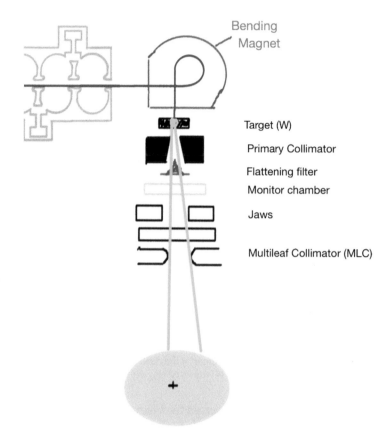

FIGURE 9.1.4
Components of the linac head. The collimation system for photon mode is shown.

The next component in the linac head is a mirror assembly system (not shown in Figure 9.1.4). The mirror is a thin mylar window oriented at 45 degrees to the beam. Off to the side is a small light source that reflects off the mirror and down onto the patient. By carefully orienting the components, the light can be placed at a virtual location that is the same as the source in the target. This allows the treated field to be visualized on the patient.

The next component in the head depends on the design of the linac. In some linacs the next components are the "jaws." These are thick metal collimators that can be moved independently across the beam to provide blocking. One set of jaws moves in the "gun-target" direction (the top jaws in Figure 9.1.4) while the other set of jaw moves in the perpendicular direction (bottom jaws in Figure 9.1.4). The jaws can be moved to define a rectangular irradiation field as observed from the beams-eye point of view. The final component is the multileaf collimator (MLC), a series of numerous metal blades, each of which can be adjusted to provide the field shaping desired. The MLC is an important component of the linac and will be described in much more depth in Section 9.1.5.

9.1.3 Linac Electron Beams

The previous section focused on the components of the linac head when operating in photon mode. However, many linacs are also designed to provide electron beams. These are

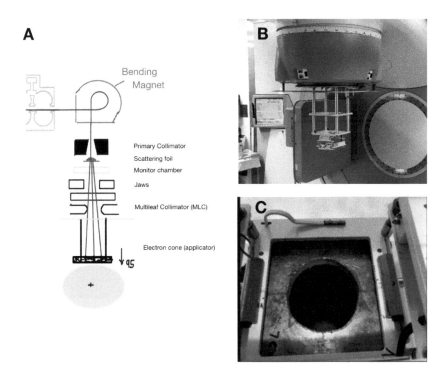

FIGURE 9.1.5
(A) Components of the linac head in electron mode. (B) Applicator attached to the head of the linac. (C) Custom block poured from Cerrobend inserted in the applicator.

useful for treating patients with a dose restricted to more superficial depths. In electron mode the flattening filter is removed. Instead, a scattering foil is introduced. This foil, made of low-Z metal, scatters the beam to provide an approximately flat fluence across the face of the beam. The other components of the beam remain in place, although the jaws and MLCs are pulled back to a predefined location out of the beam path. Finally an electron applicator (or "cone") is attached to the head of the machine (Figure 9.1.5B). The distal edge of the cone is usually 95 cm from the source and is placed close to the patient (typically 5 cm). It is important to have collimation close to the surface of the patient because electrons scatter a great deal in air and if the collimator is not placed close the patient the beam edge can be very blurry. The applicator holds a block in place which is poured from Cerrobend to the shape needed for the patient. Figure 9.1.5C shows the example of a circular cutout.

9.1.4 Beam-Shaping Devices

Here we consider beam-shaping devices for photon mode. Figure 9.1.6 shows an example lateral field from a whole-brain treatment for a patient. The jaws can be used to define the superior, inferior, anterior, and posterior edges. However, these can only define straight edges. In this case, one would like to block the eyes and anterior anatomy shown in the red-hatched region.

One tradition way to accomplish this is through the use of blocks poured from a low-melting point metal such as Wood's metal (tradename Cerrobend) which has a density of 9.7 g/cm^3 and a melting point of 70°C (Figure 9.1.7). The blocks can be constructed by

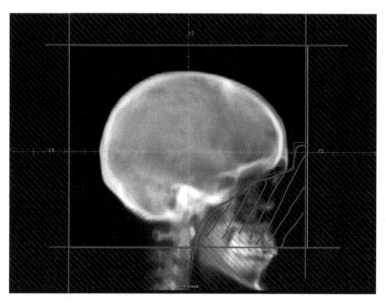

FIGURE 9.1.6
Example collimation for a whole-brain treatment.

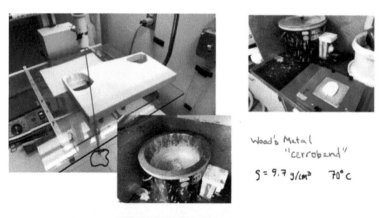

Wood's Metal
"cerrobend"
$\varsigma = 9.7 \ g/cm^3$ $70°C$

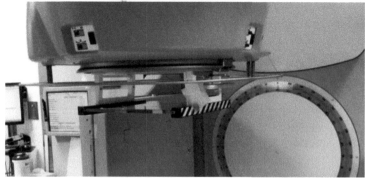

FIGURE 9.1.7
Poured blocks. Block shapes are cut from Styrofoam and then filled with metal (Cerrobend). After cooling they are mounted on a tray in the head of the linac.

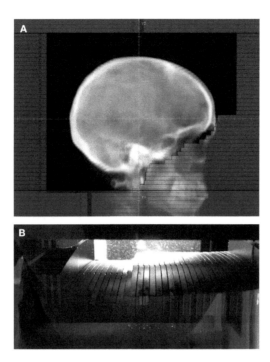

FIGURE 9.1.8
(A) Beam shaping with MLCs and (B) the MLC as seen looking up toward the source in the linac head.

first cutting the shape in a Styrofoam block using a heated wire cutter and tracing out the required shape from a film. The wire cutter is designed to have a divergence which exactly matches the beam. The metal is then poured in or around the Styrofoam block on a cooling table. Finally the block is mounted on a plastic tray in the head of the linac. This is a labor-intensive process, requires dedicated facilities and staff, and introduces toxic metals into the clinic. Also, importantly, it does not allow for dynamic shaping of the beam and requires that the blocks be replaced between each beam which makes the treatment delivery much more time consuming. For these reasons, poured blocks have largely been replaced by MLCs. Figure 9.1.8 shows the potential of an multileaf collimator (MLC) for beam shaping. The MLCs are thin blades of tungsten arranged one next to the other as seen in Figure 9.1.8B. This particular linac head has 80 MLC leaves (40 per side), each of which can be moved to the desired location by tiny motors. This provides the beam shaping as needed.

9.1.5 MLC Design

The design of MLC leaves represents a key concept and is important to understand in some detail. MLCs are the subject of AAPM Task Group #50 (Boyer et al. 2001) which provides a wealth of detail, though is somewhat dated. Figure 9.1.9A shows an example MLC leaf. It is constructed of tungsten which has a high density and favorable mechanical properties. In Figure 9.1.9A, the beam direction is top to bottom. The thickness of the leaf in the beam direction is 5.5 cm to 7.5 cm, which provides attenuation of the beam. Each leaf is driven by a motor, and the motor drive-screw goes through the central channel shown. The typical leaf width is 0.5 cm to 1 cm projected at isocenter. This width determines, in

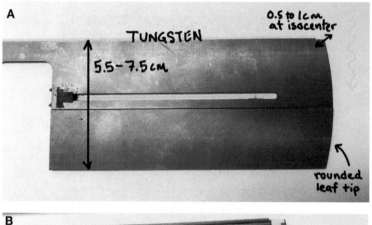

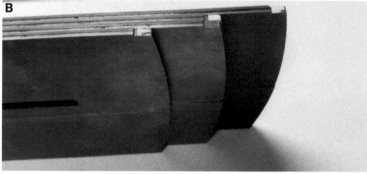

FIGURE 9.1.9
MLC leaf design. (A) A single MLC leaf. (B) Multiple MLC leaves next to each other.

part, how accurately the field can be shaped. Another property is the rounded leaf tip which will be discussed in more detail below. The MLC leaves are stacked together and slide past each other (Figure 9.1.9B). For an illustration of this see the accompanying video.

9.1.6 Penumbra

Penumbra is a key concept underlying the properties of radiation therapy beams, and considerations of penumbra drive many of the design choices of medical linacs including the MLC. Penumbra is illustrated in Figure 9.1.10 where it is seen that the dose profile across a beam does not have an extremely sharp falloff at the edge, but rather there is some "smearing" of the dose gradient. This is a penumbra.

The first component of penumbra is a simple geometric one. As shown in Figure 9.1.11 if the source has some finite size, s, it will project down to some finite size at the patient. Another way to think of this is to imagine standing at a point in the patient just at the nominal edge of the field and look up toward the source. The entire source is not visible; some of it is obscured by the collimator. This makes the overall flux at that point lower. As you move in closer to the central axis, however, more of the source is visible and the dose increases until it finally reaches its full value when the whole source is visible. Conversely, moving away from the central axis means less of the source is visible and the dose drops until it is nominally zero.

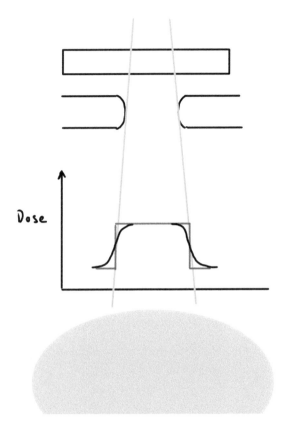

FIGURE 9.1.10
Penumbra results in dose falloffs that are less sharp than ideal.

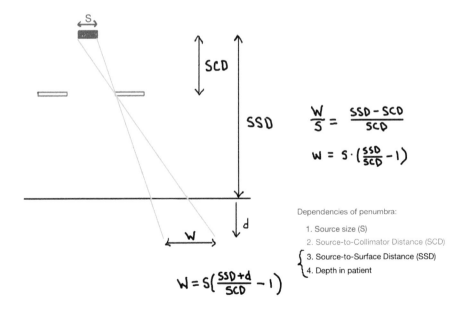

$$\frac{W}{S} = \frac{SSD - SCD}{SCD}$$

$$W = S \cdot \left(\frac{SSD}{SCD} - 1\right)$$

Dependencies of penumbra:

1. Source size (S)
2. Source-to-Collimator Distance (SCD)
3. Source-to-Surface Distance (SSD)
4. Depth in patient

$$W = S\left(\frac{SSD + d}{SCD} - 1\right)$$

FIGURE 9.1.11
Geometric penumbra. Dependence on relevant variables.

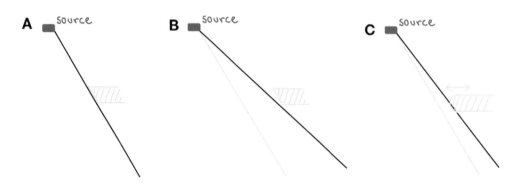

FIGURE 9.1.12
Collimators of finite thickness.

The relevant formulas for penumbra can be derived by using similar triangles (see Figure 9.1.11). We consider a source-to-collimator distance of SCD. The collimator here is considered to be extremely thin so that issues like the rounded leaf tip of the MLC do not factor in at the moment. The width, w, of the penumbra at a distance SSD is given by $w = s\left(\dfrac{\text{SSD}}{\text{SCD}} - 1\right)$. Or, if one wanted to consider the more general problem of the penumbra width at some depth, d, in the patient the formula would be:

$$w = s\left(\frac{\text{SSD} + d}{\text{SCD}} - 1\right) \tag{9.1}$$

From this formula the relevant dependencies can be appreciated. The penumbra will be larger as the source size increases and as the SSD and/or depth increases. The penumbra will be smaller as the SCD increases. If the collimation were placed on the skin then SCD = SCD and the penumbra at the skin would be zero.

A second contribution to the penumbra is the transmission through the edge of the collimator. Collimators, of course, are not infinitely thin so some consideration must be made for their thickness. As Figure 9.1.12A illustrates, an ideal collimator might have an edge that diverges exactly with the beam. This would provide a sharp edge (small penumbra width). However, as the collimator is moved away from the central axis, this edge would no longer diverge with the beam angle (Figure 9.1.12B). One solution to this is to mount collimators on circular track so their edges always match the divergence of the beam. This was the approach of the linac from Siemens Inc., but this is no longer manufactured. More recently, though, the linac from ViewRay Inc. employs this approach and achieves a relatively small penumbra width. A second commonly used approach is to employ rounded leaf ends to the MLC (Figure 9.1.12C). Although this does not produce an ideal penumbra, it does produce a penumbra that is roughly independent of the position of the MLC in the field (Figure 9.1.13).

9.1.7 MLC Interleaf Leakage and the Tongue-and-Groove Effect

In some MLCs there is a small groove along the length of the MLC near the middle. This groove matches with a "tongue" on the other side of the leaf and allows the

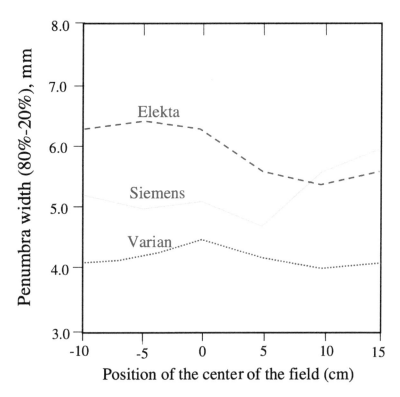

FIGURE 9.1.13
Penumbra width vs. position of the MLC for three different systems, circa 2002. (Reprinted from M.S. Huq, I.J. Das, T. Steinberg, and J.M. Galvin. 2002. A dosimetric comparison of various multileaf collimators. *Phys Med Biol* 47(12):N159–70.)

leaves to slide past each other (Figure 9.1.9B). The exact design of the tongue-and-groove varies by manufacturer, but an example is shown in Figure 9.1.14 where the leaves are viewed in cross-section, i.e. viewed looking toward the tip of the leaf. Here the source would be at the top of the page and the MLCs would slide in and out of the page. There is a small airgap between leaves (shown in red) because the leaves are not tightly bonded together but rather slide past each other. Because of this gap there is extra transmission between the leaves. This is called interleaf leakage and is illustrated in the dose profile. Note that the transmission through the leaf itself ("leaf transmission") is small, but the leakage between leaves is higher. Some actual data are shown in Figure 9.1.14C where it is seen that the leaf transmission through the full thickness of the leaf is 1 to 1.5% in these MLCs whereas the interleaf leakage is roughly double that.

In the dose profile shown in Figure 9.1.14A there is a region where the leaves are open. This might be the intended region of irradiation. Note that the tongue-and-groove edges of the MLC contribute to a dose profile that has a falloff that is less sharp than if these features were not present. This is called the tongue-and-groove effect. Some treatment planning systems can model this but many do not. It does have some effect on the overall dose distribution. Note that Figure 9.1.14A is a simplification of the actual MLC. In reality the edges of the MLC are designed to be divergent with the field (Figure 9.1.14B). Nevertheless, the tongue-and-groove effect is still present.

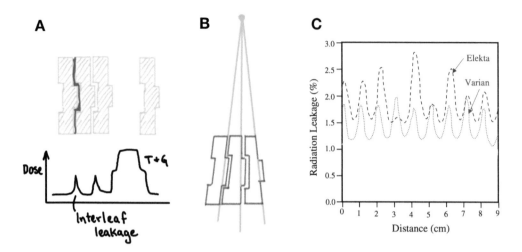

FIGURE 9.1.14

Tongue-and-groove effect and interleaf leakage. The source is at the top of the figure, and the leaves slide in and out of the page. (A) Schematic of transmission and leakage. (B) Divergent leaves. (C) Data from two different MLCs. (Reprinted from AAPM Task Group 50, Boyer et al. 2001.)

9.1.8 C-Arm Linac Collimation Systems

Here we discuss collimation systems from actual C-arm linac systems, focusing on systems from two particular manufacturers as examples. This will provide a practical example of the principles discussed above and will illustrate some of the design trade-offs involved.

Figure 9.1.15 shows a scale illustration of the Millennium MLC (Varian Inc.) and the Agility MLC (Elekta Inc.). In the former, the MLC is placed as the last component of the collimation system, making the SCD larger and the penumbra smaller. In the latter the

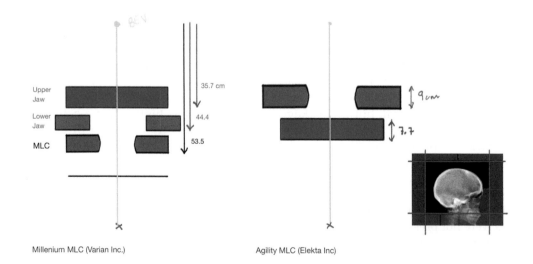

FIGURE 9.1.15

MLC and collimation designs from two manufacturers. Source is shown in green and the isocenter is shown as with a red x. Drawing is to scale.

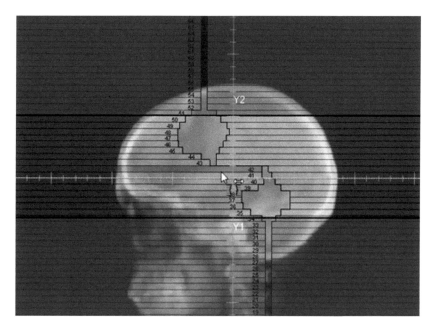

FIGURE 9.1.16
MLC leaf interdigitation.

MLC (9 cm thick and 0.5 cm width at isocenter) is placed as the first component with another jaw below it moving perpendicular to the MLC direction. This results in a more compact head.

Other design features of the MLC include the maximum speed of the leaves which can be important for dynamic deliveries (Chapter 14) and whether the leaves can "interdigitate." This latter point is illustrated in Figure 9.1.16 where it is seen that two immediately adjacent leaves from either side move past each other (red highlight). This is called interdigitation. Some older designs of MLC (e.g. MLCi, Elekta Inc.) could not interdigitate which meant that it was not possible to block regions like that at the center of Figure 9.1.16 in a single field.

9.2 Linear Accelerator Systems

This section considers some of the available commercial linac solutions. This is meant to be practical, though not comprehensive. It is useful to begin with a brief historical overview since many of the technologies used today are a direct outgrowth of these historical systems.

Prior to the widespread use of linear accelerators, external beam radiation therapy was delivered with ^{60}Co-teletherapy units. These machines used ^{60}Co produced in nuclear reactors and had the advantage of being much simpler and easier to maintain. These units are currently in use in many underserved areas of the world. Like many modern linacs they employ the C-arm design (e.g. Figure 9.1.1) in which a gantry rotates around the patient. The main disadvantage of treatment with ^{60}Co is the steeper depth-dose falloff due to the

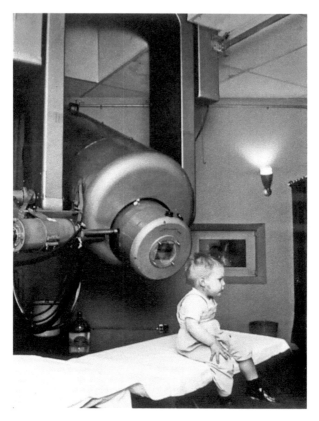

FIGURE 9.2.1
Gordan Isaacs, two years old in 1957, was treated for retinoblastoma by Dr. Henry Kaplan using a linac. The first linac treatments were four years prior at London Hammersmith Hospital in England.

lower energy photons (1.17 MeV and 1.33 MeV vs. for example a 6 MV linac beam) and also the larger sources size (2-cm diameter) which results in worse penumbra. Though many of these problems can be obviated with the use of multiple beams and IMRT modulation, ^{60}Co-teletherapy does not see widespread use anymore.

The first treatment with a linear accelerator was performed on August 19, 1953, at London Hammersmith Hospital in England. This was accomplished with an 8 MV linac manufacturer by Metropolitan-Vickers Inc., a company which was later sold. Around this same time work was ongoing in the San Francisco Bay area by a pioneering oncologist Dr. Henry Kaplan in collaboration with physicists and engineers. In 1957 Dr. Kaplan treated his first patient with a linac, Gordan Issacs, a two-year-old boy with a retinoblastoma tumor of the eye (Figure 9.2.1). Dr. Kaplan later pioneered the use of radiation therapy for Hodgkins lymphoma.

Linac technology developed quickly in the subsequent decade and in 1968 Varian Associates Inc. released the "Clinac 4" treatment unit, which was a C-arm linear accelerator system. It was not the first such device, but it would become the first one that was widely available. In the 50-plus years since, linac technologies have advanced with the addition of beam shaping using MLCs, image guidance, and other features, but the basic C-arm design is still widely used at present. More recently in 2017, Varian Inc. released the

Halcyon™ device which is the first major departure for this company from the 50-year-old C-arm design. This is a single energy (6 MV) ring-based device with an integrated kV imaging system.

The next section considers other non-C-arm linac solutions. Though the market-share of these devices is smaller at present, they are important technologies.

9.2.1 TomoTherapy

The TomoTherapy® system (Accuracy Inc.) is a ring-based system that delivers modulated radiation therapy (Figure 9.2.2). From outside the unit appears much like a CT scanner (Figure 9.2.2A), but inside it contains a compact X-band linac operating at 6 MV (treatment mode) and a detector to acquire MV CT images with the linac operating in imaging mode at 3.5 MV (Figure 9.2.2B). This technology was originally pioneered by Dr. T. Rock Mackie and colleagues at the University of Wisconsin–Madison. The system treats in a helical mode. That is, the gantry rotates around continuously as the patient slides through the bore and, as such, long treatment lengths (up to 135 cm) are possible. The first human patients were treated with TomoTherapy in 2002. This device was one of the first to allow for volumetric imaging at the time of treatment via megavoltage CT. This helped to usher in the era of image-guided radiation therapy, IGRT (Chapter 21).

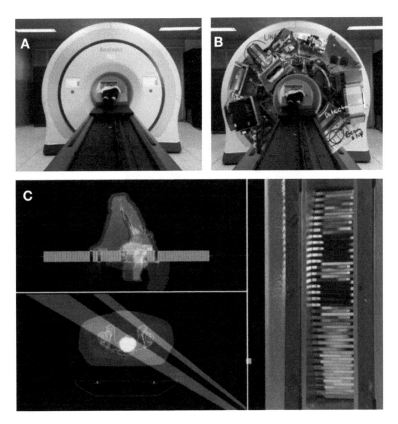

FIGURE 9.2.2
TomoTherapy unit. (A) The patient slides through the bore as the gantry rotates. (B) Inside is a low-energy linac. (C) Beam shapes are modulated by a binary MLC with fast opening/closing. (Images courtesy of Accuray Inc.)

The TomoTherapy unit employs an 85-cm SAD design which is smaller than the typical 100 cm in C-arm linac. The smaller SAD results in an increased output. Also this unit does not have a flattening filter (FFF beam) which also results in increased output. The MLC employs a simple but unique design (Figure 9.2.2C). At any given time each MLC leaf can be either open or closed (i.e. a "binary" MLC). As the gantry rotates around, the MLCs open and close very quickly to irradiate different regions of tissue. In this way modulated delivery is achieved. For an animation of this process see the video.

The newer version of the system (RadiaExact®) provides for kV CT imaging through a system mounted orthogonally to the treatment beam (Figure 9.2.2B). A further detailed description of the TomoTherapy device and associated QA considerations circa 2010 can be found in AAPM Task Group #148 (Langen et al. 2010).

9.2.2 CyberKnife

The CyberKnife® unit (Accuray Inc.) is yet another type of linac design. This system was developed by the neurosurgeon Dr. John Adler and colleagues at Stanford for SRS and SBRT treatments (see Chapter 22) and was first cleared in the United States by the FDA in 1999. A further detailed description of the CyberKnife device and associated QA considerations circa 2010 can be found in AAPM Task Group #135 (Dieterich et al. 2011).

The CyberKnife employs a compact X-band linac that is mounted on a robotic manipulator arm adopted from the automotive industry (Figure 9.2.3A). The beam (green in Figure 9.2.3) can be aimed downward from any direction over a roughly hemispherical region (2π steradians) over the top of the patient. As such, the system is non-isocentric, i.e.

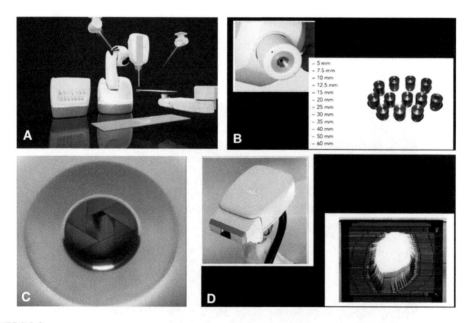

FIGURE 9.2.3
CyberKnife unit for SRS and SBRT. (A) X-band linac on the robotic manipulator arm can be aimed in any direction over the upper roughly hemispherical region. (B) Collimation of the original system using cones. (C) Collimation of the second-generation system with a tungsten iris. (D) Collimation and head of the newest system using an MLC. (Images courtesy of Accuray Inc.)

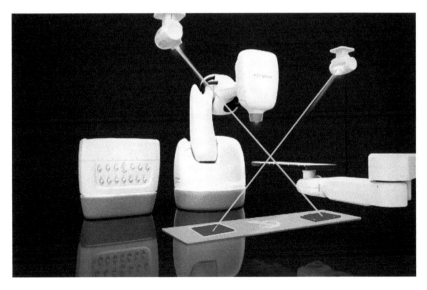

FIGURE 9.2.4
CyberKnife imaging system. Two kV X-ray tubes (ceiling) are paired with two X-ray imaging panels (floor). By simultaneously imaging from orthogonal directions features such as bony anatomy can be located in 3D space.

the various beams used to treat a patient need not intersect at a single point. For an animation of the CyberKnife in action see the video.

The linac is coupled with an imaging system consisting of two kV X-ray tubes paired with two X-ray imaging panels in the floor (Figure 9.2.4). This system acquires two images simultaneously from two orthogonal directions which allows features like bony anatomy to be localized in 3D space. This is called stereoscopic imaging (see Chapter 21 for further details). This can also be used for respiratory tracking.

The original design for the collimation system consisted of 12 different-sized collimators ranging in size from 5 mm to 60 mm (Figure 9.2.3B). The robotic arm automatically mounted these collimators from an exchange table. The second-generation system uses an iris consisting of six tungsten blades which can be adjusted to any aperture size up to 60 mm (Figure 9.2.3C). The newest collimation system uses an MLC which allows more control over beam shaping and field sizes up to 10×11 cm (Figure 9.2.3D).

9.2.3 MR-Guided Linacs

The most advanced linac-based treatment units to emerge in recent years are MR-guided linacs in which a linear accelerator is combined with an MRI unit (Figure 9.2.5). This allows high-resolution MR imaging during treatment.

The first commercially available unit, MRIdian® (ViewRay Inc.), uses a 0.35 T magnet and began treating patients at Washington University–St. Louis in 2014. The original design used ^{60}Co-based beams, but a linac version has since become available. A second commercial system became available in 2019, Unity® (Elekta Inc.), which uses an MRI unit with a field strength of 1.5 T. Further background on MRI can be found in Section 20.1, and we will return to a more detailed discussion of these units and the MR-guided radiotherapy approach in Chapter 21 on image guidance.

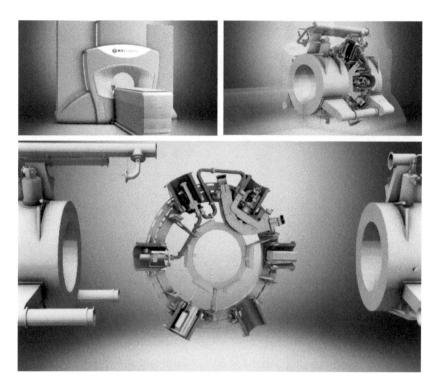

FIGURE 9.2.5
MR-guided linac unit from ViewRay Inc. Between the two magnet poles is a linac unit. (Images courtesy of ViewRay Inc.)

Further Reading

Boyer, A., et al. 2001. Basic applications of multileaf collimator: Report of Task Group No. 50 Radiation Therapy Committee. American Association of Physicists in Medicine (AAPM TG-50, 10) (AAPM Report No. 72).

Dieterich, S., et al. 2011. Report of AAPM TG 135: Quality assurance for robotic radiosurgery. *Med Phys* 38(6):2914–2936.

Klein, E.E., et al. 2009. Quality assurance of medical accelerators: Report of AAPM Task Group 142. *Med Phys* 36(9):4197–4212.

Langen, K.M., et al. 2010. Quality assurance for helical tomotherapy: Report of the AAPM Task Group 148. *Med Phys* 37(9):4817–4853.

McDermott, P.N. and C.G. Orton. 2010. *The Physics of Radiation Therapy*. Chapter 13. Madison, WI: Medical Physics Publishing.

Metcalfe, P., T. Kron and P. Hoban. 2007. *The Physics of Radiotherapy X-rays and Electrons*. Chapter 1. Madison, WI: Medical Physics Publishing.

Chapter 9 Problem Sets

Note: * indicates harder problems.
See Section 5.3 for references to data of mass attenuation coefficients.

1. Which devices in the beam path are changed when the mode of the linac is switched from 6 MV photon to 6 MeV electron? Indicate what the change is.

   a. Microwave frequency in the waveguide

   b. Electron current in linac pulses

   c. Strength of bending magnet

   d. Primary collimator

   e. Flattening filter

   f. Scattering foil

   g. Monitor chamber

   h. Jaws

   i. MLC

   j. Collimator cone

   k. Isocenter

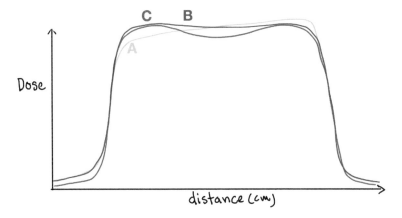

FIGURE PS9.1
MV beam profiles.

2. Match the scenarios below with the 6 MV photon beam profiles shown in Figure PS9.1.

   a. Beam spot not aligned with the flattening filter _____

   b. Beam energy too low _____

   c. Correct beam at depth of d_{max} _____

3. What is the shape of the flattening filter in an 18 MV photon beam vs. a 6 MV photon beam?
   a. Thicker at center
   b. Thinner at center
   c. Wider at the base
   d. Narrower at the base

4. What is the geometric penumbra at 100 SAD for an MLC at 37 cm from the source? Assume an infinitely thin collimator and source diameter of 2 mm.
   a. 1.3 mm
   b. 2.7 mm
   c. 3.4 mm
   d. 5.4 mm

5. What is the geometric penumbra at 100 SAD for a jaw at 51 cm from the source? Assume an infinitely thin collimator and source diameter of 2 mm.
   a. 1.0 mm
   b. 1.9 mm
   c. 2.5 mm
   d. 3.9 mm

6. What is the geometric penumbra at 100 SAD for a stereotactic cone at 70 cm from the source? Assume an infinitely thin collimator and source diameter of 2 mm.
   a. 0.9 mm
   b. 1.4 mm
   c. 2.1 mm
   d. 2.9 mm

7. What are the key choices of linac operation that allow for the compact systems such as TomoTherapy, CyberKnife, or Mobetron? Check all that apply.
   a. High repetition rate
   b. High frequency
   c. Lower required MU
   d. Low energy

8. What is the transmission through a 9-cm thick tungsten leaf for a monoenergetic beam of 2 MeV? What will be different in an actual measurement?

   a. 0.045%

   b. 0.48%

   c. 4.7%

   d. 42.5%

9. * What is another design for jaws or MLC leaves that overcomes the problems for a non-ideal penumbra from a rounded end and also a constant penumbra vs. field size? What are the disadvantages of this design?

10. * Describe how the physical segments in a monitor chamber allow for steering of the electron beam in a linac.

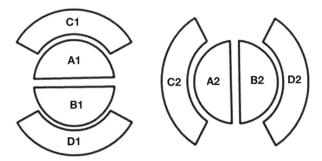

FIGURE PS9.2
Monitor chamber in a linac.

10

Megavoltage Photon Beams

10.1 Basic Properties of Megavoltage Photon Beams

Previous chapters considered the production of megavoltage (MV) beams, the acceleration of electrons to energies >6 MeV and the subsequent production of bremsstrahlung photons, then the collimation of the beam to a shape that suits the clinical requirements. Here we consider the properties of the MV photon beam in a patient. Figure 10.1.1 shows the dose distribution of one such beam entering straight down onto a water phantom. (Note that "phantoms" are devices which are used to make measurements of the beam. They can be tanks of water, pieces of plastic, or other devices.) Figure 10.1.1 shows two plots of data through the dose distribution: a percent depth dose (PDD) (Figure 10.1.1A) and a profile (Figure 10.1.1B). These are useful ways to characterize a beam and will be used extensively in this and subsequent chapters.

10.1.1 Introduction to Percent Depth Doses (PDDs)

The concept of PDD is illustrated in Figure 10.1.1. Here a patient is aligned at some source-to-surface distance (SSD) (say, for example, SSD of 90 cm). The collimation system (see Figure 9.1.4) is used to create a field (say, for example, a square field 10×10 cm at isocenter). The goal is to find the dose along the central axis at any depth in the patient. Figure 10.1.1B shows actual PDD data from this example beam. These data were acquired with a water tank. That is, a water tank was set up under the beam and dose measurements were acquired with an ionization chamber as it was scanned down through the water (this process is presented in more detail in Section 16.1). PDDs are always rescaled so that the dose is 100% at a depth of d_{max} (i.e. depth where the dose is maximum).

One might predict that PDD should follow an exponential falloff. Recall from Chapter 6 and Equation 6.1, the form of attenuation in matter, $I = I_0 e^{-\mu x}$. However, this is not the case for such a beam. The actual dose (blue line in the figure) is larger than that predicted by exponential falloff (dashed line in figure). This is due mostly to scatter. That is, photons are scattered from the sides of the beam in toward the center of the beam, making the dose higher than predicted by simple exponential attenuation.

PDDs can be used for practical dose calculations. The dose at any depth d is given by $D(d) = D(d = d_{max}) \cdot PDD(d)$. Say we know that the dose at a depth of 10 cm is 100 cGy. To find the dose at a depth of d_{max} we look up the $PDD(d = 10)$ which is 0.60 for the example beam in Figure 10.1.1B. We can then calculate that the dose at d_{max} is 167 cGy ($D = 100/0.60$). In practice, we will see that tissue-maximum ratios (TMRs) are more useful for isocentric treatments. TMR is closely related to PDD.

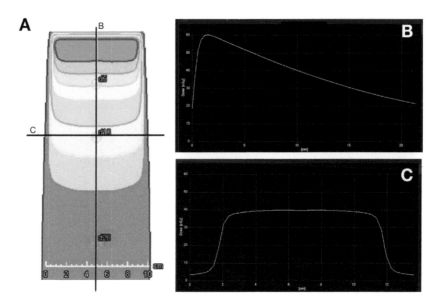

FIGURE 10.1.1
(A) Dose distribution from a 6 MV 10×10 cm beam at 100 SSD. (B) The percent depth dose (PDD) along the line through the center of the beam. (C) Dose profile across the beam along the line shown.

10.1.2　Buildup and d_{max}

Another important property of MV beams is that the dose at superficial depths is relatively low. This "skin-sparing" effect allows for the treatment of deep-seated tumors while limiting the dose to normal tissues in the superficial regions. Note that this is also a deviation from a pure exponential falloff with depth (blue vs. red dashed lines in Figure 10.1.2). This process is described in detail in Section 6.2.2 and discussed further in the patient context in Section 10.2.1. Recall that it is actually electrons that deposit dose and there are relatively few electrons at the surface. At deeper depths the contributions of electrons build up until equilibrium is reached at a depth of d_{max}.

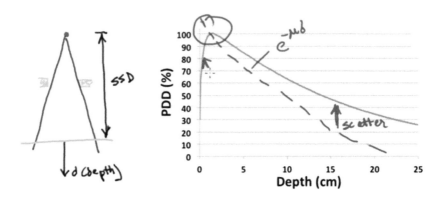

FIGURE 10.1.2
Percent depth dose (PDD) for a megavoltage photon beam. The PDD deviates from a pure exponential form (red dashed line).

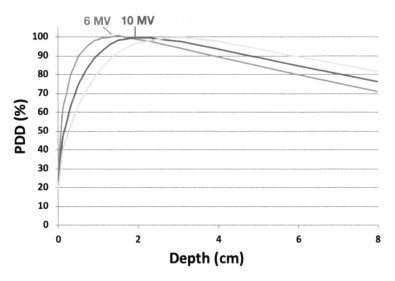

FIGURE 10.1.3
Depth d_{max} increases with beam energy.

Note that at higher energies the range of the electrons is longer and so the build up to equilibrium happens at a deeper depth. This is illustrated in Figure 10.1.3. **This is a key concept: that the depth of d_{max} depends on beam energy, with higher energy beams having a deeper depth of d_{max}.** Some approximate values are 6 MV $d_{max} = 1.5$ cm, 10 MV $d_{max} = 2$ cm, 18 MV $d_{max} = 3$ cm, which are useful values to remember.

10.1.3 PDD: Energy and Field Size

The PDD of a beam depends on the energy and field size. Not only does the depth of d_{max} depend on energy (Figure 10.1.3), but the falloff beyond d_{max} also depends on the energy and field size. Figure 10.1.4A shows that the PDD is larger for higher energy beams. This is due to the fact that a high-energy beam is attenuated somewhat less, or, in other words, the mass attenuation coefficient is slightly smaller at higher energies (see Section 5.2 and Figure 5.2.5). Note that the depth of d_{max} also depends somewhat on the field size.

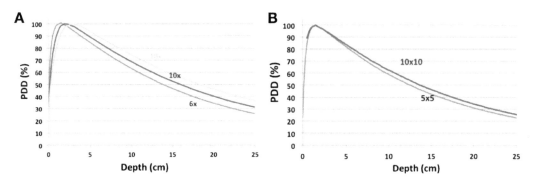

FIGURE 10.1.4
PDD dependence on (A) beam energy and (B) field size.

The nominal field size for d_{max} is 10×10 cm, but at field sizes below approximately 5×5 cm the depth of d_{max} decreases due to phantom scatter effects, and also at larger field sizes the depth of d_{max} decreases due to scatter effects in the linac head.

The shape of the PDD also depends somewhat on field size (Figure 10.1.4B) with larger field sizes having larger PDDs. This is due to a scatter effect. There is more scatter toward the center for larger fields since there is a larger region for the scatter contribution to originate from. The dose is therefore higher. Note though that the field size effect is not large.

10.1.4 Profiles and Penumbra

We now discuss the profiles of the beam, that is the dose as a function of the position across the beam at a fixed depth (Figure 10.1.1B). One important aspect of this profile is the dose falloff (gradient) at the edge of the field. The width of this region is called the "penumbra" and is important for dose shaping in treatment planning. A quantitative metric is often useful, and one that is in common use is the "80–20% penumbra" which is defined as the distance from the location of 80% dose to the position of 20% dose in the profile (Figure 10.1.5).

There are several effects which contribute to penumbra. First is geometric penumbra. This was described in Section 9.1.6. Geometric penumbra increases with source size, SSD, and depth and decreases with source-to-collimator distance (SCD). Recall from Equation 9.1 that the dependency is $w = s\left(\dfrac{SSD+d}{SCD} - 1\right)$. A second contribution is the shape of the beam collimation device (e.g. MLC) and the transmission through the edge of this device. Finally, there is another contribution to penumbra arising from the motion of electrons in the medium. Photons undergo Compton scattering in the medium and produce high-energy electrons (Figure 10.1.6). Some of these electrons scatter out away from the field edge. If the scattering event happens close to the edge (compared to the range of electrons in tissue) then the electron can scatter out of the field and contribute extra dose outside the field edge. Similarly, some of the electrons that normally would have contributed dose near the field edge have scattered out and so the dose in this region is reduced. The result is a "smearing out" of the dose falloff at the edge, i.e. an increase in the penumbra. There is

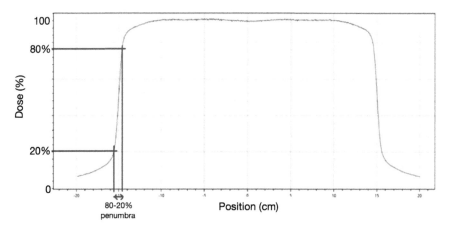

FIGURE 10.1.5
Beam profile and penumbra definition.

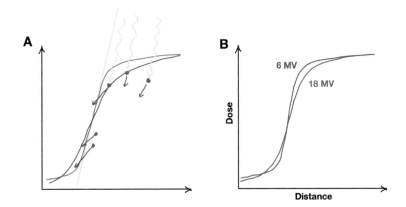

FIGURE 10.1.6
Radiological penumbra. (A) Electrons (blue) are scattered by photons and travel some distance before depositing their energy. (B) Beam data show a larger penumbra for an 18 MV beam vs. a 6 MV beam.

an energy-dependence to this effect. At higher energies the range of the electrons in tissue is larger and so the smearing out is larger. The penumbra of a high-energy beam is larger than a low-energy beam. This can be appreciated in Figure 10.1.6B which shows beam data from a 6 MV vs. 18 MV beam.

10.1.5 Profile Flatness

As can be appreciated in Figure 10.1.5 at this depth (10 cm) the profile of the beam is relatively flat. That is, the dose is relatively uniform across the field except at the extreme edges. This is a desirable property of the beam since it provides a uniform dose to the target tissue in this region. Figure 10.1.7 shows, however, that the beam flatness depends strongly on depth. In order for the dose to be flat at a depth of 10 cm the dose at shallower depth like d_{max} must be very unflat. This can be seen in the profile where, at d_{max}, the dose near the edges of the field is substantially higher than in the center. These are the so-called "horns" of the beam. At deeper depths, the beam becomes flatter because there is more

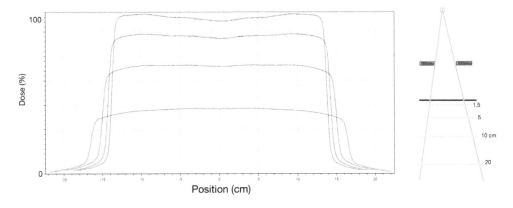

FIGURE 10.1.7
Beam profile dependence on depth. Profiles for a 6 MV beam are shown for four different depths in tissue as labeled.

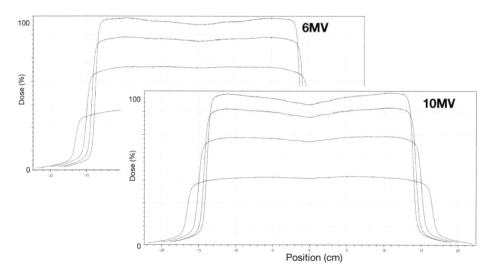

FIGURE 10.1.8
Beam profiles: dependence on energy.

scatter contribution to the center of the beam. That is, photons and electrons are scattered in toward the center of the beam and contribute a higher dose there. Also the spectrum of the beam is somewhat harder at the center of the beam since it has passed through the thicker part of the flattening filter, resulting in a faster falloff with depth. One can imagine there is an energy dependence to these effects, since scattering and the spectrum depend on energy. This is shown in Figure 10.1.8. The horns at d_{max} are larger in a 10 MV beam vs. a 6 MV beam.

10.2 Megavoltage Photon Beams: Effects in Patients

This section considers the properties of megavoltage photon beams in the context of patient treatments. This will move the description beyond the theoretical considerations outlined above to practical considerations in the clinical context.

10.2.1 The Physics of Skin-Sparing

Of the many layers of skin, the basal layer is the most important in the context of radiation therapy. This layer is home to rapidly growing and dividing cells and so is especially sensitive to radiation. The basal layer is 0.05 mm to 0.4 mm deep, and physicists commonly use 0.1 mm deep as a reference depth. ICRU Report #38 recommends the specific depth of 0.07 mm as a reference point. Figure 10.2.1 shows the dose at these depths acquired through careful measurement. Note that the dose at the surface is low (20–40% of that at d_{max}). This is the effect of "skin-sparing" and buildup described in Section 6.2.2 and Section 10.1.2. Recall that it is actually electrons that deposit dose and there are relatively few electrons at the surface. However, the contribution is not zero. There is some dose. This is due to electrons created by components in the head of the linac, such as the flattening filter or the

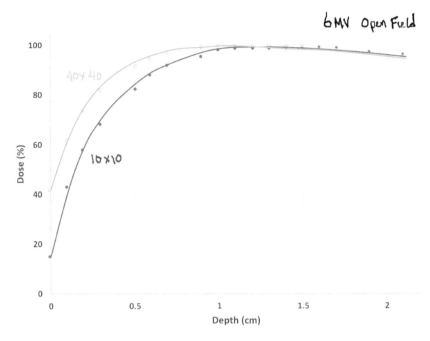

FIGURE 10.2.1
Superficial dose from a 6 MV field. (Adapted from Meydanci, T. and Kemikler, G. *Radiat Med*, 26, 539–544, 2008.)

primary collimator (Section 9.1). Skin-sparing is a key effect that allows for the treatment of deep-seated tumors while limiting toxicity.

10.2.2 Skin-Sparing Dependencies

The skin-sparing effects depend on field size, beam energy, the devices in the beam, and the angle of the beam relative to the patient. A wealth of information on this topic can be found in AAPM Report #176 (Olch et al. 2014). This section provides a brief summary of the essential points.

As seen in Figure 10.2.1 the skin dose depends strongly on field size. For a 40×40 cm field the skin dose is approximately twice that of the 10×10 cm field. The enhancement at large field sizes is due to the increased number of electrons that are scattered from the head of the linac collimation system at larger field sizes.

Figure 10.2.2 shows the dependence on the beam energy. Clearly the dose in the superficial region is lower with the higher energy beam. However, at the skin (0.1 mm reference depth) the dose from the two beams is roughly similar. Some medical physics texts have promulgated a canonical view that the skin dose is lower for a higher energy beam, but while this is true the effect is not large (c.f. Figure 10.2.2). While the superficial dose in the first cm or so is certainly lower, the dose at the skin is similar in most cases.

One of the parameters which affects skin dose most strongly is the presence of devices in the beam path. This topic is presented clearly and in great detail in AAPM Task Group #176 (2014). Figure 10.2.3 shows an example of one such device, the table or "couch" or patient support assembly. If a beam is directed through the table, then the table has two effects on the dose distribution. First, the dose at deeper depths is attenuated somewhat, typically approximately 2%, though the actual magnitude depends on the situation. Second the skin dose is increased.

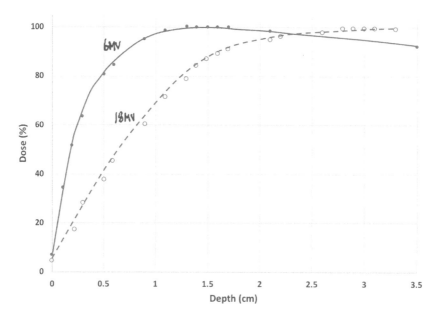

FIGURE 10.2.2
Superficial dose: Energy effect. (Adapted from Meydanci, T. and Kemikler, G. *Radiat Med*, 26, 539–544, 2008.)

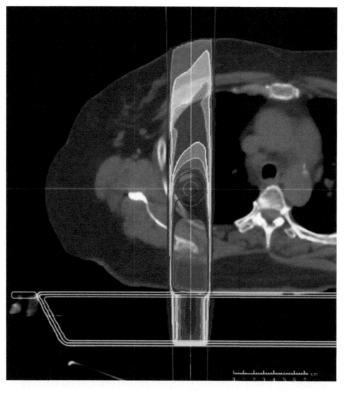

FIGURE 10.2.3
Effects of the table (patient support assembly) on dose distributions.

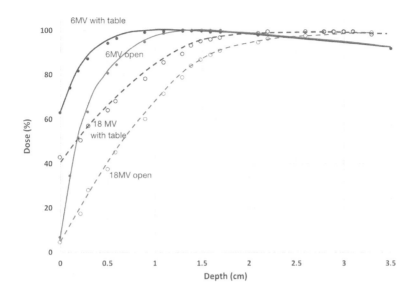

FIGURE 10.2.4
Effect of the table on superficial doses. (Adapted from Meydanci, T. and Kemikler, G. *Radiat Med*, 26, 539–544, 2008.)

The increase in skin dose can be substantial. Figure 10.2.4 shows PDD data for superficial doses from beams with a table in the beam (red) vs. without a table in the beam (blue). These data for a 6 MV beam show an increase of skin dose from approximately 10% to 60% with the table in place. The magnitude of the increase depends sensitively on the type of couch used and other details of the field design. More details can be found in the AAPM Task Group #176 Report. The magnitude of the increase also depends on the energy of the beam. The increase in skin dose is more modest for an 18 MV beam (dashed lines in Figure 10.2.4) vs. a 6 MV beam (solid lines in Figure 10.2.4).

The table is not the only type of device in the beam. Immobilization devices are often used to hold the patient in place during treatment and to ensure an accurate setup from day to day. One such device is a bag that fits to the form of the patient (e.g. a vacuum-sealed bag with beads inside, Figure 10.2.5). As with the table itself this bag also increases skin dose. Values in TG176 range from 10% to 50% for a 6 MV beam depending on the field size and thickness of the bag.

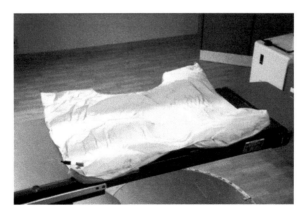

FIGURE 10.2.5
Vacuum-sealed bag for patient immobilization. (Image courtesy of Civco Inc.)

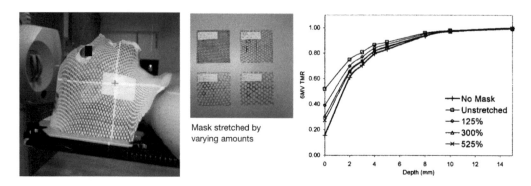

FIGURE 10.2.6
Thermoplastic masks for patient immobilization and the effect on surface dose. Masks can be stretched by different amounts which affects the surface dose. (From Hadley et al. *J Appl Clin Med Phys*, 6(1), 2005.)

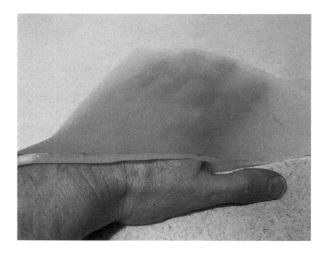

FIGURE 10.2.7
Bolus placed on top of the hand. Here, a 5-mm thick Superflab.

Another common example of an immobilization device is the thermoplastic mask shown in Figure 10.2.6. As with the table, this mask increases the skin dose even though there are holes present to minimize this effect. These masks are heated up and stretched over the patient. The more the mask is stretched, the thinner the material is, and the lower the skin dose is (Figure 10.2.5).

Another device that is sometimes place in the beam is a tray made of plastic (lucite), fixed just under the linac head (see Figure 9.1.7). This is most often used to hold blocks for beam shaping. The tray also increases skin dose, again due to electrons produced in the device that reach the patient. The amount of skin dose increase depends on how close the device is placed to the patient; the closer to the patient the higher the skin dose.

10.2.3 Bolus

The devices outlined in the previous section increase skin dose in a way that is usually unwanted. However, sometimes the goal is to increase the superficial dose. Disease may

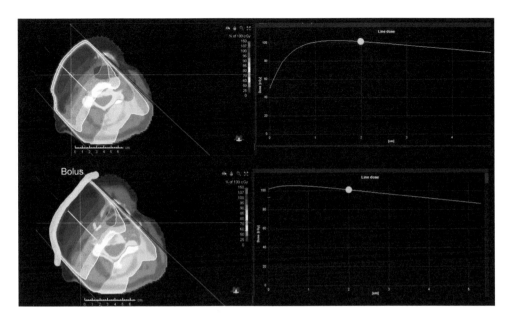

FIGURE 10.2.8
Effect of bolus on skin dose. A plan without bolus (top) vs. a plan with 5 mm bolus (bottom). PDDs along the central axis line are shown. Both cases use the same prescription dose at depth of 2 cm (dot).

extend up to or include the skin, for example, or a surgical scar may need to be treated in order to eliminate possible cancer cells there. In these cases bolus is used.

"Bolus" in this context means material placed on the skin of the patient in order to increase skin dose. Examples include synthetic materials such as the oil gel "Superflab" (Figure 10.2.7) or natural materials such as bees wax or paraffin. The material is often tissue equivalent but not always. For example, metal meshes are sometimes used to provide bolus.

The impact of bolus can be appreciated in the treatment plan shown in Figure 10.2.8. This shows two plans, one without bolus (top) and one with 5-mm bolus (bottom). In both cases the prescription dose is set to be the same at a depth of 2 cm (dot). Clearly, with the bolus the surface dose is much higher (100%) than without the bolus (50%).

Further Reading

Kahn, F.M. and J.P. Gibbons. 2014. *Kahn's The Physics of Radiation Therapy*. 5th Edition. Chapters 9 and 13. Philadelphia, PA: Wolters Kluwer.

McDermott, P.N. and C.G. Orton. 2010. *The Physics of Radiation Therapy*. Chapters 9, 10 and 14. Madison, WI: Medical Physics Publishing.

Metcalfe, P., T. Kron and P. Hoban. 2007. *The Physics of Radiotherapy X-rays and Electrons*. Chapter 4. Madison, WI: Medical Physics Publishing.

Olch, A.J., et al. 2014. Dosimetric effects caused by couch tops and immobilization devices: Report of AAPM Task Group 176. *Med Phys* 41(6):061501. doi:10.1118/1.4876299.

Chapter 10 Problem Sets

*Note: * indicates harder problems.*
See below for tables of useful data.

TABLE 10.1

6 MV, 10×10 cm, SSD 100

Depth (cm)	PDD (%)
0	65
1.5	100
5	86
10	67
15	52
18.5	43
20	39

1. What is the dose delivered at d_{max} for a single field for a prescribed dose of 100 cGy at a depth of 10 cm for a single 6 MV beam 10×10 cm field size? Use Table 10.1 for the PDD data.

    a. 1.49 cGy

    b. 67 cGy

    c. 149 cGy

    d. 222 cGy

2. Rank the following beams in order of smallest to largest dose delivered at a depth of d_{max} from a single field for a prescribed dose of 100 cGy at a depth of 10 cm.

    a. 6 MV, 10×10 cm

    b. 18 MV, 10×10 cm

    c. 6 MV, 30×30 cm

    d. 18 MV, 30×30 cm

3. * What is the dose delivered to a point at a depth of d_{max} for a plan with two fields, assuming a total dose of 100 cGy delivered to the midline point through equally weighted fields? Assume a separation of 20 cm, field size of 10×10 cm, energy of 6 MV and each field is set up at 100 SSD. Use PDD table below.

    a. 74.6 cGy

    b. 96.1 cGy

    c. 106.7 cGy

    d. 118 cGy

4. Rank the following beams in order of increasing dose to the skin.
   a. 10 MV, 30×30 cm field, open field
   b. 6 MV, 10×10 cm field, 1 cm bolus
   c. 15 MV, 30×30 cm field, carbon fiber tabletop in beam

5. What might you do to reduce skin reaction in a patient receiving treatment for a sarcoma on the upper posterior thigh (select all that apply)?
   a. Change the patient orientation to prone
   b. Remove the immobilization device in the field
   c. Decrease the beam energy
   d. Increase the weighting of the posterior beams

6. What is the impact of not including the patient support assembly (couch) in the treatment plan (select all that apply)?
   a. Delivered dose low at isocenter
   b. Delivered dose high at isocenter
   c. Delivered dose low at skin
   d. Delivered dose high at skin

7. List at least three ways to increase the superficial and skin dose in a breast cancer treatment with photon beam tangents.

11

Megavoltage Photon Beams: TMR and Dose Calculations

11.1 PDD and TMR

The previous chapter described the percent depth dose (PDD) and its dependence on energy, field size, and other factors. Unfortunately, the PDD is not directly useful for isocentric treatments. Instead a more useful, and closely related, quantity is the tissue-maximum ratio (TMR). **The TMR is a key concept and is described in this section**.

To understand the behavior and limitations of PDD consider the setup in Figure 11.1.1 where the PDD is acquired at one SSD setup (blue, "SSD_1"). As described in Chapter 10, the PDD falloff includes the following effects: attenuation in tissue (described by the $e^{-\mu x}$ formalism), scatter, and inverse square falloff. For a discussion of the inverse square effect see Section 4.2.3. If the PDD data were acquired for a different SSD (red, "SSD_2" in the figure) then the attenuation in tissue and scatter effects would be approximately the same. However, the falloff due to the inverse square effect would be substantially less. To understand this, remember that the PDD is the dose-relative to the dose acquired at a depth of d_{max}. If the SSD were very large, say 100 meters, then the inverse square falloff relative to the depth of d_{max} would be very small. Conversely, if the SSD were very small, say 10 cm, then the inverse square falloff relative to the depth of d_{max} would be very large. **This represents a key concept, namely that the relative dose falls off more quickly with depth for smaller SSDs than larger SSDs.**

From the above considerations it is clear that the PDD depends on the SSD used. It is possible to convert the PDD from one SSD setup to that of another SSD setup. The formula is as follows:

$$\text{PDD}_2 = \left[\left(\frac{\text{SSD}_2 + d_m}{\text{SSD}_2 + d} \right)^2 \cdot \left(\frac{\text{SSD}_1 + d}{\text{SSD}_1 + d_m} \right)^2 \right] \text{PDD}_1 \tag{11.1}$$

The term in square brackets is known as the "Mayneord F factor." If SSD_2 is larger than SSD_1, then the PDD_2 should be larger than PPD_1 since the inverse square falloff has less of an effect. This can be helpful in remembering the formula in the brackets.

To understand the limitations of PDD in routine clinic use consider the situation presented in Figure 11.1.2. Here a patient is treated isocentrically. That is the isocenter is a fixed point in the patient (black plus symbols in the figure) and the gantry is rotated around. As a result, the SSD is 80 cm for the AP beam but is 95 cm for the PA beam. Because the SSDs are not the same for the two beams it would not be possible to use a single PDD table for performing calculations. Rather the PDD would have to be converted to the appropriate

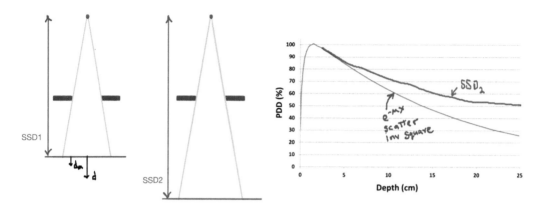

FIGURE 11.1.1
PDD depends on SSD. Although tissue attenuation and scatter effects are roughly the same, the inverse square falloff is less at the larger SSD.

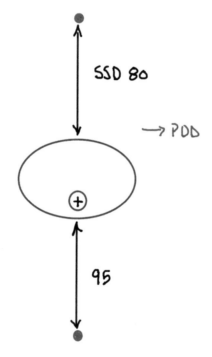

FIGURE 11.1.2
Variable PDD for isocentric treatments.

SSD value using Equation 11.1 and then used for calculation. This is a cumbersome and error-prone process. PDD, therefore, is not a convenient quantity for isocentric treatments.

11.1.1 TMR Definition

Here we consider a quantity called the tissue-maximum ratio (TMR) which is more useful for isocentric treatments. TMR is measured and defined by fixing the isocenter and changing the depth in water or tissue. This is different than the PDD where the surface is fixed

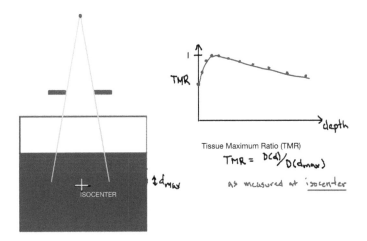

FIGURE 11.1.3
Tissue-maximum ration (TMR) measurements. The isocenter point is kept fixed and the depth is varied.

(fixed SSD) and the dose is measured at different depths. TMR is defined as the ratio of doses at two depths:

$$TMR(\text{depth} = d) = \frac{D(\text{depth} = d)}{D(\text{depth} = d_{\max})} \qquad (11.2)$$

TMR measurements are illustrated in Figure 11.1.3 (for an animation see the video). The isocenter is fixed and the depth is varied. This might be done, for example, in a water tank by progressively filling the tank with water as measurements are taken. Figure 11.1.3 shows a plot of the dose vs. depth curve under this scenario. The TMR is the ratio of the measurements at each depth scaled to a depth of $d_{\max}$. Note that because of the way that TMR is defined and measured the point of measurement is not moving. The detector is fixed at isocenter. Therefore, there is no inverse square effect in the TMR. The only effects at play are attenuation and scatter. A quantity that one sometimes encounters is tissue-phantom ratio (TPR). This is simply defined as the ratio of dose at depth, d, scaled to some reference depth, d_{ref}: $TPR(d) = D(d)/D(d_{\text{ref}})$. Note that this is similar to TMR except that the reference depth may be something other than $d_{\max}$.

Like PDD, the TMR also depends on beam energy and field size, since the same physical effects are at work. As the beam energy increases, TMR falls off less quickly with depth due to reduced attenuation at higher energy. Also the depth of $d_{\max}$ increases. The TMR also increases with field size due to scatter effects, just like PDD.

11.2 Monitor Unit (MU) Calculations

This section describes the methodology for dose calculations in megavoltage beams. It relies on the concepts of TMR described above. **These calculations are fundamental to the practice of radiation oncology and as such form a key concept.** The most effective way to learn this is to do problems. A sample selection of problems is found at the end of this chapter.

11.2.1 MU Calculation Formula

The goal for MU calculations is to find the number of monitor units (MU) required to deliver the desired dose at some depth in tissue. Recall that the monitor unit chamber is the ionization chamber in the head of the linac that tracks dose (Section 9.1.2). One MU is essentially one unit or one "tick" in the chamber and represents a certain amount of dose delivered. The linac is typically calibrated to deliver 1 cGy per 1 MU under standard calibration conditions (at isocenter at a depth of d_{max} with 10×10 cm field size).

The remainder of this chapter will describe the formulae used to calculate MU, presented in increasing complexity. Monitor unit calculations are the subject of AAPM Task Group #71 (Gibbons et al. 41(3), 2014) and further details can be found there. In its simplest form the formula for calculating the number of MU given the dose at a depth, d, given the dose, D, is as follows:

$$MU = \frac{D(cGy)}{REF \cdot S_c \cdot S_p \cdot TMR(d, FS)} \qquad (11.3)$$

Here TMR depends on depth, d, and field size, FS, so it written as $TMR(d, FS)$. S_c and S_p are collimator and phantom scatter factors respectively, and will be described momentarily. The term REF here represents the reference calibration, that is, the dose per MU under reference conditions which is typically 1 cGy/MU for a 10×10 cm field.

Consider a simple example where we want to calculate the number of MU required to deliver 100 cGy at isocenter to a depth of d_{max} for a 10×10 cm field. Here the TMR=1 since the depth is d_{max}. S_c and S_p are 1 since they are defined as such at the reference field size. REF is 1. The number of MU is therefore 100. Say the problem were instead to calculate the MU required to deliver 100 cGy to isocenter at a depth of 10 cm with this field. To do this we would look up the $TMR(d = 10\ cm, 10×10)$. Say this is 0.75. The required MU, therefore, would be MU = 100/0.75 = 133. Note that more MU are needed to deliver the same dose to this point since the depth is increased and the tissue attenuation is larger.

Hopefully, it is obvious also that Equation 11.3 can also be used to find the dose delivered if the number of MU are known and if S_c, S_p, and TMR can be determined.

Before considering more complex calculations let us first consider the S_c and S_p factors in more detail. This follows the formalism of AAPM TG71. Figure 11.2.1 shows an example treatment field. Note that the outside border of the field is defined by the jaws or collimator

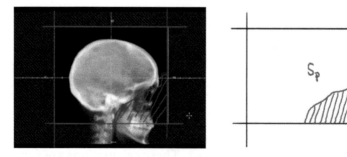

FIGURE 11.2.1
Field definition and the S_c and S_p factors used for MU calculations.

while the blocks or MLCs define a blocked field. S_c is the collimator scatter factor, a factor which accounts for the scatter radiation from the collimator system (or jaws). As such it is related to the unblocked field, i.e. the field defined by the jaws. S_p is the patient or phantom scatter factor, a factor which accounts for the scatter radiation in the field that is irradiated in the patient. As such it is related to the blocked field in the patient, i.e. the part of the irradiated field that actually intersects with the patient. Both S_c and S_p are defined to be one for a 10×10 cm field. S_c and S_p vary over a range of approximately 0.95 to 1.05 for most clinical fields. For field sizes larger than 10×10 cm the S_c and S_p factors are greater than one while for field sizes less than 10×10 cm they are less than one.

Equation 11.3 can be generalized somewhat to include three other factors that are important in some situations, namely:

- *Wedge factor (WF)*. This is included if a wedge is used. The factor is the ratio of the dose with the wedge vs. without the wedge.
- *Off-axis ratio (OAR)*. This is included if the calculation point is not at isocenter at the center of the field.
- *Tray factor (TF)*. This is included if the plastic block tray is in the beam. The factor is a ratio of the dose with the tray vs. without the tray.

$$MU = \frac{D(\text{cGy})}{\text{REF} \cdot S_c(\text{FS}_{\text{col}}) \cdot S_p(\text{FS}_{\text{eq}}) \cdot \text{TMR}(d, \text{FS}_{\text{eq}}) \cdot \text{WF} \cdot \text{OAR} \cdot \text{TF}} \tag{11.4}$$

Note that this equation is written to explicitly include the field size dependencies. That is, $S_c(\text{FS}_{\text{col}})$ indicates that the field size from the collimators or jaws is used, while $S_p(\text{FS}_{\text{eq}})$ and $\text{TMR}(d,\text{FS}_{\text{eq}})$ indicate that the field size from equivalent square of the blocked field, FS_{eq}, is used. The topic of equivalent squares is discussed in Section 11.2.5.

11.2.2 Example MU Calculations

To further understand S_c and S_p consider the example above of calculating the number of MU required to deliver 100 cGy to isocenter at a depth of d_{max}. However, instead of a 10×10 cm field as above we now consider a 20×20 cm field defined by the jaws and an effective 16×16 cm field defined by the MLC. We use Equation 11.3. Again the TMR$=1$ since the depth is d_{max} and REF$=1$. We look up S_c and S_p on a table for a 6 MV beam and find $S_c(20 \times 20) = 1.023$ and $S_p(16 \times 16) = 1.018$. Therefore, the number of MU required is $100/1.023/1.018 = 96$. Note that fewer MU are required in this case (96 vs. 100) since the S_c and S_p factors are greater than one. That is, there is more contribution from scatter radiation in the 20×20 cm and 16×16 cm fields compared to a 10×10 cm field, and therefore fewer MU are required to deliver the same dose.

To take another example problem, consider the field shown in Figure 11.2.2. The problem here is to calculate the number of MU required to deliver 150 cGy to the isocenter point using a 6 MV beam. Equation 11.3 is: $MU = \dfrac{150}{1 \cdot S_c(10 \times 10) \cdot S_p(9.5 \times 9.5) \cdot \text{TMR}(d = 10, FS = 9.5 \times 9.5)}$.

Note the blocked field is taken to have an equivalent field size of 9.5×9.5 cm (see Section 11.2.5). Looking up the numbers on a table we find $MU = \dfrac{150}{1 \cdot 1 \cdot 0.99 \cdot 0.81} = 187$. In other words, if the patient were set up as shown in Figure 11.2.2 and a 6 MV beam were used to deliver the blocked field indicated with 187 MU then the resulting dose would be 150 cGy to the isocenter point.

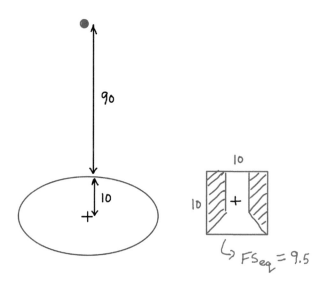

FIGURE 11.2.2
Setup for an example MU calculation. Isocenter (black plus) is at a depth of 10 cm.

11.2.3 MU Calculations at an Extended Distance

Now we consider a situation where the patient is set up at an extended distance. This is illustrated in Figure 11.2.3. Normally, the isocenter (plus symbol in the figure) is in the patient (left panel), but with extended SSD treatments the patient is set up distal to the isocenter (right panel). For the normal setup (left panel) MU calculations would be possible using Equation 11.4. However, for the extended SSD (right panel) this formalism does not work. A correction factor is needed.

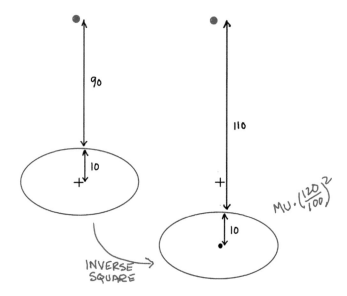

FIGURE 11.2.3
Geometry for an extended SSD setup. Normally, the isocenter (plus) is in the patient (left panel), but with extended SSD treatments the patient is set up distal to the isocenter (right panel).

To understand this, consider that the goal for extended SSD is still to calculate the dose to the point at a depth of 10 cm in the example shown. This is the same point in the patient in the right vs. left panel. The only difference is that at extended SSD this point is no longer at the isocenter. To find the MU then, we first calculate the MU in the left panel using the standard equations and then apply an inverse square factor to find the MU for extended SSD. This inverse square factor accounts for the fact that the patient is farther from the source in the right panel vs. left panel. Note that more MU will be needed in order to deliver the same dose.

The considerations above can be written as a slight modification to Equation 11.4 for treatment at extended SSD:

$$\text{MU} = \frac{D(\text{cGy})}{\text{REF} \cdot S_c(\text{FS}_{\text{col}}) \cdot S_p(\text{FS}_{\text{eq}}) \cdot \text{TMR}(d, \text{FS}_{\text{eq}}) \cdot \text{WF} \cdot \text{OAR} \cdot \text{TF} \cdot (100/(\text{SSD}+d))^2} \quad (11.5)$$

Here "SSD" is the extended SSD that is being used for treatment. Note also, that to be more accurate the FS_{eq} is the field size at the calculation point which is larger than the field size at isocenter because of divergence. Note that for an isocentric treatment (not at extended SSD), $\text{SSD}+d = 100$ and the term in the bottom becomes one and Equation 11.5 is the same as Equation 11.4.

11.2.4 MU Calculations with PDD

Though it is most often convenient to use TMRs for calculations it is also possible to use PDDs. This proves most useful in situations where the patient is set up at a particular SSD. The relevant equation is:

$$\text{MU} = \frac{D(\text{cGy})}{\text{REF} \cdot (100/(\text{SSD}+d_{\text{max}}))^2 \cdot S_c(\text{FS}_{\text{col}}) \cdot S_p(\text{FS}_{\text{eq}}) \cdot \text{PDD}(d, \text{FS}_{\text{eq}}, \text{SSD}) \cdot \text{WF} \cdot \text{OAR} \cdot \text{TF}} \quad (11.6)$$

This is very similar to Equation 11.5 except that a PDD is used instead. Note that since PDD depends on the SSD, it is important to use the correct PDD value for the SSD being used for treatment. Often an SSD of 100 is used. The extra factor in the denominator, $(100/(\text{SSD}+d_{\text{max}}))^2$, accounts for the fact that the machine is typically calibrated in an isocentric mode. That is "REF" is 1 cGy/MU at isocenter at a depth of d_{max}, so an inverse square correction is needed for a treatment with an SSD setup.

11.2.5 Equivalent Squares

For rectangular fields it is useful to consider what the radiologically equivalent square field would be. This greatly simplifies the data since values for square fields can be tabulated. Consider a rectangular field with dimensions $L \times W$ (Figure 11.2.4). There is an empirical formula, Sterling's formula, from which the equivalent square can be calculated:

$$\text{Equiv Square}, S = \frac{4A}{P} = \frac{2LW}{L+W} \quad (11.7)$$

Here A is the area and P is the perimeter. It is not possible to "prove" this formula mathematically. Rather it is based on empirical observations and fits to data.

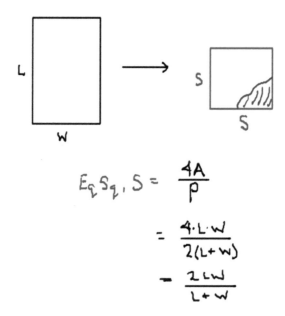

$$Eq\ Sq\ ,\ S = \frac{4A}{P}$$

$$= \frac{4 \cdot L \cdot W}{2(L + W)}$$

$$= \frac{2 LW}{L + W}$$

FIGURE 11.2.4
Equivalent square. The rectangular field (L×W) can be represented by a square field with size S×S.

To take an example consider a 2×9 cm field. The equivalent square would be 4·18/22 = 3.3 cm. That is, a 2×9 cm field is radiologically equivalent to a 3.3×3.3 cm field. In performing MU calculations for this case (e.g. Equation 11.5) an FS_{eq} of 3.3 would be used.

In some cases the field is not a simple rectangle but there may be some blocking (e.g. Figure 11.2.1). There are methods to account for blocking and to calculate an equivalent square in these cases. However, this is beyond the scope of this text and, in any case, is usually performed by computer algorithms in most clinics. However, a simple method for estimating the equivalent square in these cases is to estimate the size of the rectangle that would fit into the blocked field. A commonly used estimate for the equivalent square size of the blocked field is $FS_{Eq\text{-}blocked} = FS_{Eq\text{-}unblocked}\sqrt{1-f}$, where f is the fraction of the field that is blocked.

Further Reading

Kahn, F.M. and J.P. Gibbons. 2014. *Kahn's The Physics of Radiation Therapy*. Edition 5. Chapters 9 and 10. Philadelphia, PA: Wolters Kluwer.

McDermott, P.N. and C.G. Orton. 2010. *The Physics of Radiation Therapy*. Chapters 10, 12 and 13. Madison, WI: Medical Physics Publishing.

Metcalfe, P., T. Kron and P. Hoban. 2007. *The Physics of Radiotherapy X-rays and Electrons*. Chapters 4 and 6. Madison, WI: Medical Physics Publishing.

Chapter 11 Problem Sets

*Note: * indicates harder problems.*
See Section 11.3 below for data tables where needed.

1. What is the PDD (depth = 20 cm) of a 6 MV beam for a 90 cm setup, given that the PDD ($d = 20$ cm) for this beam is 0.350 with a 100 cm setup and the depth of d_{max} is 1.5 cm?

    a. 0.319
    b. 0.338
    c. 0.362
    d. 0.381

2. During an emergency treatment on the weekend, a PDD table is accidentally used instead of a TMR table to calculate dose at a depth of 10 cm for an isocentric setup. What is the approximate impact on the dose delivered?

    a. Dose to isocenter 20% low
    b. Dose to isocenter 10% low
    c. Dose to isocenter 10% high
    d. Dose to isocenter 20% high

3. What are the required MU to treat each field for the following setup?

    Treatment to abdomen with AP and PA beams
    Total dose 300 cGy, evenly weighted from both fields
    Field size 10 × 20 cm, no blocking
    Separation = 20 cm. Isocenter in the middle of the abdomen
    Energy = 6 MV

    a. 184
    b. 264
    c. 368
    d. 528

4. What are the required MU per field in Problem 3 if the treatment is delivery in a 100 SSD setup for each field instead of an isocentric treatment?

    a. 152
    b. 167
    c. 202
    d. 223

5. What are the required MU per field if the following parameters are changed in Problem 3? Note isocentric treatment is maintained.

Parameter Change	MU
6 MV -> 10 MV	
10×20 cm field -> 4×10 cm field	
20 cm separation -> 30 cm separation	

6. What are the required MU per field in Problem 3 if the lower left quarters of the fields are blocked?

   a.   175

   b.   181

   c.   187

   d.   195

7. What is the dose to isocenter for the following delivery?

   4 × 10 field with no blocking, 6 MV beam

   Isocenter at a depth of 10 cm

   180 MU delivered

   a.   125 cGy

   b.   130 cGy

   c.   135 cGy

   d.   140 cGy

8. What are the required MU for each field for the following treatment?

   AP/PA treatment, total dose 800 cGy, weighted 3:1 with more dose from the posterior beam

   Field size 4 × 10 cm, no blocking

   Separation = 22 cm. Isocenter at a depth of 5 cm posteriorly

   Energy = 10 MV (AP), 6 MV (PA)

   a.   AP = 218, PA = 772

   b.   AP = 330, PA = 688

   c.   AP = 678, PA = 491

   d.   AP = 847, PA = 385

9. * What is the dose at a depth of d_{max} of 1.5 cm from a single beam in Problem 3?

   a.   169 cGy

   b.   187 cGy

   c.   205 cGy

   d.   224 cGy

10. * What is the dose at the exit surface of the patient from a single beam in Problem 3?

    a. 86 cGy

    b. 95 cGy

    c. 104 cGy

    d. 113 cGy

Section 11.3 Data Tables

6 MV Scatter Factors

Field Size	Total S	S_c	S_p
4×4	0.930	0.949	0.980
6×6	0.958	0.972	0.986
8×8	0.984	0.986	0.998
10×10	1.000	1.000	1.000
12×12	1.015	1.006	1.009
16×16	1.034	1.016	1.018
20×20	1.047	1.023	1.024
24×24	1.060	1.028	1.031
30×30	1.077	1.031	1.045
40×40	1.085	1.032	1.051

6 MV TMR

Depth [cm]	Equivalent Square Field Size [cm]														
	-3×3-	-4×4-	-5×5-	-6×6-	-7×7-	-8×8-	-9×9-	-10×10-	-12×12-	-15×15-	-18×18-	-20×20-	-25×25-	-30×30-	-35×35-
0.0	41.4	42.2	43.3	44.6	45.7	46.8	47.7	48.6	50.9	54.3	56.9	58.7	61.8	64.9	66.1
0.2	49.8	50.5	51.4	52.8	54.0	55.3	56.3	57.3	59.4	62.5	64.9	66.5	69.4	72.2	73.1
0.4	66.3	66.7	67.5	68.9	70.0	71.2	72.0	72.7	74.3	76.7	78.4	79.6	81.4	83.2	83.8
0.6	80.6	80.8	81.2	82.0	82.8	83.6	84.2	84.8	85.9	87.5	88.6	89.3	90.4	91.4	91.6
0.8	89.2	89.2	89.5	90.1	90.6	91.1	91.4	91.7	92.4	93.5	94.1	94.5	95.1	95.7	95.7
1.0	94.3	94.3	94.5	94.9	95.1	95.4	95.7	95.9	96.3	96.9	97.2	97.5	97.8	98.0	98.1
1.2	97.5	97.3	97.5	97.8	97.9	98.0	98.1	98.2	98.5	98.8	99.0	99.0	99.1	99.2	99.3
1.4	99.2	99.0	99.0	99.1	99.3	99.4	99.4	99.5	99.6	99.7	99.7	99.8	99.8	99.8	99.9
1.6	99.8	99.8	99.8	99.9	100.0	100.0	99.9	99.9	99.9	100.0	100.0	100.0	100.0	100.0	99.9
$d_{max}=1.8$	100.0	100.0	100.0	100.0	100.0	100.0	100.0	100.0	100.0	100.0	100.0	99.9	99.9	99.9	100.0
2.0	99.7	99.8	99.9	99.9	99.9	99.9	99.9	99.9	99.9	99.9	99.8	99.8	99.9	99.9	99.9
2.2	99.3	99.3	99.6	99.7	99.6	99.6	99.5	99.5	99.4	99.4	99.4	99.3	99.4	99.4	99.8
2.4	98.8	98.9	99.0	99.2	99.2	99.2	99.1	99.1	99.1	99.1	99.1	99.0	99.1	99.1	99.5
2.6	98.1	98.2	98.5	98.7	98.7	98.8	98.7	98.7	98.7	98.8	98.8	98.8	98.9	98.9	99.2
2.8	97.4	97.7	98.0	98.2	98.3	98.3	98.3	98.3	98.3	98.4	98.4	98.4	98.4	98.5	99.0
3.0	96.9	97.1	97.4	97.6	97.7	97.7	97.7	97.7	97.8	97.9	98.0	98.0	98.1	98.2	98.8
3.2	96.1	96.4	96.8	97.1	97.2	97.2	97.2	97.3	97.4	97.6	97.6	97.7	97.8	97.9	98.5
3.4	95.4	95.8	96.2	96.4	96.6	96.6	96.7	96.8	96.9	97.1	97.1	97.1	97.3	97.5	98.1
3.6	94.5	95.1	95.6	96.0	96.1	96.2	96.2	96.3	96.4	96.6	96.7	96.7	96.9	97.1	97.7
3.8	93.9	94.4	95.0	95.4	95.6	95.7	95.7	95.8	96.0	96.2	96.3	96.4	96.5	96.7	97.3
4.0	93.2	93.7	94.2	94.6	94.9	95.0	95.2	95.3	95.5	95.8	95.9	96.0	96.2	96.4	97.0
4.2	92.5	93.1	93.6	94.0	94.2	94.4	94.6	94.8	95.0	95.3	95.5	95.5	95.8	96.0	96.7
4.4	91.8	92.4	93.0	93.4	93.6	93.8	94.0	94.2	94.4	94.8	95.0	95.1	95.4	95.6	96.3
4.6	91.0	91.7	92.3	92.8	93.1	93.2	93.4	93.6	93.9	94.3	94.6	94.7	95.0	95.3	95.9
4.8	90.2	91.0	91.6	92.2	92.5	92.7	92.9	93.1	93.4	93.8	94.0	94.1	94.5	94.9	95.6
5.0	89.4	90.2	91.0	91.6	92.0	92.2	92.4	92.6	92.9	93.3	93.6	93.7	94.1	94.5	94.7
5.2	88.8	89.5	90.2	90.8	91.2	91.5	91.8	92.0	92.3	92.8	93.1	93.3	93.6	94.0	94.4
5.4	87.9	88.8	89.5	90.1	90.5	90.9	91.2	91.5	91.8	92.3	92.7	92.9	93.2	93.6	94.0
5.6	87.3	88.2	88.9	89.4	89.9	90.3	90.6	90.9	91.3	91.9	92.2	92.4	92.8	93.2	93.6
5.8	86.5	87.4	88.3	88.9	89.4	89.7	90.0	90.3	90.7	91.3	91.7	91.9	92.3	92.8	93.2

6 MV TMR

Depth [cm]	-3×3-	-4×4-	-5×5-	-6×6-	-7×7-	-8×8-	-9×9-	-10×10-	-12×12-	-15×15-	-18×18-	-20×20-	-25×25-	-30×30-	-35×35-
6.0	85.7	86.7	87.6	88.3	88.8	89.1	89.4	89.8	90.2	90.8	91.2	91.4	91.9	92.4	92.8
6.2	85.1	86.1	87.0	87.7	88.2	88.5	88.9	89.2	89.6	90.3	90.8	91.0	91.4	91.9	92.2
6.4	84.3	85.3	86.2	86.9	87.5	87.9	88.4	88.7	89.2	89.8	90.3	90.5	91.0	91.4	91.9
6.6	83.5	84.6	85.5	86.2	86.8	87.3	87.7	88.1	88.6	89.3	89.8	90.0	90.6	91.1	91.5
6.8	82.8	83.9	84.8	85.7	86.3	86.7	87.1	87.5	88.0	88.8	89.3	89.5	90.1	90.6	91.1
7.0	82.0	83.2	84.2	85.1	85.7	86.1	86.6	87.0	87.5	88.3	88.8	89.1	89.6	90.2	90.7
7.2	81.4	82.5	83.5	84.3	85.0	85.4	85.9	86.4	87.0	87.8	88.4	88.6	89.2	89.8	90.3
7.4	80.8	81.9	82.9	83.7	84.3	84.8	85.3	85.8	86.4	87.3	87.9	88.2	88.8	89.3	89.9
7.6	80.1	81.2	82.3	83.2	83.9	84.3	84.8	85.3	85.9	86.8	87.4	87.7	88.3	88.9	89.5
7.8	79.3	80.5	81.6	82.5	83.2	83.7	84.2	84.7	85.3	86.2	86.9	87.2	87.9	88.6	89.1
8.0	78.6	79.9	80.9	81.8	82.5	83.1	83.7	84.2	84.8	85.8	86.4	86.7	87.4	88.1	88.7
8.2	78.2	79.3	80.3	81.2	81.9	82.5	83.0	83.6	84.3	85.2	85.9	86.3	87.0	87.7	88.2
8.4	77.4	78.6	79.7	80.7	81.4	81.9	82.5	83.0	83.7	84.7	85.4	85.8	86.5	87.3	87.9
8.6	76.6	77.9	79.1	80.1	80.8	81.3	81.9	82.4	83.1	84.1	84.9	85.3	86.1	86.8	87.4
8.8	76.0	77.3	78.4	79.4	80.2	80.7	81.3	81.8	82.6	83.7	84.5	84.9	85.7	86.4	87.0
9.0	75.4	76.7	77.8	78.7	79.6	80.1	80.8	81.3	82.1	83.2	84.0	84.3	85.1	86.0	86.6
9.2	74.6	76.0	77.1	78.2	79.0	79.6	80.2	80.8	81.6	82.6	83.5	83.8	84.7	85.6	86.2
9.4	74.1	75.4	76.5	77.6	78.4	79.0	79.7	80.3	81.1	82.2	83.0	83.4	84.3	85.1	85.7
9.6	73.5	74.7	75.9	77.0	77.9	78.4	79.1	79.7	80.5	81.6	82.5	82.9	83.8	84.6	85.3
9.8	72.8	74.1	75.3	76.3	77.2	77.8	78.5	79.1	79.9	81.1	82.0	82.4	83.4	84.2	84.9
10.0	72.2	73.5	74.7	75.8	76.6	77.2	77.9	78.5	79.4	80.5	81.5	81.9	82.9	83.7	84.4
10.2	71.6	72.9	74.1	75.1	75.9	76.6	77.3	77.9	78.9	80.1	81.0	81.5	82.4	83.2	83.9
10.4	70.9	72.2	73.4	74.4	75.3	76.0	76.7	77.4	78.3	79.5	80.5	81.0	81.9	82.8	83.5
10.6	70.2	71.5	72.7	73.8	74.7	75.4	76.2	76.7	77.6	78.9	79.9	80.4	81.4	82.3	83.1
10.8	69.7	70.9	72.1	73.2	74.1	74.8	75.6	76.2	77.1	78.4	79.4	79.9	80.9	81.9	82.7
11.0	69.0	70.2	71.5	72.7	73.6	74.3	75.0	75.6	76.6	77.9	79.0	79.5	80.5	81.5	82.3
11.2	68.4	69.6	70.9	72.1	73.0	73.7	74.5	75.1	76.1	77.4	78.5	79.0	80.0	81.0	81.8
11.4	67.8	69.1	70.3	71.5	72.4	73.1	73.9	74.6	75.6	76.9	78.0	78.5	79.5	80.5	81.3
11.6	67.3	68.5	69.7	70.9	71.8	72.5	73.3	74.0	75.1	76.4	77.5	78.0	79.1	80.1	80.9
11.8	66.7	68.0	69.2	70.3	71.2	71.9	72.7	73.4	74.5	75.9	77.1	77.6	78.7	79.7	80.6
12.0	66.2	67.5	68.6	69.7	70.6	71.4	72.2	73.0	74.1	75.4	76.6	77.1	78.2	79.3	80.2
12.2	65.6	66.8	68.0	69.2	70.2	70.9	71.7	72.5	73.5	74.9	76.1	76.5	77.7	78.7	79.7
12.4	65.1	66.3	67.5	68.6	69.6	70.3	71.2	71.9	73.0	74.4	75.6	76.1	77.3	78.4	79.3
12.6	64.5	65.7	66.9	68.1	69.1	69.8	70.7	71.4	72.5	73.9	75.1	75.7	76.9	77.9	78.8
12.8	63.9	65.1	66.3	67.5	68.6	69.3	70.2	70.9	72.0	73.4	74.7	75.2	76.4	77.5	78.4
13.0	63.4	64.6	65.8	67.0	68.0	68.7	69.6	70.3	71.4	72.9	74.2	74.7	76.0	77.1	78.0

6 MV TMR

Depth [cm]	Equivalent Square Field Size [cm]														
	-3×3-	-4×4-	-5×5-	-6×6-	-7×7-	-8×8-	-9×9-	-10×10-	-12×12-	-15×15-	-18×18-	-20×20-	-25×25-	-30×30-	-35×35-
13.2	62.9	64.1	65.3	66.5	67.5	68.2	69.0	69.8	71.0	72.5	73.8	74.4	75.6	76.6	77.6
13.4	62.2	63.5	64.7	65.9	66.9	67.6	68.5	69.3	70.4	71.9	73.2	73.8	75.1	76.2	77.2
13.6	61.7	63.0	64.2	65.4	66.4	67.1	68.0	68.8	69.9	71.4	72.7	73.3	74.6	75.8	76.7
13.8	61.1	62.5	63.7	64.8	65.8	66.5	67.4	68.2	69.5	70.9	72.2	72.8	74.1	75.3	76.3
14.0	60.7	61.8	63.0	64.1	65.2	66.0	66.9	67.7	68.9	70.4	71.8	72.4	73.7	74.9	76.0
14.2	60.1	61.3	62.6	63.8	64.9	65.6	66.4	67.2	68.4	69.9	71.3	71.9	73.3	74.4	75.5
14.4	59.5	60.7	62.0	63.2	64.3	65.0	65.8	66.6	67.8	69.4	70.8	71.4	72.8	74.0	75.1
14.6	59.0	60.3	61.4	62.7	63.7	64.5	65.4	66.1	67.4	68.9	70.3	71.0	72.4	73.5	74.6
14.8	58.5	59.8	61.0	62.2	63.2	64.0	64.8	65.6	66.9	68.4	69.8	70.5	72.0	73.1	74.1
15.0	58.0	59.3	60.4	61.5	62.6	63.4	64.2	65.0	66.3	67.9	69.4	70.0	71.4	72.6	73.7
15.2	57.5	58.7	59.9	61.1	62.1	62.8	63.7	64.5	65.7	67.4	68.9	69.5	71.0	72.1	73.2
15.4	56.9	58.2	59.4	60.6	61.6	62.3	63.2	64.0	65.3	66.9	68.4	69.0	70.5	71.7	72.8
15.6	56.5	57.7	58.9	60.0	61.1	61.8	62.7	63.5	64.9	66.5	68.0	68.6	70.0	71.3	72.4
15.8	56.0	57.2	58.4	59.5	60.6	61.3	62.2	63.0	64.4	66.0	67.5	68.1	69.6	70.9	72.0
16.0	55.5	56.7	57.9	59.0	60.0	60.8	61.7	62.6	64.0	65.6	67.1	67.7	69.2	70.5	71.6
16.2	55.0	56.2	57.4	58.5	59.6	60.4	61.3	62.1	63.5	65.1	66.7	67.3	68.8	70.0	71.1
16.4	54.5	55.7	56.8	58.0	59.1	59.9	60.8	61.6	63.0	64.6	66.1	66.8	68.3	69.6	70.7
16.6	54.0	55.2	56.4	57.6	58.6	59.4	60.2	61.0	62.3	64.0	65.7	66.3	67.9	69.2	70.3
16.8	53.6	54.8	56.0	57.2	58.2	58.9	59.8	60.6	61.9	63.6	65.2	65.9	67.4	68.7	69.9
17.0	53.2	54.4	55.5	56.6	57.6	58.4	59.3	60.1	61.5	63.2	64.8	65.4	66.9	68.2	69.5
17.2	52.6	53.9	55.1	56.2	57.2	58.0	58.9	59.7	61.1	62.7	64.3	65.0	66.5	67.8	69.0
17.4	52.2	53.4	54.6	55.6	56.7	57.5	58.4	59.3	60.6	62.3	63.9	64.5	66.1	67.4	68.6
17.6	51.9	53.0	54.2	55.2	56.2	57.0	57.9	58.8	60.2	61.9	63.5	64.1	65.7	67.0	68.2
17.8	51.4	52.6	53.7	54.8	55.8	56.5	57.4	58.3	59.7	61.4	63.0	63.7	65.3	66.6	67.8
18.0	51.0	52.1	53.2	54.3	55.3	56.1	56.9	57.8	59.3	60.9	62.5	63.2	64.9	66.2	67.4
18.2	50.5	51.6	52.7	53.9	54.9	55.7	56.5	57.4	58.8	60.4	62.1	62.8	64.5	65.8	67.0
18.4	50.2	51.2	52.3	53.4	54.5	55.3	56.2	57.1	58.4	60.1	61.7	62.4	64.1	65.4	66.6
18.6	49.7	50.7	51.8	53.0	54.1	54.9	55.8	56.6	57.9	59.6	61.3	62.0	63.7	65.0	66.3
18.8	49.2	50.3	51.5	52.7	53.8	54.5	55.3	56.2	57.5	59.2	60.9	61.6	63.2	64.6	65.9
19.0	48.8	49.9	51.1	52.2	53.2	54.0	54.9	55.7	57.1	58.8	60.5	61.2	62.9	64.2	65.5
19.2	48.5	49.6	50.7	51.8	52.8	53.6	54.4	55.3	56.7	58.4	60.1	60.8	62.4	63.8	65.0
19.4	48.1	49.1	50.3	51.3	52.3	53.1	53.9	54.8	56.2	58.0	59.8	60.4	62.0	63.4	64.6
19.6	47.7	48.7	49.8	50.9	51.9	52.6	53.4	54.3	55.8	57.5	59.3	60.0	61.6	63.0	64.2
19.8	47.2	48.2	49.4	50.5	51.4	52.2	53.0	53.9	55.4	57.1	58.8	59.5	61.2	62.5	63.9
20.0	46.9	47.8	48.9	50.1	51.0	51.8	52.6	53.5	54.9	56.6	58.3	59.0	60.7	62.1	63.5
20.2	46.5	47.4	48.5	49.6	50.5	51.3	52.1	53.0	54.5	56.2	58.0	58.7	60.5	61.8	63.1

6 MV TMR

Depth [cm]	-3×3-	-4×4-	-5×5-	-6×6-	-7×7-	-8×8	-9×9-	-10×10-	-12×12-	-15×15-	-18×18-	-20×20-	-25×25-	-30×30-	-35×35-
20.4	46.1	47.0	48.0	49.1	50.1	50.9	51.7	52.6	54.1	55.8	57.5	58.3	60.1	61.4	62.6
20.6	45.7	46.6	47.6	48.7	49.7	50.5	51.3	52.2	53.7	55.4	57.1	57.9	59.7	61.0	62.3
20.8	45.3	46.2	47.2	48.2	49.2	50.0	50.8	51.8	53.3	55.0	56.7	57.4	59.1	60.5	61.8
21.0	44.9	45.8	46.8	47.9	48.8	49.6	50.4	51.3	52.9	54.5	56.2	57.0	58.7	60.1	61.4
21.2	44.4	45.3	46.4	47.5	48.5	49.3	50.1	51.0	52.4	54.1	55.8	56.5	58.3	59.7	61.1
21.4	44.2	45.0	46.0	47.1	48.2	49.0	49.8	50.6	52.0	53.7	55.4	56.2	57.9	59.3	60.6
21.6	43.8	44.7	45.7	46.8	47.7	48.5	49.3	50.2	51.6	53.3	55.1	55.8	57.5	58.9	60.3
21.8	43.4	44.3	45.4	46.4	47.3	48.1	48.9	49.8	51.2	52.9	54.6	55.4	57.2	58.6	59.9
22.0	43.0	43.9	45.0	46.0	47.0	47.7	48.6	49.4	50.8	52.5	54.2	55.0	56.9	58.2	59.5
22.2	42.6	43.4	44.6	45.6	46.6	47.3	48.1	49.0	50.4	52.1	53.8	54.6	56.5	57.8	59.1
22.4	42.2	43.1	44.2	45.2	46.2	47.0	47.7	48.6	50.0	51.7	53.4	54.2	56.1	57.4	58.7
22.6	42.0	42.7	43.8	44.9	45.8	46.6	47.3	48.2	49.6	51.3	53.0	53.9	55.7	57.1	58.4
22.8	41.7	42.4	43.5	44.4	45.4	46.1	46.9	47.8	49.2	50.9	52.6	53.5	55.3	56.7	58.0
23.0	41.4	42.1	43.1	44.1	45.0	45.8	46.5	47.4	48.9	50.5	52.2	53.0	54.8	56.3	57.6
23.2	41.0	41.8	42.7	43.7	44.6	45.4	46.2	47.0	48.4	50.2	51.9	52.7	54.5	55.9	57.3
23.4	40.6	41.4	42.3	43.3	44.3	45.1	45.8	46.6	48.0	49.8	51.5	52.3	54.1	55.5	56.9
23.6	40.3	41.0	41.9	43.0	43.9	44.7	45.4	46.2	47.7	49.4	51.1	51.9	53.7	55.1	56.5
23.8	39.9	40.7	41.6	42.6	43.5	44.3	45.0	45.9	47.3	49.0	50.7	51.5	53.4	54.9	56.2
24.0	39.6	40.3	41.2	42.2	43.1	43.9	44.6	45.5	46.9	48.6	50.3	51.2	53.1	54.5	55.8
24.2	39.2	39.9	40.9	41.8	42.7	43.5	44.3	45.1	46.6	48.3	50.0	50.8	52.7	54.1	55.4
24.4	38.9	39.6	40.5	41.5	42.5	43.3	44.0	44.8	46.2	47.9	49.6	50.4	52.3	53.7	55.0
24.6	38.6	39.3	40.3	41.3	42.2	42.9	43.7	44.5	45.9	47.5	49.2	50.1	51.9	53.3	54.6
24.8	38.3	39.0	40.0	40.9	41.8	42.6	43.3	44.1	45.5	47.2	48.9	49.7	51.6	52.9	54.2
25.0	37.9	38.7	39.7	40.6	41.4	42.2	43.0	43.8	45.2	46.9	48.5	49.4	51.2	52.6	53.9
25.2	37.6	38.3	39.3	40.3	41.1	41.9	42.6	43.4	44.8	46.5	48.2	49.1	50.9	52.2	53.5
25.4	37.4	38.0	39.0	39.9	40.8	41.5	42.2	43.0	44.4	46.1	47.8	48.7	50.6	51.9	53.2
25.6	37.1	37.7	38.6	39.5	40.4	41.2	41.9	42.7	44.0	45.7	47.4	48.3	50.1	51.5	52.8
25.8	36.7	37.3	38.3	39.2	40.1	40.8	41.5	42.3	43.7	45.4	47.0	47.9	49.6	51.0	52.4
26.0	36.3	37.0	37.9	38.9	39.8	40.5	41.2	42.0	43.4	45.0	46.7	47.6	49.4	50.8	52.1
26.2	36.1	36.7	37.6	38.5	39.4	40.1	40.8	41.6	43.0	44.7	46.4	47.3	49.1	50.5	51.8
26.4	35.8	36.4	37.3	38.2	39.1	39.8	40.5	41.3	42.6	44.3	46.0	46.9	48.7	50.1	51.5
26.6	35.5	36.1	37.0	37.9	38.8	39.5	40.2	41.0	42.4	44.0	45.6	46.5	48.3	49.7	51.1
26.8	35.2	35.8	36.7	37.6	38.4	39.1	39.8	40.7	42.1	43.7	45.3	46.2	48.0	49.4	50.8
27.0	34.8	35.5	36.3	37.2	38.1	38.8	39.5	40.3	41.7	43.4	45.0	45.8	47.7	49.1	50.5
27.2	34.7	35.2	36.0	36.9	37.7	38.4	39.2	40.0	41.4	43.0	44.7	45.6	47.4	48.8	50.2

6 MV Photon Wedge Factors

FS [cm]	Depth @ 10 cm	Depth @ d_{max}
5	0.255	0.241
6	0.256	0.242
8	0.258	0.245
10	0.260	0.248
15	0.266	0.256
20	0.269	0.262
25	0.272	0.266
30	0.274	0.268

10 MV Scatter Factors

Field Size	Total S	S_c	S_p
4×4	0.925	0.963	0.961
6×6	0.959	0.980	0.979
8×8	0.981	0.990	0.991
10×10	1.000	1.000	1.000
12×12	1.013	1.006	1.007
16×16	1.033	1.013	1.020
20×20	1.048	1.020	1.028
24×24	1.057	1.024	1.032
30×30	1.070	1.030	1.039
40×40	1.078	1.037	1.040

10 MV TMR

Depth [cm]	-3×3-	-4×4-	-5×5-	-6×6-	-7×7-	-8×8-	-9×9-	-10×10-	-12×12-	-15×15-	-18×18-	-20×20-	-25×25-	-30×30-	-35×35-
0.0	30.6	31.7	33.0	34.5	35.8	37.2	38.4	39.6	42.5	46.8	50.3	52.6	56.7	59.5	60.7
0.2	38.1	39.0	40.1	41.8	43.4	45.0	46.3	47.7	50.5	54.8	58.1	60.3	64.1	66.7	67.6
0.4	54.0	54.4	55.2	56.8	58.4	60.1	61.3	62.5	64.9	68.5	71.2	73.1	76.0	77.8	78.4
0.6	68.8	69.1	69.8	71.0	72.2	73.5	74.5	75.4	77.3	80.0	82.1	83.5	85.7	86.7	86.9
0.8	79.0	79.0	79.6	80.5	81.4	82.3	83.0	83.8	85.3	87.4	88.9	89.9	91.4	92.1	92.1
1.0	86.2	86.1	86.5	87.0	87.7	88.4	89.0	89.6	90.7	92.3	93.4	94.1	95.1	95.5	95.4
1.2	91.1	91.0	91.2	91.7	92.2	92.7	93.1	93.5	94.3	95.4	96.2	96.7	97.3	97.5	97.5
1.4	94.6	94.5	94.5	94.8	95.2	95.6	95.9	96.2	96.7	97.5	97.9	98.3	98.6	98.8	98.8
1.6	97.0	96.8	96.8	97.0	97.2	97.5	97.7	97.9	98.3	98.8	99.0	99.2	99.4	99.5	99.5
1.8	98.5	98.3	98.3	98.4	98.6	98.8	98.9	99.0	99.2	99.5	99.6	99.7	99.9	99.9	99.8
2.0	99.3	99.2	99.2	99.3	99.4	99.5	99.6	99.7	99.8	99.9	99.9	99.9	100.0	100.0	100.0
2.2	99.8	99.7	99.8	99.8	99.8	99.9	99.9	99.9	100.0	100.0	100.0	100.0	100.0	100.0	100.0
$d_{max} = 2.4$	100.0	100.0	100.0	100.0	100.0	100.0	100.0	100.0	100.0	100.0	100.0	99.9	99.9	99.8	99.8
2.6	100.0	100.0	99.9	99.9	99.8	99.8	99.9	99.9	99.9	99.8	99.7	99.6	99.7	99.6	99.6
2.8	99.7	99.7	99.7	99.7	99.6	99.6	99.6	99.7	99.7	99.6	99.5	99.5	99.5	99.4	99.5
3.0	99.4	99.4	99.6	99.5	99.5	99.4	99.4	99.4	99.4	99.3	99.2	99.1	99.2	99.2	99.2
3.2	99.1	99.1	99.2	99.2	99.1	99.1	99.1	99.1	99.1	99.0	98.9	98.8	98.9	98.9	98.9
3.4	98.5	98.6	98.7	98.7	98.7	98.7	98.7	98.7	98.7	98.6	98.6	98.5	98.6	98.6	98.7
3.6	97.9	98.0	98.2	98.2	98.2	98.2	98.2	98.3	98.3	98.2	98.2	98.2	98.2	98.3	98.4
3.8	97.3	97.5	97.7	97.7	97.7	97.7	97.8	97.8	97.9	97.8	97.8	97.8	97.9	98.0	98.0
4.0	96.8	97.1	97.2	97.3	97.3	97.3	97.4	97.5	97.5	97.5	97.4	97.4	97.5	97.6	97.7
4.2	96.3	96.5	96.7	96.7	96.8	96.8	96.9	97.1	97.1	97.1	97.1	97.1	97.2	97.3	97.3
4.4	95.6	95.9	96.2	96.2	96.3	96.4	96.5	96.6	96.7	96.7	96.8	96.8	96.9	97.0	97.0
4.6	94.9	95.3	95.6	95.7	95.8	95.9	96.0	96.1	96.2	96.3	96.3	96.3	96.5	96.6	96.7
4.8	94.3	94.7	94.9	95.1	95.3	95.4	95.5	95.6	95.7	95.8	95.9	95.9	96.2	96.3	96.4
5.0	93.7	94.1	94.4	94.6	94.8	94.9	95.0	95.2	95.3	95.4	95.5	95.6	95.8	95.9	96.1

10 MV TMR

Depth [cm]	-3×3-	-4×4-	-5×5-	-6×6-	-7×7-	-8×8-	-9×9-	-10×10-	-12×12-	-15×15-	-18×18-	-20×20-	-25×25-	-30×30-	-35×35-
											Equivalent Square Field Size [cm]				
5.2	93.0	93.4	93.9	94.2	94.3	94.4	94.5	94.7	94.8	94.9	95.1	95.2	95.4	95.5	95.7
5.4	92.4	92.9	93.3	93.5	93.7	93.9	94.1	94.3	94.5	94.5	94.7	94.7	95.0	95.2	95.3
5.6	91.7	92.3	92.8	93.0	93.3	93.5	93.7	93.9	94.0	94.0	94.2	94.3	94.6	94.8	94.9
5.8	91.1	91.7	92.2	92.5	92.7	92.9	93.1	93.3	93.5	93.7	93.8	93.9	94.3	94.5	94.6
6.0	90.6	91.1	91.5	91.8	92.1	92.4	92.6	92.9	93.1	93.2	93.4	93.5	93.9	94.2	94.3
6.2	89.8	90.5	91.1	91.4	91.6	91.8	92.1	92.4	92.6	92.8	93.0	93.1	93.5	93.7	93.9
6.4	89.2	89.9	90.5	90.8	91.1	91.3	91.6	91.8	92.1	92.4	92.7	92.8	93.1	93.3	93.6
6.6	88.5	89.2	89.7	90.1	90.5	90.8	91.1	91.4	91.6	91.9	92.2	92.3	92.7	93.0	93.2
6.8	88.0	88.7	89.2	89.5	89.9	90.2	90.6	90.9	91.2	91.5	91.7	91.9	92.3	92.6	92.8
7.0	87.3	88.0	88.6	89.1	89.5	89.8	90.1	90.4	90.7	91.0	91.3	91.5	91.9	92.2	92.4
7.2	86.7	87.5	88.1	88.6	89.0	89.3	89.6	89.9	90.2	90.6	90.9	91.1	91.5	91.8	92.0
7.4	86.2	86.9	87.5	88.0	88.4	88.7	89.1	89.4	89.8	90.2	90.5	90.7	91.1	91.5	91.7
7.6	85.5	86.3	87.0	87.5	87.9	88.2	88.6	89.0	89.4	89.8	90.1	90.2	90.8	91.2	91.4
7.8	84.9	85.8	86.5	87.0	87.4	87.7	88.1	88.5	88.9	89.3	89.7	89.9	90.4	90.8	91.1
8.0	84.4	85.2	85.9	86.4	86.9	87.2	87.6	88.0	88.4	88.9	89.3	89.5	90.0	90.5	90.7
8.2	83.7	84.6	85.3	85.8	86.3	86.7	87.2	87.6	88.0	88.5	88.9	89.1	89.7	90.1	90.3
8.4	83.1	84.1	84.9	85.4	85.9	86.3	86.8	87.1	87.6	88.0	88.5	88.7	89.3	89.7	90.0
8.6	82.7	83.5	84.3	84.9	85.4	85.8	86.2	86.6	87.1	87.7	88.1	88.4	88.9	89.4	89.6
8.8	82.1	82.9	83.7	84.3	84.8	85.2	85.6	86.1	86.6	87.2	87.7	87.9	88.5	89.0	89.3
9.0	81.5	82.4	83.1	83.7	84.2	84.7	85.2	85.6	86.2	86.8	87.3	87.5	88.2	88.7	89.0
9.2	80.9	81.8	82.6	83.2	83.7	84.1	84.7	85.1	85.7	86.3	86.9	87.1	87.8	88.3	88.5
9.4	80.3	81.2	82.0	82.6	83.2	83.6	84.1	84.6	85.1	85.8	86.4	86.7	87.4	87.8	88.1
9.6	79.7	80.6	81.4	82.1	82.7	83.1	83.7	84.1	84.7	85.4	86.0	86.3	87.0	87.5	87.8
9.8	79.1	80.1	80.8	81.5	82.1	82.6	83.2	83.7	84.3	84.9	85.5	85.9	86.6	87.1	87.5
10.0	78.6	79.6	80.4	81.0	81.6	82.1	82.6	83.2	83.8	84.4	85.0	85.3	86.1	86.7	87.0
10.2	78.0	79.0	79.9	80.5	81.1	81.6	82.2	82.7	83.3	83.9	84.6	84.9	85.7	86.3	86.6
10.4	77.4	78.5	79.3	80.0	80.6	81.1	81.7	82.2	82.7	83.5	84.1	84.5	85.3	86.0	86.3
10.6	77.0	77.9	78.7	79.5	80.1	80.6	81.2	81.7	82.3	83.0	83.7	84.0	84.9	85.5	85.9
10.8	76.4	77.3	78.2	78.9	79.6	80.1	80.7	81.2	81.9	82.6	83.3	83.6	84.4	85.0	85.5
11.0	75.8	76.7	77.6	78.3	79.0	79.5	80.1	80.6	81.3	82.1	82.8	83.2	84.1	84.7	85.1
11.2	75.2	76.1	77.0	77.7	78.4	78.9	79.6	80.1	80.9	81.7	82.4	82.7	83.6	84.3	84.7
11.4	74.6	75.6	76.5	77.2	77.9	78.4	79.1	79.7	80.4	81.2	82.0	82.3	83.2	84.0	84.4

10 MV TMR

Depth [cm]	-3×3-	-4×4-	-5×5-	-6×6-	-7×7-	-8×8-	-9×9-	-10×10-	-12×12-	-15×15-	-18×18-	-20×20-	-25×25-	-30×30-	-35×35-
11.6	74.2	75.2	76.0	76.8	77.4	78.0	78.6	79.2	79.9	80.8	81.6	82.0	82.9	83.6	84.0
11.8	73.6	74.6	75.5	76.3	77.0	77.5	78.2	78.7	79.5	80.3	81.2	81.5	82.5	83.2	83.6
12.0	73.1	74.1	75.0	75.7	76.5	77.0	77.7	78.3	79.1	79.9	80.8	81.1	82.1	82.8	83.2
12.2	72.7	73.7	74.5	75.3	76.0	76.5	77.2	77.8	78.6	79.6	80.4	80.8	81.7	82.4	82.8
12.4	72.2	73.1	74.1	74.9	75.7	76.1	76.8	77.4	78.2	79.1	80.0	80.4	81.4	82.1	82.5
12.6	71.6	72.6	73.5	74.3	75.0	75.6	76.3	76.8	77.6	78.5	79.4	79.8	80.9	81.7	82.2
12.8	71.0	71.9	72.9	73.8	74.5	75.1	75.8	76.4	77.2	78.1	79.0	79.4	80.5	81.4	81.9
13.0	70.5	71.5	72.5	73.3	74.0	74.5	75.2	75.9	76.7	77.7	78.6	79.0	80.1	81.0	81.5
13.2	70.0	71.0	72.0	72.8	73.6	74.1	74.8	75.5	76.4	77.3	78.3	78.7	79.8	80.6	81.1
13.4	69.6	70.6	71.5	72.3	73.1	73.6	74.4	75.0	75.9	76.9	77.9	78.3	79.4	80.2	80.7
13.6	69.2	70.1	70.9	71.8	72.6	73.2	73.9	74.6	75.5	76.5	77.4	77.9	79.0	79.8	80.3
13.8	68.6	69.5	70.5	71.3	72.0	72.7	73.4	74.1	75.0	76.1	77.0	77.5	78.6	79.4	80.0
14.0	68.1	69.1	70.1	70.9	71.6	72.1	72.9	73.6	74.5	75.5	76.5	77.0	78.2	79.1	79.6
14.2	67.5	68.6	69.5	70.3	71.1	71.6	72.4	73.0	74.0	75.0	76.1	76.5	77.8	78.7	79.3
14.4	67.0	68.0	69.0	69.9	70.6	71.2	71.9	72.6	73.6	74.6	75.7	76.1	77.4	78.3	78.8
14.6	66.6	67.6	68.5	69.4	70.2	70.7	71.5	72.2	73.2	74.2	75.3	75.8	77.0	77.8	78.4
14.8	66.2	67.1	68.0	68.8	69.7	70.3	71.1	71.7	72.7	73.8	74.9	75.4	76.6	77.5	78.0
15.0	65.6	66.5	67.4	68.3	69.1	69.8	70.6	71.3	72.3	73.4	74.4	75.0	76.2	77.1	77.7
15.2	65.2	66.1	67.0	67.9	68.7	69.3	70.1	70.8	71.8	72.9	74.0	74.5	75.8	76.7	77.3
15.4	64.8	65.7	66.6	67.4	68.2	68.8	69.6	70.3	71.4	72.5	73.6	74.1	75.4	76.3	76.9
15.6	64.2	65.1	66.0	66.9	67.8	68.4	69.2	69.9	71.0	72.0	73.1	73.6	74.9	75.9	76.5
15.8	63.7	64.6	65.6	66.4	67.3	67.9	68.7	69.4	70.5	71.6	72.7	73.2	74.5	75.5	76.1
16.0	63.3	64.2	65.2	66.0	66.9	67.5	68.2	68.9	70.0	71.1	72.2	72.8	74.1	75.1	75.8
16.2	62.8	63.8	64.7	65.5	66.4	67.0	67.8	68.5	69.5	70.7	71.9	72.4	73.8	74.7	75.4
16.4	62.2	63.2	64.2	65.1	66.0	66.6	67.4	68.1	69.1	70.3	71.5	72.1	73.4	74.4	75.0
16.6	61.8	62.8	63.8	64.6	65.4	66.1	66.9	67.7	68.7	69.9	71.1	71.6	72.9	74.0	74.7
16.8	61.4	62.4	63.3	64.2	65.0	65.7	66.5	67.2	68.3	69.5	70.6	71.2	72.6	73.6	74.3
17.0	61.0	61.9	62.9	63.7	64.6	65.2	66.0	66.8	67.9	69.1	70.2	70.8	72.2	73.3	74.0
17.2	60.5	61.5	62.5	63.4	64.3	64.9	65.6	66.3	67.4	68.7	69.9	70.5	71.8	72.8	73.6
17.4	60.1	61.0	62.0	62.9	63.7	64.4	65.1	65.9	67.0	68.2	69.5	70.1	71.5	72.5	73.2
17.6	59.7	60.6	61.5	62.5	63.3	63.9	64.6	65.4	66.6	67.8	69.1	69.7	71.1	72.2	72.9
17.8	59.3	60.2	61.1	62.0	62.9	63.4	64.1	64.9	66.1	67.5	68.8	69.3	70.7	71.8	72.5

Equivalent Square Field Size [cm]

10 MV TMR

Equivalent Square Field Size [cm]

Depth [cm]	-3×3-	-4×4-	-5×5-	-6×6-	-7×7-	-8×8-	-9×9-	-10×10-	-12×12-	-15×15-	-18×18-	-20×20-	-25×25-	-30×30-	-35×35-
18.0	58.9	59.7	60.7	61.6	62.5	63.1	63.8	64.6	65.7	67.0	68.3	68.9	70.3	71.3	72.0
18.2	58.4	59.3	60.3	61.2	62.1	62.7	63.5	64.3	65.4	66.6	67.9	68.5	70.0	71.0	71.7
18.4	58.1	59.0	60.0	60.9	61.7	62.3	63.1	63.9	65.0	66.3	67.6	68.1	69.7	70.7	71.4
18.6	57.8	58.6	59.5	60.4	61.2	61.9	62.6	63.4	64.6	65.9	67.2	67.8	69.3	70.3	71.1
18.8	57.3	58.2	59.1	60.0	60.8	61.5	62.2	63.0	64.1	65.5	66.8	67.4	69.0	70.1	70.8
19.0	57.0	57.9	58.8	59.6	60.4	61.1	61.9	62.6	63.8	65.0	66.3	66.9	68.5	69.7	70.4
19.2	56.6	57.4	58.3	59.1	60.0	60.7	61.4	62.2	63.4	64.7	66.0	66.6	68.1	69.3	70.0
19.4	56.1	56.9	57.8	58.8	59.6	60.2	60.9	61.7	62.9	64.3	65.6	66.2	67.8	68.9	69.6
19.6	55.8	56.6	57.4	58.3	59.2	59.8	60.5	61.3	62.5	63.8	65.2	65.8	67.4	68.6	69.4
19.8	55.2	56.0	56.9	57.9	58.8	59.4	60.2	60.9	62.2	63.5	64.8	65.4	67.1	68.2	69.0
20.0	54.8	55.7	56.5	57.5	58.4	59.0	59.7	60.5	61.8	63.1	64.4	65.1	66.7	67.8	68.6
20.2	54.5	55.4	56.2	57.1	58.0	58.6	59.3	60.1	61.4	62.7	64.1	64.7	66.2	67.4	68.2
20.4	54.1	54.9	55.8	56.7	57.5	58.2	58.9	59.8	61.0	62.3	63.7	64.3	65.9	67.1	67.8
20.6	53.7	54.5	55.3	56.3	57.1	57.8	58.5	59.4	60.6	61.9	63.3	63.9	65.5	66.7	67.4
20.8	53.4	54.0	55.0	55.9	56.7	57.4	58.1	59.0	60.2	61.5	62.9	63.5	65.2	66.4	67.1
21.0	53.0	53.7	54.6	55.5	56.2	56.9	57.7	58.6	59.8	61.1	62.5	63.1	64.8	66.0	66.8
21.2	52.6	53.4	54.3	55.1	55.9	56.5	57.3	58.2	59.4	60.7	62.1	62.7	64.4	65.6	66.4
21.4	52.2	52.9	53.8	54.7	55.5	56.2	56.9	57.8	59.0	60.3	61.7	62.3	64.0	65.2	66.1
21.6	51.8	52.6	53.4	54.3	55.2	55.9	56.5	57.3	58.6	60.0	61.4	62.0	63.6	64.9	65.7
21.8	51.5	52.2	53.1	54.0	54.8	55.4	56.2	57.0	58.2	59.6	61.0	61.6	63.3	64.5	65.4
22.0	51.1	51.9	52.8	53.5	54.4	55.0	55.7	56.6	57.9	59.2	60.6	61.3	63.0	64.2	65.0
22.2	50.8	51.5	52.4	53.2	54.0	54.7	55.4	56.2	57.5	58.9	60.3	61.0	62.7	63.9	64.6
22.4	50.4	51.2	52.0	52.9	53.7	54.3	55.0	55.8	57.2	58.6	60.0	60.6	62.4	63.5	64.3
22.6	50.1	50.8	51.6	52.5	53.3	54.0	54.7	55.5	56.8	58.2	59.6	60.3	62.0	63.1	64.0
22.8	49.7	50.4	51.2	52.1	52.9	53.6	54.4	55.2	56.4	57.8	59.2	59.9	61.6	62.8	63.6
23.0	49.4	50.1	50.9	51.8	52.6	53.3	54.0	54.8	56.0	57.4	58.8	59.5	61.2	62.5	63.3
23.2	49.0	49.7	50.6	51.4	52.2	53.0	53.7	54.5	55.6	57.0	58.4	59.2	60.9	62.1	63.0
23.4	48.7	49.4	50.2	51.0	51.8	52.5	53.2	54.1	55.3	56.7	58.1	58.8	60.6	61.8	62.6
23.6	48.3	49.0	49.9	50.6	51.4	52.1	52.8	53.7	54.9	56.3	57.8	58.5	60.2	61.4	62.3
23.8	48.0	48.7	49.5	50.3	51.1	51.8	52.5	53.4	54.6	56.0	57.3	58.0	59.8	61.0	61.9
24.0	47.6	48.3	49.1	49.9	50.8	51.5	52.2	53.1	54.3	55.6	57.0	57.7	59.5	60.7	61.7
24.2	47.2	47.9	48.7	49.6	50.4	51.1	51.8	52.6	53.9	55.3	56.7	57.4	59.1	60.4	61.3

12

Photon Beam Treatment Planning: Part I

The previous chapters presented methods for calculating doses to a point. While this is useful for learning purposes it is not often used in modern clinical practice except by computer programs to calculate point doses as a second check. Most treatments, rather, are calculated and planned in a treatment planning system. These software packages calculate the dose in 3D space. They most often use a CT scan acquired at the time of "simulation" which is typically some days before treatment starts. This chapter will explore treatment planning, the associated dose calculation methods, and techniques that are frequently employed to achieve high-quality plans.

12.1 Dose Calculation Algorithms and Inhomogeneities

12.1.1 Dose Calculation: TERMA and Kernels

The goal in treatment planning dose calculations is to determine the dose delivered at each point in the patient. The patient may be represented as a 3D grid of "voxels" (a 3D pixel), tiny boxes that are typically 2 to 4 mm square in most systems. The goal is to calculate the dose in each voxel of this "dose grid." This is represented in Figure 12.1.1. The first step is to calculate the total energy released in matter (TERMA) at each voxel. To do this we draw lines (green) from the source (red) to each voxel and find the radiological pathlength to that voxel. This is the pathlength through the voxel grid; not just the physical pathlength but includes the effects of density. The density information is gathered from the CT scan. The black squares in Figure 12.1.1, for example, might be lung, and the attenuation of a ray passing through these voxels would be less. Based on the effective pathlength one can calculate the attenuation of the beam, or the photon fluence, and from that the TERMA. This is the simplest form of dose calculation and is used in pencil beam algorithms.

The next step in dose calculation accounts for the fact that the dose is not delivered just in the one voxel but rather is spread out over a larger region due to the scattering of the photons and the creation of subsequent high-energy electrons which have a long range in tissue. This effect can be accounted for using a "kernel" (blue in Figure 12.1.1). The kernel describes the contribution of the TERMA in a particular voxel to the neighboring voxels. Mathematically, this is treated with a convolution operation (written with the symbol $\otimes$) which is a sort of moving integral over the grid, i.e. repeated steps of multiply-and-move, multiple-and-move. The size and shape of the kernel depend on the energy of the beam, and it is calculated using Monte Carlo methods of numerical simulation. The size of the kernel is also scaled according to the density in the patient. That is, in regions of lower density the kernel is larger to account for the fact that the range of the electrons is larger. This algorithm for dose calculation is called superposition convolution and has been one of the main methods used in treatment planning systems for many years.

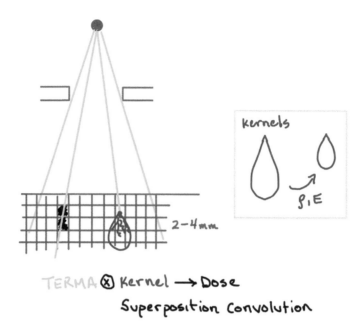

FIGURE 12.1.1
Dose calculation method in a treatment planning system.

12.1.2 TPS Beam Model

Using the dose calculation algorithm a model for the beam can be built in the treatment planning system (TPS). Figures 12.1.2 and 12.1.3 show examples from such a model from a TPS. The plots show measured data acquired in a water tank (red) compared to the model in the TPS which uses the superposition convolution algorithm (blue). Figure 12.1.2 shows the PDD for a 20×20 cm field. Numerous parameters control the shape of this curve. For example, the energy spectrum of photons in the beam. A harder spectrum will, of course, have a falloff that is less steep. In this particular system there are other parameters as well such as the flux of electrons from the head of the linac. This contributes superficial dose. If the flux of electrons is turned off (Figure 12.1.2B) the calculated dose is much lower in

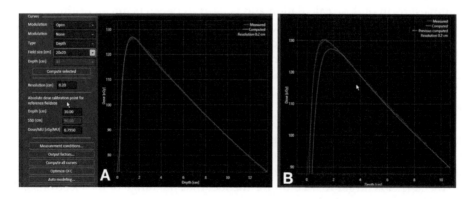

FIGURE 12.1.2
Example PDD data, measured (red) vs. calculated (blue). (B) The same model but with the contribution from electron in the linac head turned off. (Data are extracted from RayStation v. 6.1.)

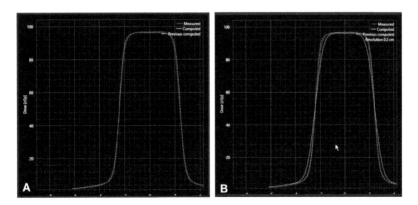

FIGURE 12.1.3
Example profile data, measured (red) vs. calculated (blue). (A) 1-mm wide source. (B) 3-mm wide source. Data are extracted from RayStation v. 6.1.

the superficial region. Another parameter that can be controlled in this system is the size of the photon source. The source is not modeled as a point source, but rather as a gaussian function whose width can be varied. Figures 12.1.3A and B show the impact of this parameter on the profile of the beam. The larger source size (3 mm, B) has a penumbra that is larger than the small source size (1 mm, A). This is the expected behavior for penumbra (see Section 9.1.6). For an animated tour of this modeling process, see the video.

There may be tens or even hundreds of parameters that control the behavior of a model in a TPS. The way this is controlled and presented to the user varies by manufacturer. Some TPS systems require the user to adjust all of these parameters themselves with the goal of fitting measured data (e.g. RayStation, RaySearch Inc.). Other systems in which the manufacturer is responsible for both the TPS and delivery system often have models that are more "locked down," offering only a very few parameters which the user can change (e.g. Eclipse, Varian Inc.).

12.1.3 TPS Dose Algorithms

Figure 12.1.4 lists dose calculation algorithms in terms of increasing accuracy. The simplest is a point dose calculation (Chapter 11). Somewhat better is a simple pencil beam

Accuracy

Simple calculations Clarkson, TAR, etc.	Pencil beam Convolution	Convolution with invariant kernels (eg. FFT)	Convolution with a source model	Collapsed Cone Convolution	Boltzmann Equation solver	Monte Carlo
	Varian PBC	CMS XiO	Varian AAA	RayStation Philips Pinnacle SCC TomoTherapy	Varian Acuros XB	CyberKnife RayStation

FIGURE 12.1.4
TPS dose calculation algorithms in order of increasing accuracy.

convolution similar to Section 12.1.1 but without kernels (used in an early version of the Varian TPS). Next is convolution with a kernel but one where the kernel is invariant, i.e. does not scale with the density in the patient. This method is fast, since a fast Fourier transform (FFT) can be used to do the convolution calculations and was useful when computing power was more limited (e.g. CMS XiO TPS from Elekta Inc.) Convolution with a source model the next most accurate algorithm. This accounts for the effect that the source of photons is not a point but has some spatial extent. Collapsed cone convolution, or superposition convolution, uses the method described in Section 12.1.1 above (e.g. Pinnacle TPS from Philips Inc., TomoTherapy from Accuray Inc. or RayStation from RaySearch Inc.). Next is Acuros XB algorithm (Varian Inc.). This algorithm solves the Boltzmann equation, which is an equation that describes diffusion through matter, in this case the diffusion of charged particles through tissue. This was developed in Lawrence Livermore National Labs and beginning around 2013 has been employed in the TPS from Varian Inc.

The final algorithm which is the most accurate is Monte Carlo. This method basically follows the propagation of all the particles through matter (photons and electrons mainly). This propagation is simulated particle-by-particle. As the particle steps through matter, each interaction is simulated using the fundamental physics interactions. To determine which interaction happens the algorithm draws randomly from the probability distributions given the physics cross-sections for each interaction. Hundreds of millions or billions of particle histories are simulated and eventually a dose distribution is built up. The main challenge with this currently is the speed of calculation. However, as computational power increases Monte Carlo algorithms are becoming available in more TPS systems. **Monte Carlo is considered the "gold standard" for dose calculation and is often used to benchmark other algorithms. This is a key concept.** For more information on Monte Carlo see Chapter 10 in Metcalfe, Kron, and Hoban (2007) and AAPM TG-105 on Monte Carlo (Chetty et al. 2007).

12.1.4 Inhomogeneities: Lung

Patient tissue is not homogeneous (i.e. uniform density everywhere) but rather highly variable in density. For example, Figure 12.1.5 shows a patient CT where density can be visualized as grayscale, light areas being high density (e.g. compact bone, $\rho = 1.85$ g/cm^3)

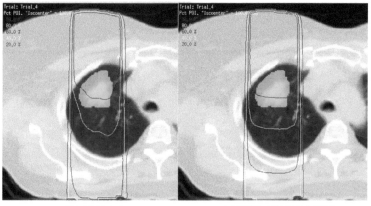

INHOMOG   WATER EQUIVALENT

FIGURE 12.1.5
Effect of density inhomogeneities in treatment planning calculations.

and dark areas being low density (e.g. lung at the center, $\rho \approx 0.3$ g/cm^3). More information on CT scans can be found in Section 19.2. It is important to include the effects of these inhomogeneities in dose calculations. Figure 12.1.5 illustrates the impact. Panel A shows a calculation with the effect of inhomogeneities included vs. panel B where the calculation is performed as if the patient is homogenous water. Note that the attenuation of the beam is much less with the inhomogeneous calculation (e.g. blue 60% isodose line extends much farther into the patient). Also the shape of the dose distribution is different. **The penumbra is wider in the presence of the inhomogeneity due to the longer range of the electrons, and this effect is larger in a higher energy beam because the electron range is larger. This is a key concept.**

To understand these effects in more detail, consider Figure 12.1.6 which shows a percent depth-dose curve for a 10×10 cm field with measured data (black dots) vs. calculations using various dose algorithms. With no correction for the inhomogeneity (green) the dose falls off more rapidly with depth, i.e. more attenuation due to the assumed higher density in the lung region. Monte Carlo is the most accurate predictor of the measured dose, but the convolution algorithm with a source model (AAA) is also accurate. The other algorithms shown are more similar to a TMR calculation but with empirical corrections applied.

There is a field size effect to this as well. Figure 12.1.7 shows similar data for a small field (2×2 cm). Here the measured dose decreases in the lung compared to the homogeneous case (no correction). The reason for this is that electrons are scattered out from this small field and there is no balance of in-scattered electrons vs. out-scattered electrons, i.e. no electronic equilibrium. Again the Monte Carlo algorithm is the most accurate predictor of the measured dose. The convolution algorithm (AAA) does not predict the dose accurately in certain regions, but note that this is an extreme situation of a small field. Figure 12.1.7 illustrates an important effect. When the beam enters from the lung back into soft tissue there is another buildup region. **The "rebuildup" region is often present when going from a low-density back to a high-density region and is a key concept.**

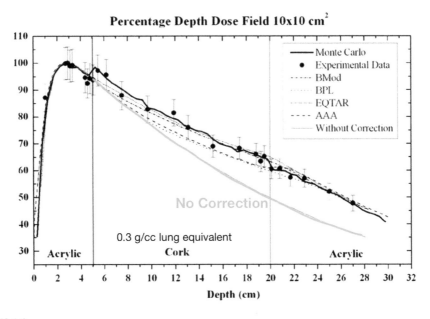

FIGURE 12.1.6
Effect of inhomogeneity on depth-dose curve. (From de Rosa et al. *J Appl Clin Med Phys*, 11(1), 2010.)

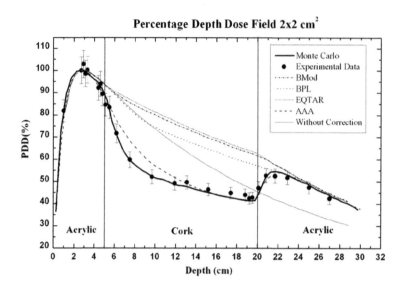

FIGURE 12.1.7
Inhomogeneity effects in a small field (2×2 cm). (From de Rosa et al. *J Appl Clin Med Phys*, 11(1), 2010.)

12.1.5 Inhomogeneities: Bone

The effects of bone are somewhat more complex. Because the bone density is higher than soft tissue (e.g. compact bone, $\rho = 1.85$ g/cm^3) the beam attenuation will increase in going through bone compared to a case without the bone. In the small region of soft tissue just distal to the bone there is an increase in dose due to backscattered electrons. Within the bone itself the dose depends on energy. At megavoltage energies of approximately 6 MV the dose in the bone itself is very similar to the dose that would be deposited to soft tissue. However, at higher energies pair production starts to become important. The dose increases because of the higher effective atomic number in bone.

12.2 Treatment Planning with Megavoltage Photon Beams

12.2.1 Treatment Planning with Multiple Fields

The previous chapters and sections have considered single photon beam incidents on a patient. This was useful for illustrating the basic physics. However, in reality, treatments rarely use a single photon beam but rather employ a combination of beams from multiple angles. Figure 12.2.1 shows an example of two parallel-opposed beams, i.e. beams that enter from opposite sides of the patient. The dose profile for these beams is shown in Figure 12.2.1 (right). This is a plot along the central axis for the composite dose distribution. It is a combination of the PDDs from two individual beams (see Figure 11.1.1), one entering from the left in the figure and one from the right.

One key feature of Figure 12.2.1 is the dose hot spots in the superficial region (the so-called "tissue lateral effect"). In this case, in order to deliver a dose of 100 cGy at the center with two 6 MV beams in this example, you must deliver a dose of 210 cGy to the superficial

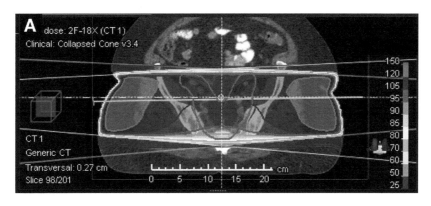

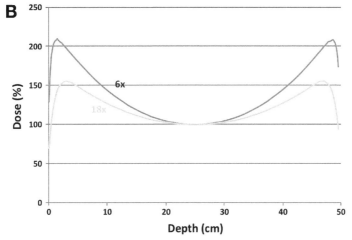

FIGURE 12.2.1

Dose distribution in two parallel-opposed beams. (A) Isodose distribution. (B) Example dose profiles horizontally along the central axis. The beam energies are 6 MV (blue) and 18 MV (green).

region at a depth of 1.5 cm. These dose hot spots can be reduced in various ways. One way is to increase the energy of the beam. Figure 12.2.1B (green) shows a treatment with two 18 MV beams. This results in reduced hot spots compared to the 6 MV beam treatment. This is because the PDD falls off less quickly for an 18 MV beam vs. a 6 MV beam. The same effect plays out with the other factors which affect PDD (see Section 10.1.3). Recall, for example, that for larger field sizes PDD falls off less quickly with depth. Therefore, the dose hot spots will be less for larger field sizes.

The magnitude of the dose hot spots depends also on the patient thickness or "separation." Figure 12.2.2 shows dose profiles along the central axis for parallel-opposed 6 MV beams for three different patient separations: 10 cm, 20 cm, and 50 cm. The dose hot spot in the superficial regions is higher for larger patient separations. This means, for example, that treatment of the pelvis or abdomen will result in larger hot spots than treatment of say the neck which is thinner. Similarly, treatment with lateral beams (vs. AP/PA, anterior–posterior beams) will also result in larger hot spots. In cases where the separation is small (e.g. brain or head-and-neck region) lower energy beams can readily be used.

The magnitude of the effect described above depends greatly on the number of beams used. For illustration purposes the above examples used an extreme case where

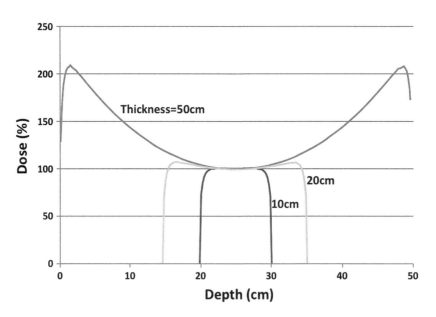

FIGURE 12.2.2
Dose hot spots for different patient separations.

only two fields were employed. However, as more fields are used the dose hot spot is reduced. This is illustrated in Figure 12.2.3 which shows a four-field plan (B) vs. a two-field plan (A). The dose profile across the center line (C) shows much smaller doses for the superficial regions in the four-field plan (dotted) vs. the two-field plan (solid). The more fields are used the lower the hot spot will be in the superficial regions. An extreme example is some of the SRS or SBRT systems which may use hundreds of beams (Chapter 22).

12.2.2 Treatment Planning with Wedges

Consider the treatment shown in Figure 12.2.4 with the goal of achieving a uniform dose in the spherical target. Treatment with parallel-opposed beams would result in a high dose to superficial tissues on the contralateral side of the patient (Figure 12.2.4A). An alternative approach might be to treat with two perpendicular beams as shown in Figure 12.2.4B (one beam from the AP direction, one laterally). This results in a much lower superficial dose. However, here we observe that the dose in the target region is not uniform. There is a hot region in the corner where the two beams intersect toward the surface and a cold region in the opposite corner. This is a case where wedges are useful.

Wedges can be used in beams to achieve differential attenuation: more attenuation on one side vs. the other. One form of a wedge is a metal device inserted into the field (Figure 12.2.5). The beam is attenuated more through the "heel" (thicker region) vs. the "toe" (thinner region). Figure 12.2.6 illustrates the effect of wedges on the dose distribution. Figure 12.2.5A shows an open beam while Figure 12.2.5B shows the same beam with a wedge in the field. Note that the MU are increased by about a factor of 4 in B vs. A in order to achieve the same dose in the isocenter point. This is because the wedge attenuates the beam by about a factor of four.

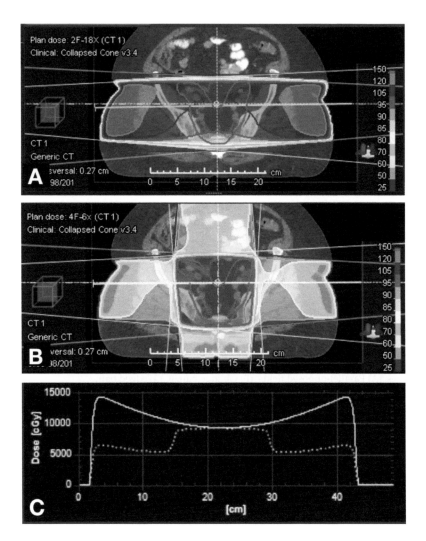

FIGURE 12.2.3
Treatment with a two-field plan (A) vs. a four-field plan (B). The dose profile across the center line (C) for the four-field plan (dotted) vs. the two-field plan (solid).

Wedges can be made with different angles to achieve more or less differential attenuation. The wedge angle is defined not by the physical angle of the metal object but rather by the angle that the isodose line makes with the horizontal (Figure 12.2.6C). Figure 12.2.5B shows a 45-degree wedge vs. Figure 12.2.5C which shows a 60-degree wedge.

Various ways have been developed to implement wedges by the manufacturer. One system uses physical wedges which are manually mounted on a holder on the head of the linac (Varian Inc., Figure 12.2.7). Various physical wedges are available with angles of 15, 30, 45, and 60 degrees. Another manufacturer uses a motorized wedge in the head of the linac (Elekta Inc.). This wedge has only one angle (60 degrees) but angles of 0 to 60 degrees are achieved by using combinations of open and wedge beams. Another method for creating wedged fields involves moving a jaw dynamically across the field to achieve a wedged dose distribution. This is called a "dynamic wedge" and does not use a physical

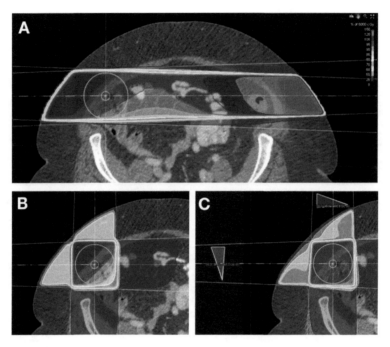

FIGURE 12.2.4
Plan with wedges. (A) Parallel-opposed fields. (B) Oblique pair. (C) Wedged pair.

FIGURE 12.2.5
Wedge (Elekta Inc.). When inserted in the beam this device attenuates the beam more in the "heel" side (thicker region) vs. the "toe" (thinner region).

wedge in the beam. The regions of the field that are covered longer by the jaw receive lower doses and correspond to the heel of the wedge.

This background should provide further clarity on the use of wedges in the example in Figure 12.2.4C. Note that the wedges are oriented such that the heels are together. This is a general feature of planning with wedges for multiple fields: "heels together, toes apart."

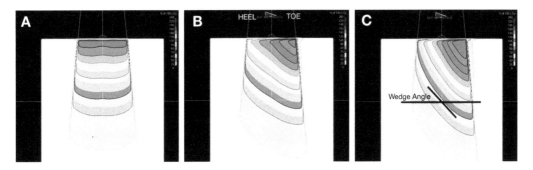

FIGURE 12.2.6
Dose distributions with (A) open field, (B) 45-degree wedge, and (C) 60-degree wedge.

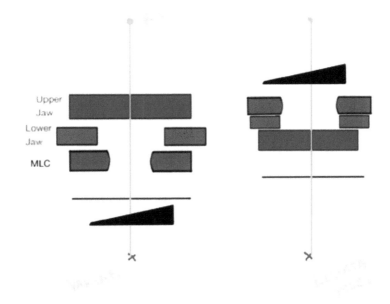

FIGURE 12.2.7
Wedge systems.

In this case 50-degree wedges were used, an angle that is achieved through a combination of open and 60-degree wedge fields. In general the angle of wedge that is required depends on the angle between the beams (Figure 12.2.8). If two oblique beams have an angle between them of ϕ (sometimes called the "hinge angle") then the required angle of the wedges is $90° - \phi/2$. This is a useful empirical formula to remember. To take an example, if the angle between the two beams is 120°, then the wedges would have an angle of 30°.

Wedges are sometimes used for other purposes as well, in particular to compensate for the fact that some sections of anatomy may be thinner than others. A classic example is in the treatment of the breast with tangent fields (parallel-opposed fields at an oblique angle that skim the chest wall interface). In this case the thickness of tissue is larger in the medial aspect of the field vs. the outer region of the breast. In this case wedges can be used with the heels oriented outward. The heel then attenuates the beam in the region where the tissue is thinner to produce a more even dose distribution in the tissue.

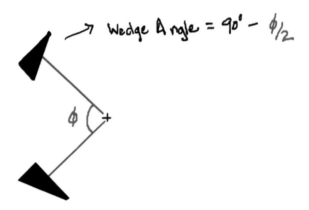

FIGURE 12.2.8
Estimation of the wedge angle given the angle between two beams.

Further Reading

Chetty, I.J., et al. 2007. Report of the AAPM Task Group No. 105: Issues associated with clinical implementation of Monte Carlo-based photon and electron external beam treatment planning. *Med Phys* 34(12):4818–4853.

Kahn, F.M. and J.P. Gibbons. 2014. *Kahn's The Physics of Radiation Therapy*. 5th Edition. Chapter 12. Philadelphia, PA: Wolters Kluwer.

McDermott, P.N. and C.G. Orton. 2010. *The Physics of Radiation Therapy*. Chapter 3. Madison, WI: Medical Physics Publishing.

Metcalfe, P., T. Kron and P. Hoban. 2007. *The Physics of Radiotherapy X-rays and Electrons*. Chapters 9 and 10. Madison, WI: Medical Physics Publishing.

Chapter 12 Problem Sets

*Note: * indicates harder problems.*
See Section 11.3 for tables of data where needed.

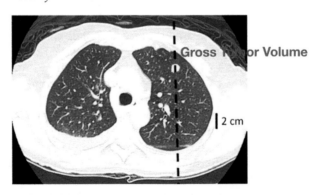

FIGURE PS12.1
CT cross-section in lung.

1. For the CT image shown in Figure PS12.1 which algorithm will produce the largest predicted penumbra in the lung around the GTV? (The 2 cm bar shows the scale of the image and the dashed line shows the beam direction.)
   a. Clarkson TAR method
   b. Pencil beam algorithm
   c. Collapsed cone convolution
   d. Jefferson Airplane

2. Which beam energy will produce the largest penumbra in the lung around the GTV?
   a. 6 MV
   b. 10 MV
   c. 15 MV
   d. 18 MV

3. For the CT image shown Figure PS12.1 how does the calculated dose (i.e. dose per number of MU) at the center of the tumor change in a heterogeneous algorithm vs. a homogeneous algorithm? Assume a 10×10 cm field size, but discuss how this answer might depend on field size.
   a. Decreases
   b. Increases
   c. Stays the same
   d. Only changed for Monte Carlo algorithms

4. * For the CT image shown Figure PS12.1, what will the ratio of predicted dose be at the center of the tumor for a homogeneous dose calculation to a heterogeneous calculation? Assume a 5×5 cm 6 MV AP beam and the same number of monitor units in both cases. Use a TMR-based calculation and refer to the data tables in Section 11.3.
   a. 0.75
   b. 0.95
   c. 1.05
   d. 1.25

5. * How will the dose difference found in Problem 4 depend on field size? Discuss from the mathematical and physical points of view.
   a. Increase with increasing field size
   b. Decrease with increasing field size
   c. Stays the same
   d. Depends on depth

6. Match the beams below with the correct image in Figure PS12.2.

    a. 6 MV, homogeneous calculation

    b. 6 MV, heterogeneous calculation

    c. 18 MV, homogeneous calculation

    d. 18 MV, heterogeneous calculation

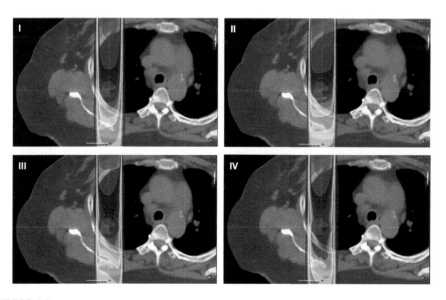

FIGURE PS12.2
Dose distributions with and without heterogeneity corrections.

7. For a treatment of breast cancer with two tangent field (i.e. opposed oblique beams), rank the following treatments in order of the smallest to largest hot spot in the breast.

    a. 6 MV, separation 15 cm

    b. 6 MV, separation 25 cm

    c. 15 MV, separation 15 cm

    d. 15 MV, separation 25 cm

8. Match the beam arrangement with the corresponding dosimetric effect if the entire vertebral body is to be treated in Figure PS12.3.

    a. 6 MV AP/PA   i. Most uniform dose to vertebral body

    b. 18 MV AP/PA + 6 MV R/L lat   ii. Lowest bowel dose

    c. 18 MV R/L lat   iii. Lowest kidney dose

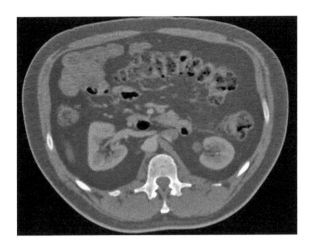

FIGURE PS12.3
CT cross-section through the abdomen.

9. Describe how a wedged pair might be arranged to treat the vertebral body shown in Figure PS12.3 and discuss the disadvantages of this approach.

10. A spine treatment is planned to deliver 300 cGy to isocenter with AP/PA 18 MV beams but instead the patient is accidentally treated with 6 MV beams (MU as planned for 18 MV are used). What is the effect on the dose to the isocenter prescription point and the hot spots near the superficial aspect?

  a. Higher isocenter dose, hotter hot spots

  b. Higher isocenter dose, cooler hot spots

  c. Lower isocenter dose, hotter hot spots

  d. Lower isocenter dose, cooler hot spots

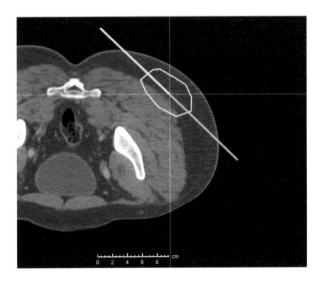

FIGURE PS12.4.1
Tangent beams.

11. Draw the wedge directions required to treat the PTV in white shown in Figure PS12.4.1 to a uniform dose using the two beams shown in yellow and blue.

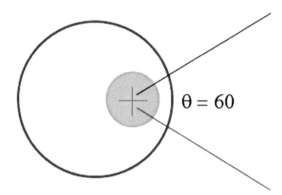

FIGURE PS12.5.1
Oblique beams.

12. Draw the wedge directions required for the treatment beams shown in Figure PS12.5.1 in order to get a uniform dose to the target in blue. Calculate the wedge angle assuming $\theta = 60$ degrees.

13. * List at least three advantages and three disadvantages of the "dynamic wedge" design (dynamically moving the jaw across the field) as compared to physical wedges.

13

Photon Beam Treatment Planning: Part II

13.1 Volume Definitions and DVHs

13.1.1 ICRU Volume Definitions

An essential part of the nomenclature radiation oncology is the terms and definitions for volumes of interest, that is, regions of interest (ROIs) on volumetric imaging (typically CT and/or MRI) for use in treatment planning. One standard reference for this is the reports from the International Commission on Radiation Units and Measurements (ICRU), namely ICRU Report No. 50 (Allisy et al. 1993), ICRU Report No. 62 (Allisy et al. 1999), and the most recent update ICRU Report No. 83 (Gregoire et al. 2010). These reports define and describe the target and critical structure volumes used in treatment planning and provide a basis for comparing and evaluating treatment outcomes.

The following target volumes are defined (Figure 13.1.1):

- **Gross Tumor Volume (GTV).** Can be palpated or visualized radiographically.
- **Clinical Target Volume (CTV).** Expansion to account for the extension of occult disease that is not observable on imaging.
- **Internal Target Volume (ITV).** Expansion to account for physiological motion of the target.
- **Planning Target Volume (PTV).** Expansion to account for variability in patient setup. This expansion ensures that the intended target is treated.
- **Treated Volume (TV).** Volume enclosed by some isodose surface defined by the radiation oncologist as necessary to achieve the purpose of treatment.
- **Irradiated Volume.** Volume irradiated to a "significant dose" relative to normal tissue tolerance, typically taken to be the 50% isodose line.

The following volumes are defined for normal tissues (Figure 13.1.2):

- **Organ at Risk (OAR).** This is a structure to be avoided.
- **Planning Risk Volume (PRV).** An expansion of the OAR to account for patient setup variability. Especially useful for serial organs like the cord, this expansion ensures that the dose to the OAR will remain within tolerance even in the setting of patient motion and other uncertainties.

The special case of tubular organs is discussed in the ICRU reports. An example of this is the rectum, which is a tubular structure (rectal wall) with inner contents (stool). ICRU

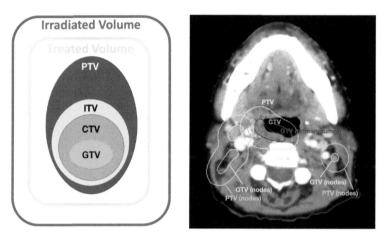

FIGURE 13.1.1
Left: Target volumes defined by the ICRU (Allisy et al. 1999). Right: Example head-and-neck cancer case showing two GTVs (one for primary disease and one for nodal disease).

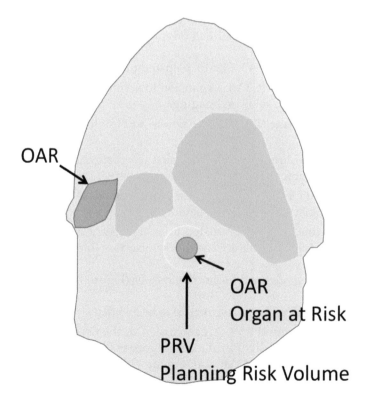

FIGURE 13.1.2
Normal tissue volumes defined by the ICRU. (From Allisy et al. 1999.)

recommends that that tubular structures should consider the lumen (wall) and not the inner contents. In practice many planners will use the full structure to optimize and create plans but then evaluate the dosimetry metrics on the tubular (wall) structure.

13.1.2 Nomenclature from AAPM Task Group 263

It is helpful to have standard naming conventions for regions of interest (ROIs). This supports communication within the clinic, data collection, clinical trials, automation, and potentially error checking and the enforcement of standards. ICRU reports do not provide detailed guidance on naming conventions for ROIs, and many possibilities exist for naming. In 2018, the AAPM Task Group 263 was published (Mayo et al. 2018) which provides recommended standards for the naming of structures and reporting of dose metrics. Examples from TG263 include the following: "Lung_L" (or "L_Lung") for left lung, "Lung_R" (or "R_Lung") for right lung, "Lungs" for both lungs combined, "Lungs-PTV" for the algebraic combination of two ROIs: total lung minus the PTV. Note that TG263 allows for two possible conventions to accommodate the various IT systems. The TG263 report and downloadable spreadsheets are available at the following site: www.aapm.org/pubs/reports/RPT_263_Supplemental/.

More recently the American Association of Radiation Oncology (ASTRO) has provided recommendations on which ROI structures should be included for each disease site for radiation oncology treatment plans. These recommendations employ the TG263 nomenclature.

13.1.3 Margins

The expansion of CTV (or ITV) to PTV is intended to account for variability in targeting of the beam relative to the patient, and one can evaluate what is the appropriate expansion, or "margin." A rational basis for margins requires an understanding of uncertainty in patient setup variability. This is illustrated in Figure 13.1.3 which shows the two types of variability in setup: **(1) random errors (denoted by the symbol, σ).** This is the variability of the targeting in the patient from day-to-day around an ideal position, indicated by the "+" in the figure. The symbol σ is the standard deviation (spread) of a Gaussian probability

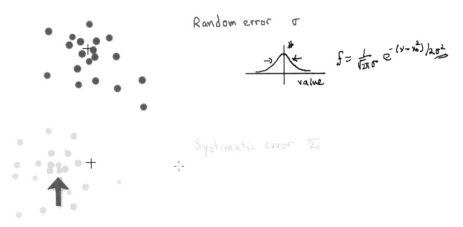

FIGURE 13.1.3
Two types of variability in the patient setup around an ideal position, indicated by the "+." (1) Random errors (σ), i.e. a day-to-day variability centered around the intended location, and (2) systematic errors (Σ), an overall offset of the collections of all days.

distribution. **(2) Systematic error (denoted by the symbol, Σ).** This is the overall offset of the average position averaged over all days compared to an ideal position, indicated by the "+" in the figure. Note that a systematic error effects every treatment in the same direction and so has a larger impact than random errors.

If the values of σ and Σ are known, it is possible to derive a "recipe" for CTV-PTV margin expansion which would guarantee the full dosimetric coverage of the CTV target in the presence of variability (see Table 4.4 in ICRU Report No. 83). One commonly used recipe is a PTV margin of $2.5\Sigma + 0.7\sigma$ from the group of Marcel van Herk et al. (2000).

Of course, since this is a stochastic statistical process, one can't guarantee that every single patient is covered. There will be statistical outliers. Rather one specifies the percentage of patients in a population that should receive full dose (e.g. 99%) and this then determines the margin. Assumptions about this population coverage are built into the margin formulae. Note that in the formula above the systematic error has a larger impact on the margin than the random error, i.e. Σ is multiplied by a larger value than σ in deriving the margins by this formalism. The margin formulae discussed here are intended to illustrate general principles and should be used with caution since these formulae rest on many assumptions about the patient population, intended coverage, and dose distribution.

13.1.4 Standards for Prescriptions

Standards for dose prescription have appeared in the ASTRO white paper on dose prescription which advocates that the following information be present in all radiation prescriptions: treatment site, delivery method, dose per fraction, number of fractions, and total dose. Furthermore the report call for the use of cGy units (instead of Gy) largely to avoid potential issues with decimal points. An example prescription then would be: L_Lung, Photons, 200 cGy × 32 to 6400 cGy. Specifying the prescription with these units and in this order avoids potential confusion of dose per fraction with the number of fractions (e.g. 54 Gy in 3 which could mean either 1800cGy×3 fractions or 3Gy×18 fractions).

In addition to the prescription, ASTRO recommendations also call for a minimum collection of data elements that should be included in a radiation oncology treatment record. These include the following 11 data elements: diagnosis (ICD-10 code), treatment site, dose per fraction (with units), radiation modality (e.g. photons), treatment technique (e.g. 3-DCRT), number of fractions planned, number of fractions delivered, total dose planned for primary target, total dose delivered for primary target, start date of treatment, end date of treatment.

13.1.5 Dose Volume Histogram (DVH)

Dose Volume Histograms (DVHs) provide a convenient way to assess and report dose. Figure 13.1.4 shows a simplified example dose distribution (left) with three different dose levels: red (100 cGy), green (50 cGy) and blue (10 cGy). Here the red point on the DVH is the volume of the ROI receiving at least 100 cGy, green is the volume receiving at least 50 cGy, and blue is the volume receiving at least 10 cGy. Each value defines a point on the DVH curve.

Figure 13.1.5 shows an example where the dose coverage of the PTV (red) is more uniform. In this case the DVH for the PTV drops off more sharply, indicating a more uniform dose. There are some modest hot spots in the target, which produce the high-dose "tail" on the DVH curve.

Example values can be read off the DVH, e.g. for the PTV $D_{95\%}$ (the dose receiving at least 95% of the prescription). The figure also shows an OAR (blue). The DVH values for this structure can be read off as well, e.g. V_{30Gy} (the volume receiving 30Gy or more).

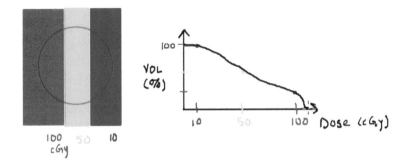

FIGURE 13.1.4
Example of dose distribution and corresponding dose volume histogram (DVH).

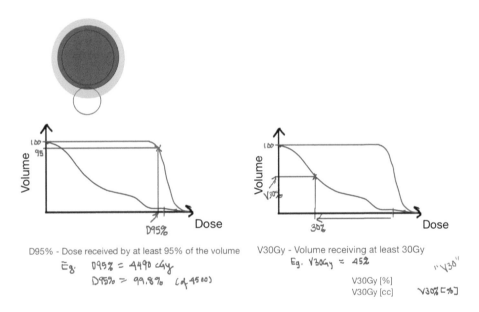

D95% - Dose received by at least 95% of the volume

Eg. D95% = 4490 cGy
D95% = 99.8% (of 4500)

V30Gy - Volume receiving at least 30Gy

Eg. V30Gy = 45%

"V30"

V30Gy [%]
V30Gy [cc]   V30%[%]

FIGURE 13.1.5
Dose distribution and DVHs for an IMRT plan.

13.1.6 Conformity Index

The conformity index (CI) is a useful metric that quantifies how tightly a dose distribution wraps around the PTV. CI is defined as the ratio of the total volume receiving the prescription dose to the volume of the PTV. Ideally the value of this would be one, though it is often larger. Sometimes the quantity CI_{95} is used, which is the ratio of the volume receiving 95% of the prescription dose to the volume of the PTV.

13.1.7 Point Dose vs. Volumetric Dose Prescription

The method of specifying the prescription dose for plans is important and has an impact on the dose delivered. For 3D conformal treatment plans, it is suitable to prescribe dose to a point. This is a rational method in these cases because the dose distributions in and around a tumor are relatively uniform. That is, if the dose is known to one point then it is known to all the other points. The ICRU Report No. 83 does specify that the point be in a region where dose can be calculated accurately (e.g. not in the buildup region, near a steep gradient, or a tissue interface), and in a central part of the PTV, most often at the isocenter or point where the beams intersect unless there is some reason for the point to be elsewhere.

For IMRT plans the situation is different. Dose distributions can be more heterogeneous. There may be hot spots in a tumor for example. If one then prescribes the dose to a point which happens to be in a hot region then the overall dose in the plan will be lower. Conversely, if the prescription point is in a cold region, the overall dose will be higher. This effect can be large. For an illustration and data see the video.

Due to these effects, ICRU Report No. 83 advocates for using a dose-volume prescription and not a point dose. Furthermore, the report advocates that the $D_{50\%}$ value of the PTV be at least reported, since this value is relatively reproducible across systems and plans.

13.2 Plan Quality, TCP, and NTCP

What makes a "good" treatment plan? One answer to this question is that the best plan maximizes tumor control probability (TCP) while balancing it with normal tissue complication probability (NTCP). Theoretical models have been developed for these quantities based on biological models of cell killing and Poisson statistics. These are presented in detail in the video.

A related topic is automated planning or "knowledge-based planning." This is an emerging area of study. Various metrics are employed, but the essential goal is to decide whether a plan is optimal. One approach is to use patient's anatomy as input and then to compare the plan to previous similar plans. Studies which have examined this have shown that plans generated in clinical practice or even in cooperative group trials are often not the optimal plans but could be improved. This is presented in much more detail in the video.

Further Reading

Allisy, A., et al. 1993. ICRU report no. 50, Prescribing, recording and reporting photon beam therapy. pp. 1–72.
Allisy, A., et al. 1999. ICRU report no. 62, Prescribing, recording and reporting photon beam therapy (Supplement to ICRU Report 50). pp. 1–62.
Bushberg, J.T., J.A. Seibert, E.M. Leidholdt Jr., and J.M. Boone. 2012. *The Essential Physics of Medical Imaging*. Chapters 4, 5, 7, and 10. Philadelphia, PA: Lippincott Williams & Wilkins.
Gregoire, V., et al. 2010. ICRU Report No. 83, Prescribing, recording, and reporting photon-beam intensity-modulated radiation therapy (IMRT). 10(1):1–106.
Mayo, C.S., et al. 2018. AAPM Task Group 263: Standardizing nomenclatures in radiation oncology. *Int J Radiat Oncol Biol Phys* 100(4):1057–1066. doi:10.1016/j.ijrobp.2017.12.013.

Chapter 13 Problem Sets

*Note: * indicates harder problems.*

1. What is the ratio of the volume of a spherical PTV of diameter 3 cm to the volume of the same PTV expanded by 3 mm in all directions?
   a. 0.58
   b. 0.75
   c. 1.33
   d. 1.72

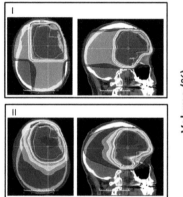 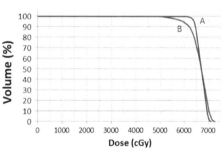

FIGURE PS13.1
Plans and DVHs.

2. Match the plans shown in Figure PS13.1 with the corresponding DVH.
   a. Plan I _____
   b. Plan II _____

3. In the DVH "B" in Figure PS13.1, estimate the following values from the plot. Assume a reference prescription dose of 6660 cGy.
   a. D95% [cGy]
   b. D99% [cGy]
   c. D50% [%]
   d. D2% [%]

4. Which of the plans in Figure PS13.1 has the largest conformity index? How is this reflected in the DVHs shown?

5. For a typical prostate cancer treatment plan of 180 cGy × 44 = 7920 cGy, indicate the structure in which the dose statistic is largest (rectum, rectal wall, or same).
   a. D_{1cc} [cGy] rectum/rectal wall/same
   b. V_{70Gy} [%] rectum/rectal wall/same
   c. V_{2Gy} [cc] rectum/rectal wall/same

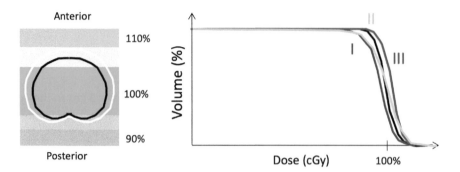

FIGURE PS13.2
Tumor coverage and DVHs.

6. Given the dose distribution (Figure PS13.2 left), the GTV (black) and PTV (white), match the anterior–posterior shift of the GTV with the corresponding DVH curve. Comment on the potential impact of these shifts on tumor control.

   a. Systematic shift anteriorly
   b. Systematic shift posteriorly
   c. Random shifts each fraction

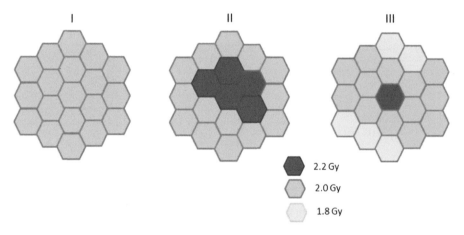

FIGURE PS13.3
Dose distributions.

7. What would be the impact of planning with a larger margin posteriorly?

8. Sketch the DVH for each of the three tumor dose distributions shown in Figure PS13.3.

9. What is the TCP for the uniform dose distribution I in Figure PS13.3? Use the LQ model with $\alpha/\beta=6$ Gy and $\alpha=0.3$ for a 30 fraction treatment. 1×10^8 clonogens in each of the 19 pixels.

   a. 0.81

   b. 0.87

   c. 0.93

   d. 0.99

10. * What is the TCP for dose distribution II and III Figure PS13.3? Use the LQ model with $\alpha/\beta=6$ Gy and $\alpha=0.3$ for a 30 fraction treatment. 1×10^8 clonogens in each of the 19 pixels.

14

IMRT and VMAT

14.1 IMRT and VMAT Delivery

14.1.1 Rationale for IMRT

Perhaps the most concrete way to understand the rationale for IMRT is to consider a case study. Figure 14.1.1 shows a patient in which the goal is to treat the target volume (blue) to a high dose while sparing the nearby organ at risk (green). One approach is to use open fields from various angles (Figure 14.1.1A). This is called 3D conformal radiation therapy (3D-CRT) because the shape of the beam conforms to the shape of the target as viewed from the particular angle being treated. Each of these beams provides a uniform dose fluence in the region being treated. However, a disadvantage is that the high dose may not conform well around the target and also the organ at risk may not be spared well. This case is a particularly challenging one since the target presents a concave shape with the organ located inside the concave region. In this case intensity modulated radiation therapy (IMRT) can provide a benefit by allowing for more complex dose distributions. In IMRT the fluence of each beam is no longer uniform but is varied or "modulated" across the face of the beam to produce higher or lower intensities where needed. In the example, for the posterior beam high intensities are required at the peripheral regions but lower intensities are required for the region of the beam that goes through the organ at risk. IMRT allows for complex dose distributions since there are many parameters which may be varied.

This highlights the main rationale for IMRT, namely:

- Conforms the high dose region to the target
- Allows for sparing of normal tissues
- Intentionally introduces inhomogeneous dose distributions when needed

Figure 14.1.2 shows another case study where IMRT is useful. That is treatment of head-and-neck cancer. Here the goals are to treat the gross disease (red-shaded region) to 6800 Gy, the intermediate region (blue) to 6000 cGy, and the at-risk nodal region (green) to 5000 cGy. It is clear that this is not achieved with the 3D-CRT plan (Figure 14.1.2A). The reason is that the posterior part of the photon treatment has to stop at approximately 4500 cGy in order to not give too much dose to the spinal cord. Though this region can be treated to higher doses with an electron field, still overall the goals for the treatment are not met. IMRT provides much finer control over the dose distribution (Figure 14.1.2B), allowing for the different regions to be treated uniformly to the desired doses while still sparing the

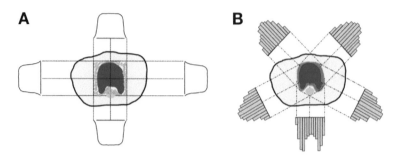

FIGURE 14.1.1
Treatment with (A) 3D conformal radiation therapy (3D-CRT) and (B) intensity-modulated radiation therapy (IMRT).

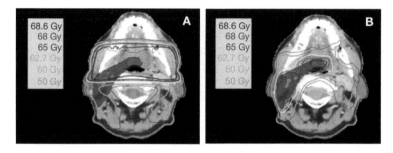

FIGURE 14.1.2
Example head-and-neck treatment plans. (A) 3D-CRT, (B) IMRT.

cord. Note also that it is possible to spare the parotid glad on the patient's left side in this case (light blue) which will preserve salivary function.

There is Level I evidence to justify the use IMRT. For example, Viani et al. 2016 report on a randomized control trial for prostate cancer treatment which shows that IMRT results in lower toxicities that 3D-CRT treatments. Also many cooperative group protocols and guideline reports require IMRT. For example, the guideline from ASTRO, ASCO and AUA on hypofractionated prostate treatment (Morgan et al. 2018) recommends that IMRT be used.

Table 14.1 shows some example sites where there is a strong rationale for the use of IMRT and the associated normal tissues where dose-sparing goals are important.

TABLE 14.1

Sites of Treatment and Normal Tissue Sparing Allowed with IMRT

Site	Normal Tissue Sparing Allowed
Central nervous system (CNS)	Brain, cord, optic structures
Head and neck	Cord, salivary glands, cochlea, other
Lung	Lung, cord, heart, esophagus, brachial plexus
Breast	Lung, heart
Abdomen	Bowel, liver, stomach, bladder
Genitourinary—prostate	Rectum, bladder, femurs
Sarcoma	Bone, lymphatic drainage

14.1.2 Delivering IMRT Treatments

To understand how IMRT treatments are delivered we must first describe what intensity modulation means. Figure 14.1.3 shows a beams-eye-view (BEV) for one particular IMRT field. BEV is the view of the beam as visualized from the location of the source. The figure on the left shows the fluence map from the treatment planning system. Some areas are bright (high photon fluence) and some areas dark (low fluence). The center plot shows the fluences plotted along the red line drawn through the BEV. One of the features of IMRT is the high dose gradients in the beam, where the dose or fluence varies rapidly over a short distance. The plot on the left is the idealized fluence distribution in the planning system. However, this distribution can be delivered on the linac using the techniques described below and verified with a detector such as film or an electronic portal imaging device (EPID). Figure 14.1.3, right, shows such a verification image.

Figure 14.1.4 illustrates the delivery of IMRT. Here the goal is to deliver the overall fluence map shown at the top. Here dark areas represent regions of high fluence while bright areas indicate regions of low fluence. For simplicity, this case just considers modulation in one dimension (i.e. long strips), but it is possible also to deliver more complex patterns like that shown in Figure 14.1.3. The delivery method shown in Figure 14.1.4 is "step-and-shoot" where the MLC is moved to form a particular shape, the beam is turned on for some amount of time and then off again. The MLC then moves on to form the next shape and that is irradiated and so on. By assembling these different shapes, any fluence map can be generated. For an animated illustration of this, see the video.

Another mode of delivery is dynamic MLC (also known as "dMLC" or sliding-window IMRT). In this method the beam remains on and the MLC moves. In regions where the MLC moves slowly more fluence is accumulated. In other words, the intensity can be modulated by controlling the leaf speed. For an illustration of this see the video.

So far the IMRT delivery considered here is performed with the beam at a fixed gantry angle. In Figure 14.1.1, for example, five separate IMRT beams were used each for a different angle roughly equally spaced around the patient. However, it is also possible to deliver IMRT while the gantry is moving. This mode of treatment goes by the generic term volumetric modulated arc therapy (VMAT) and is sometimes referred to by the trade name

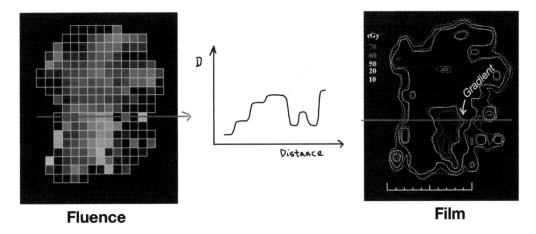

Fluence   **Film**

FIGURE 14.1.3
IMRT fluence map of a beam, profile and verification with film. (Adapted with permission from Moran et al. 2005. A dose gradient analysis tool for IMRT QA. *J Appl Clin Med Phys* 6(2) 62–73.)

FIGURE 14.1.4
Delivery of IMRT in "step-and-shoot" mode.

RapidArc® (Varian Inc.). This method is similar to dynamic MLC delivery except not only are the MLCs moving, but also the gantry is rotating while the delivery occurs. For an illustration of this see the video.

14.1.3 Other IMRT Delivery Methods

The IMRT and VMAT delivery techniques described above are the methods developed by the major manufacturers over the past 20 years. However, other methods of delivering IMRT are also possible. The TomoTherapy system (Section 9.2.1) uses a spiral delivery method similar to VMAT, but where only one slice at a time is irradiated. The system

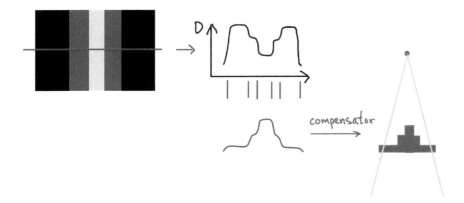

FIGURE 14.1.5
Compensator-based IMRT. A physical compensator constructed of metal has a thickness that provides the required fluence.

consists of a bank of binary MLCs that open and close quickly as the gantry rotates and the patient slides through the unit. For an animation of this process see the video.

Another method of IMRT delivery uses physical compensators. These are pieces of metal (often brass) which are machined into a shape that provides the attenuation needed for each beam. Figure 14.1.5 shows an illustration. The physical compensator map (blue) is essentially the inverse of the fluence map. Where the required fluence is higher, the compensator is thicker and vice versa. The thickness is calculated according to the attenuation law with some correction factors. A compensator is machined for each field mounted in the head of the linac (Figure 14.1.5, right). One of the main advantages of compensator-based IMRT is that it uses dose very efficiently compared to MLC-based IMRT or VMAT where the leaves are often irradiating only a thin strip. In the other areas of the field, the leaves are closed and the beam is "wasted." Some of the disadvantages of compensators are the need to produce multiple metal blocks for each patient and the need to enter the vault room to exchange blocks for each field. There are ways to address these problems, but since the 1990s little effort has been directed toward optimizing compensator-based IMRT systems since most manufacturers have pursued MLC-based systems.

14.2 Inverse Planning

The previous section described how IMRT plans were delivered, in particular, given a desired fluence map how the MLCs could be moved in order to deliver the desired fluence. Now we consider the creation of the fluence map (or leaf sequence). The question here is, given the desired dose distribution on the treatment planning computer, how is the fluence map (or leaf sequence) generated? The process for this is called "inverse planning."

14.2.1 Forward vs. Inverse Planning

To understand inverse planning it may be valuable to first understand forward planning. Figure 14.2.1 illustrates the process of forward planning with a case study for prostate

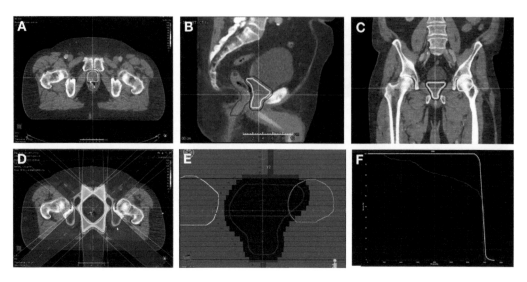

FIGURE 14.2.1
Treatment plan for prostate using a 3D-CRT forward plan with six fields.

cancer treatment. For a more detailed review of this case see the video. Here the goal is to treat the clinical target volume (cyan) to a high dose while avoiding the rectum (magenta) and bladder (contour not shown). In the process of forward planning, beam directions and MLC shapes are chosen by the planner. In this example six beams are chosen and the MLC is shaped to conform to the target volume at each angle (Figure 14.2.1E). Beam angles are selected in order to avoid the femoral heads (yellow and orange).

In the process of forward planning, the treatment planner selects the beam angles and MLC shapes and the computer calculates the dose distribution. This is a manual process. Inverse planning takes this process and inverts it. That is, the treatment planner specifies what dose distribution is desired and the computer calculates the MLC leaf shapes required.

14.2.2 Cost Functions and Optimization

Inverse planning is an optimization problem, the goal of which is to find the best planning solution given the goal of treating the target volume with a uniform dose while sparing the organs at risk (OAR). To drive this optimization, some metric for plan quality is needed. A critical measure for this is the dose volume histogram (DVH) as described in Section 13.1.5.

Figure 14.2.2A shows a cumulative DVH for an idealized dose distribution for treatment of the prostate.

In this plan the prescription isodose line (red) wraps around the target (blue) and the dose inside the target is uniform. Therefore, the DVH shows that 100% of the target volume receives at least 40 Gy. As the prescription dose is approached (79 Gy), the volume of target covered drops abruptly to zero. This is an idealized scenario of coverage of the target with a uniform dose.

Consider now the case where the dose distribution is changed somewhat in order to further spare a nearby OAR (Figure 14.2.2B) which in this case is the rectum shown in green. Now the prescription isodose (red) no longer covers the target so there is some region of the target that has a cold spot. This is reflected in the DVH. The target curve no longer

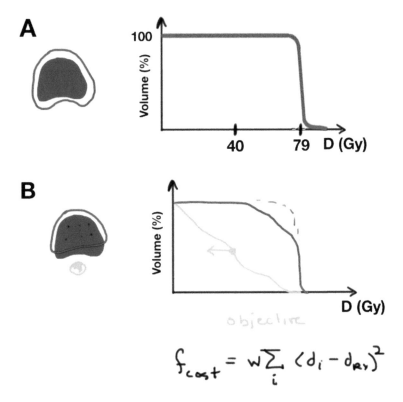

FIGURE 14.2.2
Dose volume histograms (DVH) and cost functions for optimization. (A) Ideal DVH for uniform coverage of the target volume (blue). (B) Plan which trades target coverage for dose sparing of the organ at risk (green).

drops abruptly to zero at the prescription dose as before (dotted line) but rather has a "shoulder" in the ~50–79 Gy range where 100% of the target is not covered (solid line). This represents a deviation from the ideal, and a function can be written to describe it. This so-called "cost function" penalizes solutions which deviate from the ideal dose distribution. A common form of the cost function is the following: $f_{\text{cost}} = w \sum_i (d_i - d_{Rx})^2$. The summa-

tion (Σ) is over all voxels, i, in the dose grid. At each voxel, the function $(d_i - d_{Rx})^2$ is calculated. This describes the difference between the dose at that voxel, d_i, and the intended dose at that voxel, d_{Rx}. The difference is squared in order to avoid negative values. In the simple case where only target coverage is important, all the dose values would be made to be equal to the intended dose at that voxel, d_{Rx}, and the cost function would be zero.

In more realistic scenarios there are competing factors. In Figure 14.2.2B the uniform coverage of the target volume is competing with the goal of reducing the dose to the OAR as much as possible. To account for this a second term is added to the cost function that describes the dose to the OAR. This uses an "objective" on the DVH defined by the user (green Figure 14.2.2B). An example objective for the rectum might be V_{40Gy} less than 30%. Each objective also has some weight, w, in the above cost function formula. By increasing the weight of the OAR cost function relative to the weight of the target cost function one can prioritize OAR sparing over target coverage or vice versa. More than one OAR can be included in the cost function as well as other metrics such as how tightly the dose conforms around the target or the spillage of the low-dose lines. By changing the dose distribution until the cost function is minimized one finds the ideal dose distribution that optimizes for the competing factors.

The video includes an example planning session for the prostate case shown here. IMRT and VMAT plans are created using the principles described here (see final eight-minute segment).

Further Reading

Dieterich, S., E. Ford, D. Pavord and J. Zeng. 2016. *Practical Radiation Oncology Physics*. Chapter 16. Philadelphia, PA: Elsevier.

Ezzell, G., et al. 2009. IMRT commissioning: Multiple institution planning and dosimetry comparisons, a report from AAPM Task Group 119. *Med Phys* 36(11):5359–5373.

Gregoire, V., et al. 2010. Prescribing, recording and reporting photon-beam intensity-modulated radiation therapy (IMRT). *ICRU Report 83* 10(1):1–106. doi:10.1093/jicru/ndq001.

Kahn, F.M. and J.P. Gibbons. 2014. *Kahn's The Physics of Radiation Therapy*. 5th Edition. Chapter 20. Philadelphia, PA: Wolters Kluwer.

Low, D.A., et al. 2001. Dosimetry tools and techniques for IMRT, report of AAPM Task Group 120. *Med Phys* 38(3):1313–1338.

McDermott, P.N. and C.G. Orton. 2010. *The Physics of Radiation Therapy*. Chapter 20. Madison, WI: Medical Physics Publishing.

Metcalfe, P., T. Kron and P. Hoban. 2007. *The Physics of Radiotherapy X-rays and Electrons*. Chapter 7. Madison, WI: Medical Physics Publishing.

Miften, M., et al. 2018. Tolerance limits and methodologies for IMRT measurement-based verification QA: Recommendations of AAPM Task Group No. 218. *Med Phys* e53–e83. doi:10.1002/mp.12810.

Morgan, S.C. et al. 2018. Hypofractionated radiation therapy for localized prostate cancer: Executive summary of an ASTRO, ASCO, and AUA evidence-based guideline. *Practical Radiation Oncology*, 8(6): 354–360.

Viani, G.A., B.S. Viana, J.E. Martin, G. Zuliani and E.J. Stefano. 2016. Intensity-modulated radiotherapy reduces toxicity with similar biochemical control compared with 3-dimensional conformal radiotherapy for prostate cancer: A randomized clinical trial. *Cancer*, 122(13): 2004–2011.

Chapter 14 Problem Sets

*Note: * indicates harder problems.*

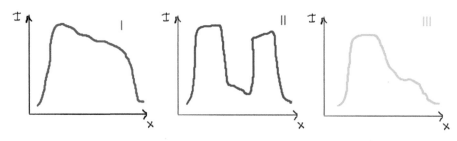

FIGURE PS14.1
IMRT fluence distributions.

1. Match the fluence profiles in Figure PS14.1 with the likely site of treatment. Each case is a profile drawn in the axial plane for one particular IMRT beam.

   a. Head and neck

   b. Prostate

   c. Breast

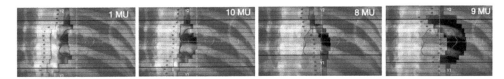

FIGURE PS14.2
IMRT leaf sequence control points.

2. Sketch the resulting fluence along a left–right line through isocenter from the step-and-shoot IMRT beam in Figure PS14.2 which has four control points (segments).

3. Rank the following treatment techniques in order from least to most MU required to deliver the same dose to a target.

   a. 3D-CRT

   b. Step-and-shoot IMRT with MLC

   c. Dynamic MLC IMRT

   d. Compensator IMRT

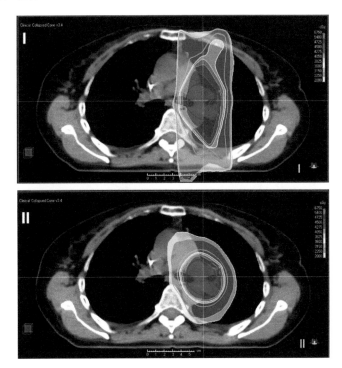

FIGURE PS14.4
Dose distributions.

4. Match the plan shown in Figure PS14.4 (I in top, II in bottom) with the technique used to plan
   a. VMAT
   b. 3D-CRT
   c. 5-field IMRT

5. Match the DVH in Figure PS14.5 for the left lung in yellow with the corresponding plan in Problem 4.
   a. A = ?
   b. B = ?

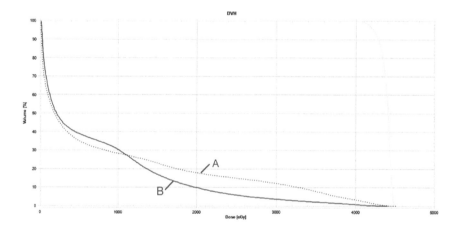

FIGURE PS14.5
DVHs.

6. What advantage is provided by rotating the collimator away from zero degrees in VMAT arcs?

7. List three advantages of VMAT vs. IMRT.

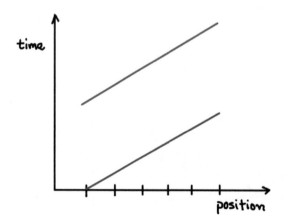

FIGURE PS14.6
IMRT leaf trajectories.

8. * The trajectory of the left MLC leaf (red) and right MLC leaf (blue) are shown in Figure PS14.6, indicating a steady speed across a 5-cm region while the beam is on. Sketch the resulting fluence distribution.

9. * Sketch the fluence distribution for Problem 8 if the left MLC is miscalibrated 2 mm to the left.

10. * Sketch the fluence distribution for Problem 8 if the left MLC is traveling at a speed 10% slower than intended. What would this look like for step-and-shoot delivery?

15

Megavoltage Electron Beams

15.1 Basic Physics and PDDs of MV Electron Beams

This chapter considers the details of megavoltage electron beams, building on previous sections which describe the production of megavoltage electron beams in a linac (Section 8.2) and the collimation system used in linacs for treatment in electron mode (Section 9.1.3). This chapter presents more details on the properties of MV electron beams and their use in patients.

15.1.1 Introduction to Electron Therapy Beams

Clinical use scenarios for MV electron beams include the following:

- Superficial lesions
- Lesions in the skin
- Nodal region (e.g. internal mammary nodes in the treatment of breast cancer)
- Boost treatment to a surgical bed or scar
- Intra-operative radiation therapy (IORT)

To understand the advantage of electrons in this context consider Figure 15.1.1 that compares a 6 MeV electron beam to a 6 MV photon beam. The 6 MeV beam has a substantially higher surface dose and is restricted to much more superficial regions.

15.1.2 Production of Beams and Spectra

Recall from Section 9.1.3 that when operating in electron mode the linac is configured as follows: the target is removed from the beam, a scattering foil is introduced, and a collimation assembly (the "cone") is attached and extends near the patient surface. See Figure 15.1.2. One important property of the beam is its spectrum, i.e. the number of electrons emerging at each energy. Figure 15.1.2 shows example spectra at three points along the beam path. As the beam emerges from the head of the linac the spectrum (gray) is relatively narrow and centered around the nominal energy of the beam. The nominal energy can be selected by controlling the energy-switch in the waveguide.

At the point where the electrons have reached the surface of the patient (blue in Figure 15.1.2) the spectrum has changed. Electrons have scattered with air molecules along the intervening path. Some electron have lost energy, but since this is a random process there are some electrons that do not undergo many scatter interactions in the air and do

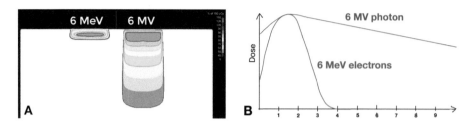

FIGURE 15.1.1
Megavoltage electron vs. photon beams. (A) Dose distribution for a 10×10 cm beam on water. (B) Percentage depth dose (PDD) along the central axis for 6 MV photons (green) vs. 6 MeV electrons (cyan).

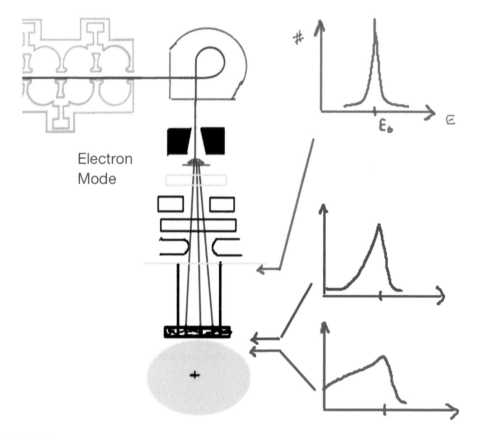

FIGURE 15.1.2
Spectra of megavoltage electron beams at various points along the beam path.

not lose much energy. The end result is a spectrum that still has a peak near the nominal energy but now has a tail extending down to low energies. As the beam enters the patient (red in Figure 15.1.2) the electrons undergo even more interactions because the material has a higher density. Again, this results in an extended low-dose tail.

The spectrum of the beam will have a profound influence on dose distribution including the percent depth doses (PDDs). This will be seen in detail in the next sections, but it is worth taking a moment to recall the physical processes at work when a charged particle like the electron interacts with matter (Chapter 7). Recall that energy is lost through two processes: (1) collisional losses which give energy to atomic electrons, and (2) radiative losses which produce bremsstrahlung X-ray photons. Recall from Figure 7.2.1 that the energy loss per unit length (stopping power) is large for collisional losses (and somewhat higher in low-Z materials like tissue vs. high-Z material like metals). The radiative energy loss is typically lower, and this is especially true for low-Z materials like water or tissue.

15.1.3 Electron PDD

Figure 15.1.3 illustrates the effects of energy loss and how these are manifest in PDDs. Figure 15.1.3A represents an idealized scenario where all the electrons have the same energy as they enter the patient (illustrated in gray). As they move through tissue they lose energy (dE/dx) and eventually stop. Because the electrons all have the same energy they stop in approximately the same place. The resulting PDD (right panel) has a sharp falloff at the depth where the electrons stop. In a more realistic scenario the electrons actually have a spectrum similar to that seen in Figure 15.1.2 (blue). Some electrons are higher energy and travel farther in tissue. Some electrons are lower energy and stop at shallower depths in tissue. This results in a PDD showing a dose falloff that is no longer sharp but is spread out in depth. Added to this is the fact that electrons do not travel in straight lines but undergo path changes due to Coulomb scattering (Figure 7.2.3). This also spreads out the depth-dose curve.

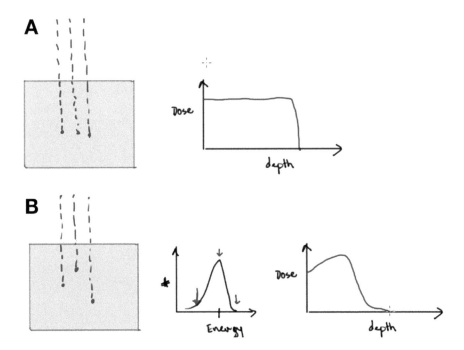

FIGURE 15.1.3
Effect of spectrum on electron percent depth dose (PDD). (A) Electrons of the same energy penetrate to approximately the same depth creating a sharp falloff in the PDD. (B) Electrons with a spectrum of energies. Higher energy electrons travel farther and lower energy electrons travel less far, resulting in a spread out of the dose falloff.

15.1.4 Energy and Field Size Effects on PDD

The energy of the electron beam has a strong effect on its properties and, in particular, on the PDD. Figure 15.1.4 shows data from two beams at two different energies (6 MeV and 12 MeV). The higher energy beam has several notable properties:

- Deeper penetration
- Higher surface dose
- More spread-out falloff
- Broader region of dose maximum

The PDD can be characterized by several metrics which are useful: the range for various doses (as a percentage of the maximum dose) including R_{50}, R_{80}, and R_{90}, and the practical range, R_p, defined by extrapolation to zero dose as shown in Figure 15.1.5. There are several

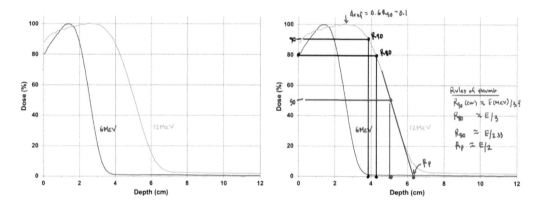

FIGURE 15.1.4
Electron PDD for two energies and associated parameters that describe the PDD.

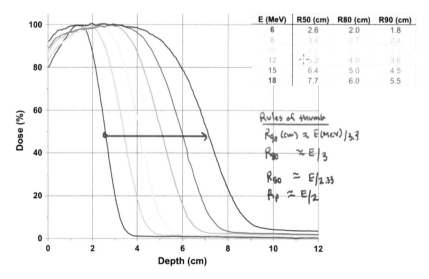

E (MeV)	R50 (cm)	R80 (cm)	R90 (cm)
6	2.6	2.0	1.8
8	3.4	2.7	2.4
12	5.2	4.0	3.6
15	6.4	5.0	4.5
18	7.7	6.0	5.5

Rules of thumb
R_{90} (cm) ≈ E(MeV)/3.7
R_{80} ≈ E/3
R_{50} ≈ E/2.33
R_p ≈ E/2

FIGURE 15.1.5
PDDs for various energy electron beams.

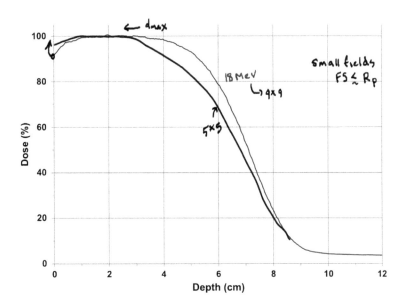

FIGURE 15.1.6
Field size effects on PDD in an electron beam.

rules of thumb for megavoltage electron beams in tissue, namely, $R_{90}(cm) \approx E(MeV)/3.3$, $R_{80}(cm) \approx E(MeV)/3$. Sometimes, $R_{50}(cm) \approx E(MeV)/2.33$ is also useful. Note that R_{50} is used in dosimetry protocols to define a reference depth for electron beams, $d_{ref} = 0.6R_{50} - 0.1$. This is the depth where the beam is calibrated and is similar to d_{max} for a photon beam. Since the dose maximum can cover a wide range of depths for an electron beam this more exact definition is often useful. A final quantity of use is R_p, the practical range. The rule of thumb is $R_p(cm) \approx E(MeV)/2$. This represents the farthest distance an electron can travel and is useful in many contexts.

Field size can also affect the PDD, though the behavior is somewhat more complex than in photon beams. Field size effects become important when the field size becomes smaller than the practical range, R_p. This is because lateral equilibrium begins to be lost. Figure 15.1.6 shows an example of the PDD from an 18 MeV beam. For this energy R_p is 9 cm so for a field size of 9×9 cm and larger the depth-dose curves look essentially identical. As the field size is reduced below 9×9 cm, however, the depth of maximum is pushed closer to the surface and the surface dose is increased. Figure 15.1.6 shows data for a 5×5 cm field.

15.1.5 Photon Contamination

It may be noted in Figure 15.1.5 that there is a non-zero dose at deep depths in the patients. This is not from electrons since they do not penetrate to these depths. Rather, it is from X-ray photons in the beam. There are several sources for these photons. One is in the head of the linac, especially the scattering foil. Recall that the radiative loss process can be very efficient in metals and, even though the scattering foil is made of low-Z metals, photons are still created. Figure 15.1.7 shows that the dose from photon contamination is roughly 1–5% of the peak dose and is larger for higher energy electron beams. This is because the radiative stopping power is larger at higher energies (recall Figure 7.2.1). Deeper in the patient, the bremsstrahlung photons that are present originate not only from the head

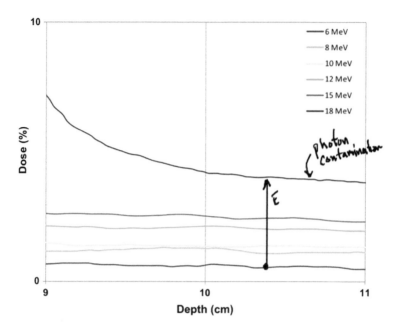

FIGURE 15.1.7
Photon contamination at deep depth in electron beams.

of the linac but from the patient themselves. Even though low-Z material like tissue is not very efficient at producing bremsstrahlung through radiative losses (recall Figure 7.2.1), at deep depths the contributions can add up to be substantial.

15.2 Properties of Treatment Beams

15.2.1 Electron Beam Penumbra

Figure 15.2.1 shows dose distributions from electron beams at two different energies. The variation of dose with depth should already be clear from the discussion in Section 15.1.4 (see e.g. Figure 15.1.5), but the beam profile depends on energy as well. Figure 15.2.1 shows dose profile plots at two different depths. At depth line A, it can be observed that the penumbra for the 18 MeV beam is smaller than for the 8 MeV beam. This is because the scatter of electrons is more forward directed at higher energies. As the depth increases, though, the penumbra becomes very wide especially for higher energy beams. This is illustrated at depth line B for the 18 MeV beam. The higher isodose lines tuck in toward the center of the field while the lower isodose lines bulge out.

15.2.2 Collimation and SSD Effects

The effects described above have an impact on how blocking is accomplished in electron fields and considerations for treatment margins. The typical setup is to treat the patient at 100 SSD; the end of the cone is at 95 cm from the source, meaning there is a 5 cm gap between the cone end and the skin. The size of the blocked area is typically drawn approximately

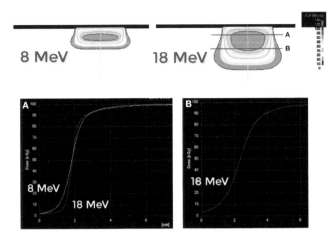

FIGURE 15.2.1

Impact of energy on penumbra in electron fields. Profiles at two different depth lines, A and B. (Dose distributions in this and other figures in this chapter are calculated with RayStation v.6.1 using a Monte Carlo algorithm with 2 million particle histories.)

1 cm wider on each side than the intended treatment area to account for the penumbra as described above, though note that this is a very approximate rule of thumb.

Patients can also be treated at extended SSDs (e.g. 105 cm or even 110 cm) as might be required to avoid some anatomical region like the shoulder. Several effects happen when treating at extended SSD compared to the standard SSD of 100 cm:

- Larger penumbra. The high-dose isodose line pinches in more and low-dose lines bulge out.
- Output decreases. An inverse square correction can be performed but here there is no "source" as such (no target); rather a "virtual source" can be defined at some distance from isocenter. Inverse square corrections using this virtual source location will be relatively accurate.
- Light field no longer accurately tracks the radiation field edge as it does in most other cases.

15.2.3 Field Matching

Some clinical scenarios require two fields to be treated next to each other. Complex dose distributions can be present in these cases. Figure 15.2.2 shows a case where a 6 MeV

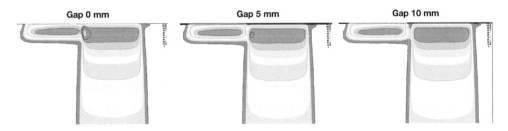

FIGURE 15.2.2

A 6 MeV electron field (left) matched with a 6 MV photon field (right).

electron field (left) is matched to a 6 MV photon field (right). When the field edges directly abut (left panel) there is a hot spot on the photon side at approximately d_{max} for the electron beam. In this case it is hot spot of approximately 30%. If a gap is introduced between fields the hot spot is reduced. At larger gaps, however, a cold spot develops at the surface (right panel). A common technique in these cases is to "feather" the gap, that is, move the position of the abutment each day on the patient in order to spread out the cold spot or hot spot.

Hot spots also occur when two electron beams are matched together. For example with two 6 MeV electron fields matched there is a hot spot (roughly 10%) where these two fields abut.

15.2.4 Obliquity and Curved Surfaces

Figure 15.2.3 shows the effect of having an electron beam incident at some oblique angle. When the beam is "en face" (perpendicular to the surface) the isodose lines are symmetric and even, but as the beam becomes more angled (20° and 40° in the figure) the isodose lines become more asymmetric. The doses are higher on the side where the angle is sharper (right side in figure). The dose maximum also gets pulled closer to the surface. This can be seen quantitatively in Figure 15.2.4 where the PDD is plotted along a line perpendicular to the surface passing through isocenter.

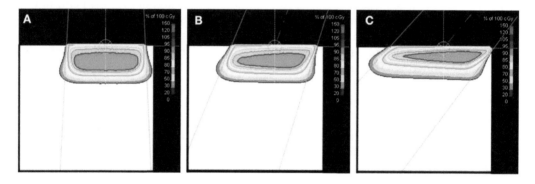

FIGURE 15.2.3
Electron fields at various angle. (A) 0 degrees or "en face," (B) 20 degrees from the normal, (C) 40 degrees from the normal.

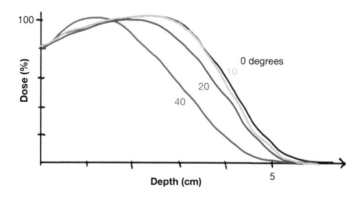

FIGURE 15.2.4
Percent depth doses (PDDs) for the beams shown in Figure 15.2.4 along a line perpendicular to the surface through the isocenter.

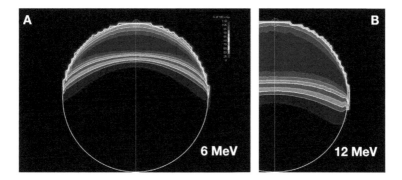

FIGURE 15.2.5
Effect of a curved surface on electron beam dose distribution. (A) 6 MeV beam. (B) 12 MeV beam.

Related to this is the dose distribution of an electron beam in the presence of a sloped surface (Figure 15.2.5). Near the edge of the beam, the isodose lines are pulled closer to the surface because in this region the beam obliquity is higher; the effect is the same as that noted above. There is also an energy dependence to this effect. Figure 15.2.5B shows data for a 12 MeV beam; the dose lines are pulled even closer to the surface for the higher energy beams.

15.2.5 Inhomogeneities in Electron Beams

Inhomogeneities have a strong effect on the dose distribution of electron beams, though the behavior is different than with photon beams. Figure 15.2.6 shows an idealized scenario to illustrate this, a tissue prominence in the center of the beam. This might be the patient's nose for example. In this case the dose under the prominence is shifted up closer to the source. Also the region on either side of the prominence has increased dose. This is due to outscatter of electrons from the prominence into this region. The hot spot is larger and higher dose for higher energy electrons due to the longer range of electrons in tissue (Figure 15.2.6B).

The same physical processes are at work around a region of a tissue defect (e.g. the "hole" in Figure 15.2.7). Here the isodose lines under the defect are shifted away from the source. Also now there is inscatter of electrons from the sides into the region underneath the defect, creating hot spots on either side just inside the edge of the defect.

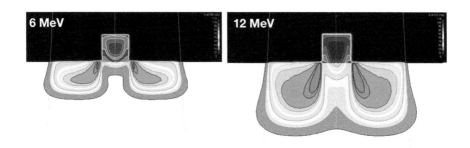

FIGURE 15.2.6
Electron dose distributions in the presence of a tissue prominence for 6 MeV and 12 MeV beams.

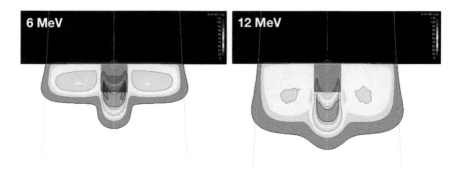

FIGURE 15.2.7
Electron dose distributions in the presence of a tissue defect for 6 MeV and 12 MeV beams.

Further Reading

Dieterich, S., E. Ford, D. Pavord and J. Zeng. 2016. *Practical Radiation Oncology Physics*. Chapter 15. Philadelphia, PA: Elsevier.

Kahn, F.M. and J.P. Gibbons. 2014. *Kahn's The Physics of Radiation Therapy*. 5th Edition. Chapter 14. Philadelphia, PA: Wolters Kluwer.

McDermott, P.N. and C.G. Orton. 2010. *The Physics of Radiation Therapy*. Chapter 15. Madison, WI: Medical Physics Publishing.

Metcalfe, P., T. Kron and P. Hoban. 2007. *The Physics of Radiotherapy X-rays and Electrons*. Chapter 5. Madison, WI: Medical Physics Publishing.

Chapter 15 Problem Sets

*Note: * indicates harder problems.*

1. What is the most appropriate cutout size and electron energy to treat a 2 cm diameter tumor whose distal aspect is 1.5 cm deep assuming the entire tumor should receive at least 90% of the dose?

   a. 6 MeV, 3 cm dia circle

   b. 6 MeV, 4 cm dia circle

   c. 10 MeV, 3 cm dia circle

   d. 10 MeV, 4 cm dia circle

2. Which electron treatment will give the highest skin dose for a given dose at maximum?

   a. 6 MeV

   b. 8 MeV

   c. 12 MeV

   d. 15 MeV

3. Which technique will provide the best dose coverage to both a scar and residual gross tumor extending to a depth of 2 cm while also sparing normal tissue?

   a. 6 MeV, no bolus
   b. 8 MeV, no bolus
   c. 10 MeV, 1 cm bolus
   d. 15 MeV, 1 cm bolus

4. For an electron treatment designed to cover the chest wall which energy will give the highest dose to the lung?

   a. 6 MeV
   b. 8 MeV
   c. 10 MeV
   d. 12 MeV

5. Which energy and cutout combination might require a special measurement to determine the PDD and MU to be used?

   a. 6 MeV, 5×5 field
   b. 8 MeV, 10×10 field
   c. 12 MeV, 5×5 field
   d. 18 MeV, 10×10 field

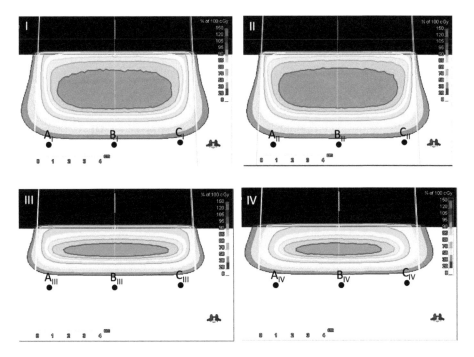

FIGURE PS15.1
Electron beam dose distributions.

6. Match the isodose plots in Figure PS15.1 with the corresponding energy and SSD.
   a. 6 MeV, 100 SSD
   b. 6 MeV, 110 SSD
   c. 12 MeV, 100 SSD
   d. 12 MeV, 110 SSD

7. When treating matched fields (electron + electron or electron + photon), what are the advantages and disadvantages of using a gap between fields? What are other ways of handling this to provide a more uniform dose?

8. When using an internal lead shield under the lip what would be the effect of reversing the position of the low-Z absorber (i.e. absorber on the distal side of the lead shield instead of the proximal side)?
   a. Decreased dose to the gum mucosa distal to the shield
   b. Increased dose to the gum mucosa distal to the shield
   c. Decreased dose to the lip
   d. Increased dose to the lip

9. * Which of the beams shown in Figure PS15.2 will result in the highest dose to the scar (shown in green)? What are other approaches to treating the tumor volume (red) and the scar (green)?

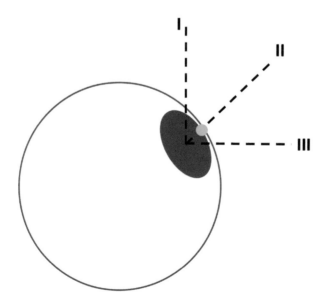

FIGURE PS15.2
Oblique beams.

10. * In Problem 6, which of the points shown has the highest dose from X-ray contamination?

16

Radiation Measurement: Ionization Chambers

16.1 Introduction to Dose Measurement

16.1.1 Operation of Ionization Chamber

One of the simplest and most accurate ways to measure radiation dose is with an ionization chamber. This is represented schematically in Figure 16.1.1 which shows a volume of air (gray) located inside some medium (red) such as water. The air in the chamber is typically vented in some way to the room and so is at the ambient temperature, T, and pressure, P. In the chamber shown in the figure there are two metal plates (electrodes), one at the top and one at the bottom of the chamber, that are held at some voltage relative to each other. A photon interacts with an electron in the medium and undergoes Compton scattering, producing an electron (blue). This electron enters the air-filled ionization chamber. At this point it experiences a force due to the electric field in the chamber. Recall that the voltage difference between the plates results in an electric field which produces a force on the electron. The electron is attracted to the electrode at positive voltage (in the figure this is the electrode on the bottom). The accumulation of many electrons on this electrode results in a charge, Q, which can be measured. Recall that charge has the units of Coulombs (C). The charges typically encountered in a therapy beam are in the range of nano-Coulomb (nC).

Once the charge is measured we need a method for converting it to dose. The formalism for doing this is called cavity theory, the most relevant form being Bragg–Gray cavity theory named after the developers of the method. This formalism will be discussed in detail in Section 16.2, but the central idea is that if the presence of the cavity does not disturb the beam then approximations can be used which make the charge-to-dose conversion much simpler. As such, small-volume ionization chambers are typically used.

16.1.2 Farmer Chamber

One commonly used type of ionization chamber is the Farmer chamber (Figure 16.1.2). This is a cylindrical (or thimble) type chamber. The active volume is the ~2 cm long cylinder the tip which is filled with air. The central electrode (red) is held at a high voltage, typically +300 V. The outer electrode (black) is constructed from a conductive plastic and is held at ground potential (0 V). An electron (blue) entering the chamber is collected on the central electrode. Also the electron ionizes molecules in the air, and the resulting electrons are collected on the central electrode. This is read out as a charge, Q. The chamber is connected to an electrometer via the connector shown in Figure 16.1.2B. The electrometer applies the voltage and also reads out the charge. The connector is a triax connector

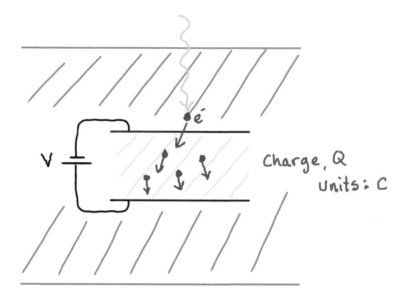

FIGURE 16.1.1
Charge measurement in an ionization chamber.

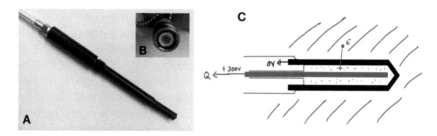

FIGURE 16.1.2
The Farmer chamber. (A) Waterproof chamber. (B) Triaxal ("triax") connector for the electrodes. (C) Chamber in water (blue). Central electrode (red) is at a high voltage and collects electrons which are read out as charge, Q.

consisting of a central electrode (pin), an inner electrode (the guard electrode, not shown in figure), and the outer ring for the ground electrode.

The chamber is placed in some medium, the standard for absolute dosimetry being water (most modern chambers are constructed to be waterproof), but it can also be placed in plastic. Plastics are available which are designed to be equivalent to water in terms of the stopping power (e.g. SolidWater®). While most electrons are created in the water surrounding the chamber, it is also possible for electrons to be generated in the wall of the chamber. Correction factors are required to account for this (see Problem Sets at the end of this chapter). If this wall of the chamber is thick and/or the energy of the beam is low, this effect can be large.

16.1.3 Parallel Plate Chamber

Another common chamber design for radiotherapy applications is the parallel plate (or "pancake") chamber (Figure 16.1.3). This design employs two circular electrodes (~2 cm diameter) separated by a small distance (~1 mm). The whole chamber is put into a

FIGURE 16.1.3
Parallel plate chamber. Inside the chamber is a thin collection area between two electrode plates. Here the chamber is shown at some depth, *d*, in the medium.

medium such as water or plastic. One of the key advantages of this design is that it has a small dimension in the beam direction and yet a relatively large collecting volume due to the large electrode area. The thin dimension in the beam direction means the measurement is collected at one depth and that depth is well-known. Many such chambers have a guard electrode, a secondary electrode ring around the central electrode. This is held at the same voltage but charge is not collected from it. This eliminates the effect of fringe fields around the outer edge of the electrode and reduces effects of in-scatter off the chamber walls. Example models of parallel plate chamber include the Markus, Spokas, and Roos chambers, each with a slightly different design of air volumes and electrode.

Applications of parallel plate chambers include depth-dose scanning in electron beams. The chamber is scanned down in depth in a water tank. Because the volume is so thin there is very little averaging in the depth dimension. This is especially useful for electron beams where there is a steep dose falloff. Another application is measuring the dose at the surface. Because of the thin window design the chamber can be brought up right to the surface. Note that measurements done in this way do not provide a direct measure of dose at the surface. Correction factors are still required.

16.1.4 Comparison of Cylindrical Chambers

Chambers come in many designs which are optimized for different uses. Figure 16.1.4 shows a comparison of various cylindrical chambers. The Farmer chamber, discussed above, stands out because of its relatively large volume (0.6 cm³). A key advantage of a larger volume chamber is that it has more air and therefore generates a higher signal since there are more air molecules to be ionized and generate electrons for collection. For this reason Farmer chambers are used for reference absolute dosimetry (establishing the number of cGy pr MU for example).

16.1.5 Applications of Small Chambers

The medium-sized chambers in Figure 16.1.4 are useful for scanning a beam to measure PDDs or profiles. The smaller volumes provide a better spatial resolution, but the volume is large enough to provide a reasonable signal. Figure 16.1.5 illustrates how chamber size affects the data acquired when measuring a beam profile. For a very large chamber like the

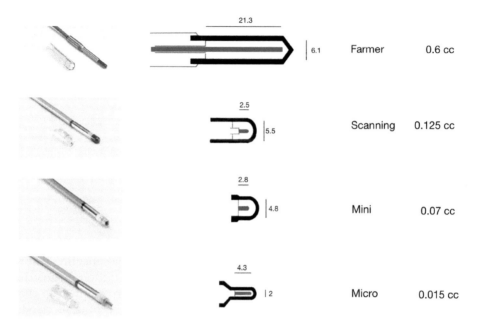

FIGURE 16.1.4
Four different cylindrical chambers. Drawings are to scale with units in mm. Applications are shown along with the approximate volume of the chamber.

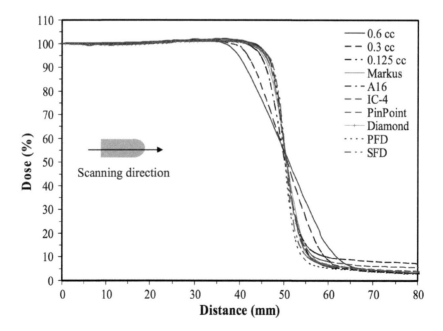

FIGURE 16.1.5
Beam profiles acquired with various chambers. (From AAPM TG106 on beam commissioning, I.J. Das, et al. 2008. Accelerator beam data commissioning equipment and procedures: report of the TG-106 of the Therapy Physics Committee of the AAPM. *Med Phys* 35(9):4186–215.)

Farmer chamber, the measured profile becomes artificially wide which does not represent the real profile. For this reason, a Farmer chamber is not used for scanning. The smaller chambers provide more accurate measurements. Note that some of the devices used to obtain the scans in Figure 16.1.5 are not chambers (e.g. diamond detector, or photon field diode [PFD]). These will be discussed in Chapter 17.

The smallest chambers (microchamber, volume ~0.015 cm³) are useful for measuring doses in very small fields. Figure 16.1.6 shows the measured charge vs. field size. The black line is the Monte Carlo calculation for the actual output. Recall that the beam output decreases as field size decreases. Note that measurements using some chambers are very low. For example, the output measured in a small field with a Farmer chamber is too low by a factor of 2. It is important to have an accurate measurement of output. If the measured output is too small then the number of MU required for a given dose will be overpredicted. This can have a major impact. The use of a large chamber was a causal factor for a series of radiation misadministration reported in 2010 which overdosed 76 patients by more than 50% (Bogdanich and Ruiz 2010).

Note that according to Figure 16.1.6 even the minichamber with volume 0.125 cm³ yields output measurements that can be deviate by more than 30% for the smallest fields. The topic of small-field dosimetry is a complex subject, and a full treatment is beyond the scope of this text. An excellent summary and description can be found in the forthcoming AAPM TG-155 report (Das et al., in preparation.) and also the 2017 TRS-483 report from IAEA and AAPM, a code of practice report for small-field dosimetry (International Atomic Energy Agency (IAEA). Dosimetry of small static fields used in external beam radiotherapy: an IAEA-AAPM International Code of Practice for reference and relative dose determination. IAEA Technical Report Series (TRS) 483. Vienna, Austria; 2017.)

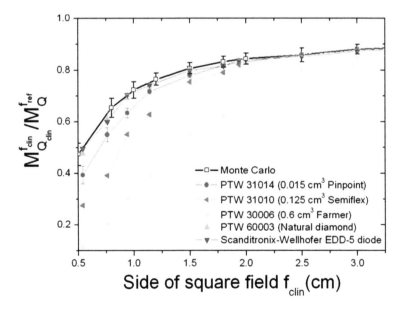

FIGURE 16.1.6
Measured charge vs. field size for various chamber sizes. (From R. Alfonso, et al. 2008. A new formalism for reference dosimetry of small and nonstandard fields. *Med Phys* 35(11):5179–86.)

16.2 Dose Measurement Protocols

16.2.1 Protocols for Dose Calibration

In order to convert the charge measured in a chamber to the dose to medium, some recipe or protocol is required. This section will describe such dose conversion using the nomenclature and procedures of AAPM TG-51 published in 1999 (Almond et al. 1999). Note though, that there are other protocols for dose calibration including TRS-398 from IAEA, Duetsch industrie Norm (DIN) in Germany, or IPEM from the UK. The nomenclature differs in these reports, but the concepts are similar. Note also that TG-51 is actually an update to an older AAPM protocol, TG-21 published in 1983 (AAPM Task Group 21, 1983). TG-51 is easier to use and less error-prone, but it is sometimes useful to refer to TG-21 for learning purposes since the underlying principles are described more thoroughly there.

16.2.2 Calibration and Quality Conversion

The setup for dose calibration measurements is shown in Figure 16.2.1. The chamber (typically Farmer chamber) is placed at isocenter (plus sign) and at a depth of 10 cm. The medium used for this measurement is water and typically the chamber would be held in

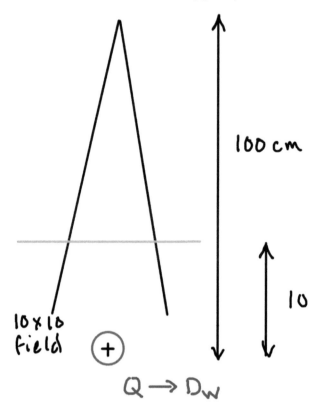

FIGURE 16.2.1
Setup for dose calibration in a common linear accelerator beam.

a water tank design for this purpose. The reference field size is 10×10 cm. The chamber is irradiated with a standard number of MU (100 typically) and the charge, Q, is measured. The goal is to convert this charge to a dose to water.

In the TG-51 protocol the dose to water is determined by the following simple formula: $D_w = M \cdot N_{D,w}$. Here M is the meter reading (the charge on the electrometer), and $N_{D,w}$ is a calibration factor that converts charge to dose. This and other symbols here may look complex with their subscripts and notation, but it is helpful to remember that it is just a number, i.e. the number of Gy per unit charge (Gy/C). To determine $N_{D,w}$ the chamber is sent to a standards lab and calibrated under a reference beam where the dose to water is known through some other independent measurement technique, typically calorimetry. This should be done at least once every two years. There are several Accredited Dosimetry Calibration Laboratories (ADCL) throughout the US, and other nations have systems as well.

One complication is that the ADCL cannot replicate the linac beam being used in the clinic. This would be an impossible task given the subtle variations in the energy and spectrum of beams from various linacs. Instead the ADCL uses a reference beam which is ^{60}Co. This is very stable, well-characterized, and relies on fundamental decay physics and not an electromechanical device. In this way the ADCL provides the user with a calibration factor, $N_{D,w}^{60\text{Co}}$, i.e. the calibration factor for dose to water in a ^{60}Co beam. It would be a mistake to apply this factor directly to the linac measurement to calculate dose (i.e. linac $D_w \neq M \cdot N_{D,w}^{60\text{Co}}$). That is because the linac beam has a very different spectrum than the ^{60}Co beam and so the chamber response to a given dose will be different. One first needs to convert the $N_{D,w}^{60\text{Co}}$ to the $N_{D,w}$ appropriate for the linac beam. This is done through a quality conversion factor, k_Q. The relevant formula is simply $N_{D,w} = k_Q \cdot N_{D,w}^{60\text{Co}}$. Putting all these formulae together then we have the following formula for determining dose to water:

$$D_w = M \cdot k_Q \cdot N_{D,w}^{60\text{Co}} \tag{16.1}$$

The k_Q factor depends on the "quality" or energy of the beam. Since the beam energy also determines the PDD, it is possible to relate the PDD to the k_Q factor. Figure 16.2.2 shows a plot of k_Q vs. PDD from TG-51 for a variety of chambers. The values in this plot were determined from Monte Carlo calculations which determine the different response of the chamber in a ^{60}Co vs. a beam from a linac with a particular PDD. Recall that the energy spectrum of the beam affects the PDD, with higher energy beams having a larger PDD (Chapter 10). The user then simply measures the PDD of the beam and then looks up the corresponding k_Q factor.

One detail to note is that the PDD is the PDD measured at a depth of 10 cm in water, %dd(10)$_x$, and includes only the X-ray component. That is, the electron component that comes from the head of the linac is filtered out. This is especially important in beams with energies >10 MV where the electron component can be large. This is accomplished by placing a thin lead foil in the beam on the head of the linac when the PDD is being measured. This filters out the electrons. The foil is removed after the PDD measurement is complete. Alternatively, there is a formula in the protocol to convert measured PDD to PDD$_x$ with the electron component removed. This is the method recommended by the updated TG51 protocol.

16.2.3 Charge Correction Factors

The meter reading, M, in Equation 16.1 indicates the reading of charge on the electrometer. However, to accurately represent the dose, several correction factors must be applied. If

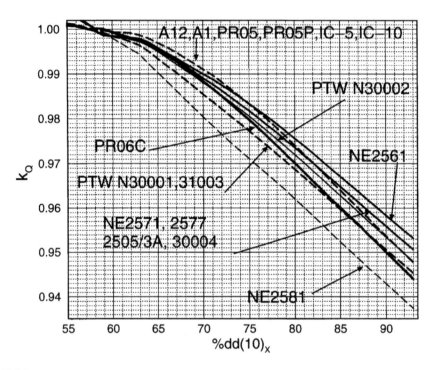

FIGURE 16.2.2
Quality conversion factor, k_Q, versus the PDD of the beam. (From AAPM Task Group 51, Almond et al. 1999.)

the actual reading on the electrometer is M_{raw} then the corrected value of M to be used in Equation 16.1 is given as follows:

$$M = M_{raw} \cdot P_{T,P} \cdot P_{ion} \cdot P_{pol} \cdot P_{elec} \tag{16.2}$$

We will discuss each of these correction factors in turn. The first is $P_{T,P}$, the correction factor for temperature and pressure. To understand this, imagine that the air in the chamber has a high pressure. This might happen naturally on a day where the ambient air pressure happens to be high. In this case there will be more air molecules in the chamber. Because there are more molecules the chamber reading will be artificially high, so some correction factor has to be applied to correct the reading back to some standard baseline.

The form of $P_{T,P}$ can be derived from ideal gas law which relates the pressure, P, volume, V, and temperature, T, of a gas: $PV = nRT$, where n is the number of molecules and R is a constant. This say that for a given volume the number of molecules will be proportional to P and inversely proportional to temperature, T, $n \propto P/T$. Therefore, to correct we multiply by T and divide by P.

$$P_{T,P} = \frac{273+T}{295} \cdot \frac{760\ \text{mmHg}}{P} \tag{16.3}$$

This corrects to standard temperature and pressure which is 22°C (or 273 + 22 Kelvin) and a pressure of 760 mmHg (or 101.3 kPa).

The next factor, P_{ion}, has to do with the efficiency of collecting charge in a chamber. Consider the parallel plate chamber shown in Figure 16.2.3. If an air molecule near the bottom of the chamber is ionized the electron will drift across the chamber toward the

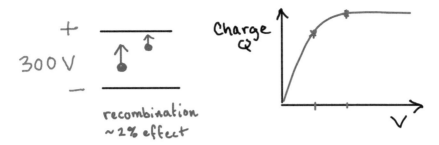

FIGURE 16.2.3
Ion collection efficiency. As electrons drift to the positive electrode they may recombine with air molecules and not be collected. This effect is less at higher voltages.

electrode at positive voltage; as it does so it may recombine with air molecules in the gas and will therefore not be collected. This reduces the charge collection efficiency. This effect will, of course, depend on the voltage. If the voltage is extremely low, the drift time is extremely large and recombination will occur a lot. The charge collected will be small. As the voltage is increased the drift time is reduced and charge collection efficiency increases until it saturates. It is possible to establish a correction factor to account for this, P_{ion}. The most common method is to measure the charge at two voltages and then apply a formula in the protocol which describes the collection efficiency. P_{ion} is typically small, well below 2%, but should be applied for accurate dosimetry.

The next term in Equation 16.2 is a polarity correction factor, P_{pol}. The polarity of a chamber can be reversed at will (e.g. the positive electrode in Figure 16.2.3 can be put at a negative voltage and the negative electrode can be made positive). When this is done the charge collection efficiency is changed because the chamber is not completely symmetric with respect to these voltages. P_{ion} is established by reversing the chamber polarity and measuring charge. The protocol contains a formula to determine the collection efficiency from these readings.

The final term in Equation 16.2 is P_{elec} which accounts for the calibration of the electrometer. In other words, if a charge of 20 nC is on the chamber is 20 nC measured on the electrometer? P_{elec} must be measured, and this is accomplished by sending the electrometer away to a calibration lab. Typically P_{elec} is thought to be very close to 1, but the experience of the calibration labs is that electrometer sensitivities often do vary substantially, or are used in such a way that they vary, and so this calibration is important.

16.2.4 Reference Depth Specification

Equations 16.1 and 16.2 provide a simple method for calibrating dose in a photon beam. Recall, though, that the protocol specifies that these measurements be performed at a depth of 10 cm in water (Figure 16.1.1). This is not, however, the reference condition of the linac. Most linacs are set to have a reference condition of 1 cGy/MU to isocenter at a depth of d_{max}. The conversion, however, is simple. We simply divide the dose at 10 cm (measured per Equation 16.1) and divide by the TMR ($d = 10$, FS $= 10 \times 10$ cm). This yields the dose at d_{max}.

16.2.5 Electron Dose Measurements

The method for calibrating electron beams is largely the same as photon beams, with a few key differences. First, the setup for electron reference dosimetry and measurements typically uses 100 SSD and not an isocentric setup as with photon beams (Figure 16.2.1). This is reflective of the way that treatments are performed (typically a 100 SSD setup). The reference depth for calibration, d_{ref}, is defined by TG-51 not as d_{max} but as $d_{ref}=0.6R_{50}-0.1$ cm. For beams with energies <10 MV this is essentially d_{max}, but for higher energy beams it is slightly deeper.

As with photon beams a quality conversion factor is required (Equation 16.1). For electron beams this is accomplished by measuring the R_{50} (Section 15.1.4) and then using this value to look up conversion factors from Monte Carlo results tabulated in the protocol.

Some subtleties are encountered in performing PDD measurement with electron beams. If one simply scans a chamber down through depth this does not provide a depth-dose curve but rather a depth-ionization curve. These are not equivalent because as the depth changes the energy spectrum of the electrons changes. This results in a change of the collisional stopping power of the electrons. Recall that stopping power changes with energy (Section 7.2). Modern water tank scanning systems have software that will convert depth-ionization curves to depth-dose curves. Note that this conversion is not necessary if the depth-dose measurements are performed with diodes (Chapter 17).

Another issue that affects electron beams is the size of the chamber and the point of measurement. To understand this consider that if there is some point in the water where the dose is deposited, then that dose is registered by the chamber at some point farther downstream due to the forward streaming of the electrons. Therefore, the center of the chamber represents the dose generated at a point upstream in the beam. Note that this effect only applies for cylindrical chambers and not parallel plate chambers which have thin collecting regions in the beam direction. The effective point of measurements for a cylindrical chamber in an electron beam is taken as 0.5 times the radius of the cavity, $0.5r_{cav}$. Thus, when acquiring PDDs the curve should be shifted upstream to shallower depths by a distance $0.5r_{cav}$. This is done before the depth-ionization to depth-dose conversion is performed. Note that this same effect is also present for photon beams (and the effective point of measurement is taken at $0.6r_{cav}$), but since the gradients are smaller the impact is minimal.

Because the size of a cylindrical chamber can be large the dose at the proximal edge of the chamber may differ from that at the distal edge. This is called the gradient effect and must be considered if cylindrical chambers are used for absolute dose measurement. A gradient correction factor can be found by measuring the charge at two depths. There is a formula in the protocol to derive the gradient correction factor from these data.

To summarize, if a cylindrical chamber is used then the following steps are required for electron beam calibration.

- PDD measurement:
  - Shift PDD curve upstream by $0.5r_{cav}$.
  - Convert depth ionization to depth dose using depth-dependent electron stopping powers.
- Measurement for absolute dose calibration:
  - Measure the gradient correction factor by taking measurements at two depths.
  - Ensure that the effective point of measurement is used, $0.5r_{cav}$ upstream from the center of the chamber.

If a parallel plate chamber is used then this process is greatly simplified. Only the conversion of depth ionization to depth dose is needed.

Further Reading

AAPM Task Group 21. 1983. A protocol for the determination of absorbed dose from high-energy photon and electron beams. *Med Phys* 10:741–771. doi:10.1118/1.595446.

Almond, P.R., et al. 1999. AAPM's TG-51 protocol for clinical reference dosimetry of high-energy photon and electron beams. *Med Phys* 26(9):1847–1870.

Bogdanich, W. and R.R. Ruiz. 2010. Radiation errors reported in Missouri. *New York Times*. Available at: https://www.nytimes.com/2010/02/25/us/25radiation.html

Dieterich, S., E. Ford, D. Pavord and J. Zeng. 2016. *Practical Radiation Oncology Physics*. Chapters 1 and 2. Philadelphia, PA: Elsevier.

Kahn, F.M. and J.P. Gibbons. 2014. *Kahn's The Physics of Radiation Therapy*. 5th Edition. Chapter 6. Philadelphia, PA: Wolters Kluwer.

McDermott, P.N. and C.G. Orton. 2010. *The Physics of Radiation Therapy*. Chapters 8 and 11. Madison, WI: Medical Physics Publishing.

McEwen, M., et al. 2014. Addendum to the AAPM's TG-51 protocol for clinical reference dosimetry of high-energy photon beams. *Med Phys* 41(4):041501-01–20. doi:10.1118/1.4866223.

Metcalfe, P., T. Kron and P. Hoban. 2007. *The Physics of Radiotherapy X-rays and Electrons*. Chapter 8. Madison, WI: Medical Physics Publishing.

Chapter 16 Problem Sets

*Note: * indicates harder problems.*

Some problems may require AAPM TG-51 or TG-21 reports.

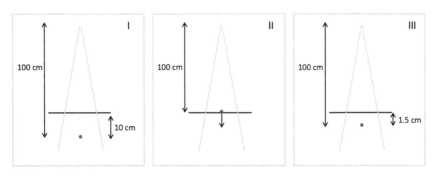

FIGURE PS16.1
Setup for measurement.

1. Match the drawing in Figure PS16.1 with the corresponding use.
   a. Electron PDD measurement
   b. Photon beam output reference 1 cGy/MU for 6 MV
   c. Photon output measurement per TG-51

2. What k_Q quality correction factor should you use for an 18 MV beam? Assume the PDD(10 cm)$_x$ is measured at 78%. Assume you are using a Farmer chamber model NE2571. See table or graph in TG-51.

   a. 0.965
   b. 0.972
   c. 1.029
   d. 1.036

3. How much does the k_Q factor in Problem 2 deviate if the incorrect PDD(10 cm) of 65% were used, i.e. the PDD for a 6 MV beam? Would the calibrated output be too low or too high?

   a. 2.2% low
   b. 3.1% low
   c. 2.1% high
   d. 3.1% high

4. What is the temperature/pressure correction factor for $T=19°C$ and $P=102.0$ kPa?

   a. 0.76
   b. 0.98
   c. 1.016
   d. 1.324

5. How much will the dose calibration (cGy/MU) deviate if a 20×20 field is accidentally used to measure output per TG-51 vs. a 10×10 field for a 6 MV beam? (Refer to the tables in Section 11.3.)

   a. 0.96
   b. 0.98
   c. 1.025
   d. 1.043

6. List three ways that the miscalibration in Problem 5 might be identified.

7. Which of the following chambers gives the highest charge reading for a delivered dose of 100 cGy?

   a. Farmer chamber, volume 0.6 cm³
   b. Mini cylindrical, volume 0.125 cm³
   c. Micro cylindrical, volume 0.07 cm³
   d. Markus pancake chamber, active volume 0.02 cm³

8. Match the task listed below with the appropriate chamber listed in Problem 7.

   Scanning PDDs and dose profiles in water tank

   Measurement of dose and/or profiles in small field

   Absolute reference dose measurements in large fields

   Electron beam PDDs and dose

9. * What fraction of electrons come from the wall of a chamber vs. from the water that the chamber sits in for a PTW Model TN30013 Farmer chamber (wall thickness 0.06 g/cm^2) in an 18 MV beam? See AAPM TG21 Section IV.F. How would this change for a lower energy beam?

   a.   0.18

   b.   0.35

   c.   0.55

   d.   0.75

10. * Describe what chamber(s) could be used to measure the PDD and output of an electron beam. Discuss how the beam quality correction factor is determined for an electron beam and the associated correction factors.

17

Other Radiation Measurement Devices

This chapter discusses three devices that are widely used to measure dose in the clinical setting: diodes (Section 17.1), luminescent dosimeters (Section 17.2), and film (Section 17.3). These devices offer many attractive features such as immediate readout (diodes) or high spatial resolution (film), but they also have limitations which are important to be aware of.

It should be noted that there are other radiation measurement devices not discussed in this chapter. These include the following:

- *Metal oxide silicon field-effect transistors (MOSFETs)*. These are similar in many ways to diodes, though are less widely used.
- *Scintillation detectors*. These low-Z detectors emit light in response to radiation. The optical signal can be calibrated to read dose. It is also possible to generate scintillation in a liquid volume. These provide immediate readout of dose.
- *Dosimetric gels*. The properties of these gels (e.g. optical opacity or MR relaxivity) change in response to radiation. They can provide a three-dimensional image of dose.
- *Calorimetry detectors*. These detectors measure temperature change in a medium (typically water) in response to radiation. They are used as a primary standard in calibration labs. New methods of calorimetry are being explored such as laser interferometry, that is, measuring the change in the index of refraction of water (which results from the change in temperature) via the interference patterns of a split laser beam.

Devices like these offer attractive features but are more complex and are not currently widely in clinical use.

17.1 Diodes

Diode detectors are rugged and small in size (e.g. 1 mm^2 in area and 30 μm thick) and produce a large signal when exposed to radiation. They have therefore found many and various applications in radiation therapy. This section explores the principles of operation of these devices and their applications.

17.1.1 Physics of Operation

Diodes for radiotherapy applications usually consist of two silicon crystals as shown in Figure 17.1.1. The n-type crystal (right) is doped with an impurity that supplies extra electrons while the p-type crystal (left) is doped with a different impurity that provides "holes" or sites which electrons may fill. When the crystals are in contact like this an

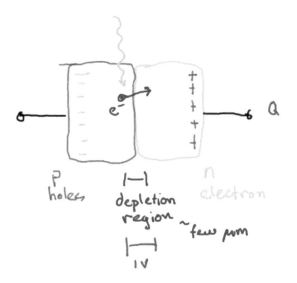

FIGURE 17.1.1
Silicon diode consisting of an n-type crystal (right) and a p-type crystal (left). The extra electrons from the n-type crystal aggregate at the left while the "holes" aggregate at the right. An electron entering the device is accelerated in the large electric field of the thin depletion region.

excess of electrons accumulates on the p-side and an excess of holes collects on the n-side as shown in Figure 17.1.1. The result is a thin "depletion region" in the middle where there are no charges. Because of the charge aggregation there is a voltage across the depletion region (typically 1 V), and because this region is so thin (e.g. few μm) the electric field is very large. This field will accelerate any charges which enter it.

In this device electrons enter from the top and are then accelerated through the high electric field toward the electrode on the n-side. This can be read out as a charge. Note that, as with all detectors in this chapter, these electrons are either from the primary beam itself (i.e. an electron therapy beam) or are created from a photon beam through Compton scattering.

17.1.2 Advantages and Limitations of Diodes

Diodes have several properties that make them appealing for certain clinical applications but one must also be aware of the limitations.

Advantages:

- **Instant readout.** Unlike OSLDs (Section 17.2) or film (Section 17.3) one gets an immediate readout of dose from a diode. This makes it possible to use the device to scan in a water tank. Beam profiles can be measured, as can PDDs (see the associated video for a demonstration of this process). Diodes are useful in QA devices and for in vivo dosimetry readings.
- **Sensitive.**
- **Compact.** See Figure 17.1.2 for a diagram of typical sizes. The size is much smaller than most of the chambers discussed in Chapter 16. The small size makes them ideal for scanning beam profiles or for measuring outputs in small fields.

- **No bias voltage.** Unlike chambers where a high voltage is applied (e.g. 300 V) no voltage bias is applied to the diode. This makes them ideal for use on patients.
- **Rugged.**

Limitations:

- **Energy.** Because of the high-Z of silicon diodes, the device will overrespond to low-energy photons.
- **Temperature dependence.** Temperature will affect the diode readout, with a change of approximately 0.5% per °C.
- **Directional dependence.** The response depends on the angle of the beam relative to the device with an overresponse of approximately 5% at 45°.
- **Damage.** The diode can experience damage leading to response changes. This effect is relatively small (0.1% at 1 k Gy), but since accumulated doses can be large it may be important.
- **Dose rate.** Response is dependent on the instantaneous dose rate.

17.1.3 Diodes for Scanning and Small Fields

Example diode devices for scanning beams are shown in Figure 17.1.2. To scan, the diode is oriented vertically in a water tank (beam coming from the top in Figure 17.1.2) and then is moved around in the beam to acquire either profiles (at different depths) or PDDs. This process is shown in the video.

The diode illustrated in the center of Figure 17.1.2 is a photon diode. The 1 mm² active area is 2.03 mm below the front surface, while in the electron diode (Figure 17.1.2, right) the active area is at a shallower depth of 0.77 mm or 1.33 mm water equivalent. The extra depth of the active area in a photon diode provides shielding from low-energy scattered photons. This is important because the diode overresponds to low-energy photons. Specialized diodes are also available for small fields. These "SRS diodes" have a higher output, e.g. 175 nC/Gy vs. 9 nC/cGy for the diodes shown in Figure 17.1.2.

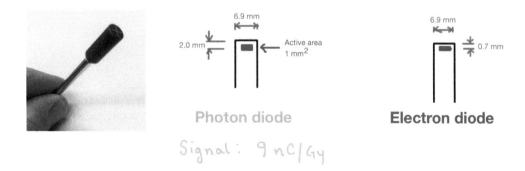

SRS diode: 175 nC/Gy

FIGURE 17.1.2
Diode devices used for scanning in a water tank. (Diagrams courtesy of PTW Inc. for the Doismetry P diode [model 60016] and the Dosimetry E diode [model 60017].)

FIGURE 17.1.3
Diodes for in vivo dosimetry, i.e. measurement of dose delivered during treatment. The active area (blue) is behind a buildup cap (red). Caps of different thicknesses are used for different energy beams.

17.1.4 Diodes for In Vivo Dosimetry

Diodes are often useful for "in vivo" dosimetry, i.e. measurement of dose delivered during treatment. For this application the diodes shown in Figure 17.1.3 are typically placed on the surface of the patient and give an instantaneous readout of dose at that point. The diagram in Figure 17.1.3 illustrates the construction of this type of device. These diode devices are cross-calibrated to read dose with a known standard as described in AAPM Task Group #62.

17.1.5 Absolute vs. Relative Dosimetry

One important point to note is that diodes are typically not used for absolute dosimetry, i.e. calibrating the number of cGy per MU in a beam. Neither are OSLDs (Section 17.2) or film (Section 17.3) used for this purpose. Instead ionization chambers are used for absolute calibration as described in Chapter 16. Diodes are useful for measuring relative dose, e.g. the profile across a beam or the PDD along a beam. They can be cross-calibrated to one of the absolute standards to provide a reading of dose, but they are not themselves used as a primary standard.

17.2 Luminescent Dosimeter

17.2.1 Principles and Operations of Luminescent Dosimeters

Luminescent dosimeters function by releasing light after exposure to radiation. The light emission is triggered either by stimulation with another light source or by heating. Figure 17.2.1 shows an example luminescent dosimeter which is a crystal made from aluminum oxide (Al_2O_3). The crystal is doped with carbon (written as Al_2O_3:C) which makes the crystal light sensitive. The Al_2O_3:C crystal is embedded in a plastic matrix, and this is read out. In these crystals the quantum mechanical energy levels are bands instead of discrete levels like in an atom. Electrons can occupy any quantum state with an energy in these bands. Figure 17.2.1 shows two bands: the valence band at low energy and the conduction band at a higher energy. When the crystal is irradiated energy absorbed, and electrons in the crystal can transition up from the valence band into the conduction band. In these crystals the electrons remain trapped in the higher energy state for a long period of time.

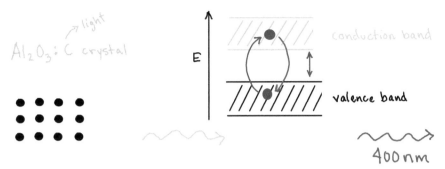

Optically Stimulated Luminescence

FIGURE 17.2.1
Optically Stimulated Luminescent Dosimeter (OSLD) made from an Al_2O_3:C crystal. During irradiation electrons in the valence band are excited into the conduction band (blue arrow) where they are trapped. Later, the crystal is read out by stimulation with a light source (red). Thermal Luminescent Dosimeters (TLDs) are similar, but the stimulation is provided by heat.

To read out the device the crystal is stimulated with a light source, which is a light-emitting diode (LED) in modern readers. This stimulation causes the trapped electrons to transition down to the valence band. As a result, they emit optical photons. The wavelength of these photons for Al_2O_3:C crystals is 400 nm. Higher radiation doses result in more trapped electrons which produce a higher optical signal during readout. Thus the device can be calibrated to read dose. This type of device is called an Optically Stimulated Luminescent Dosimeter (OSLD).

An alternative type of luminescent device is the Thermal Luminescent Dosimeter (TLD). TLDs are made from a different type of crystal, typically LiF, but the principle of operation is the same. To read out this device heat is applied up to about 180°C. As the crystal is heated trapped electrons are released and the output optical signal increases. The signal reaches a maximum and then decreases. This "glow curve" can be calibrated to yield a dose reading.

17.2.2 Advantages and Limitations of Luminescent Dosimeters

Like diodes, luminescent dosimeters have several properties that make them appealing for certain clinical applications, but one must also be aware of the limitations.

Advantages:

- **Small and passive.** OSLD and TLD crystals can be small. They can be encased in a holder and placed on the patient's skin or in a cavity. Or they can be placed inside a cavity in a phantom to provide dose readings. No wires or other attachments are needed.
- **Sensitive.** Doses as low as 0.1 mrem can be accurately read with OSLDs. This makes them ideal for monitoring personnel exposure.
- **Ease of use.**
- **Minimal dependence on temperature, dose rate, or direction of the beam**. This is in contrast to diodes whose response depends on all of these factors.

Limitations:

- **Non-linear dose dependence.** The output of an OSLD or TLD does not increase linearly with dose. There is a superlinear response, which is especially strong above approximately 2 Gy. In principle, this can be accounted for in the calibration, but care must be taken.
- **Fading.** The signal from the OSLDs fades dramatically in the first few minutes after exposure. Dose readings taken in this time will have a large uncertainty.
- **Reuse.** OSLDs can be reused by resetting them via long exposure to intense light. However, the process is complex and care must be taken.
- **Energy.** Like diodes, these devices overrespond dramatically to low-energy photons.
- **No immediate readout.**

17.3 Film

Film provides a readout of dose with a high spatial resolution that is similar to a photograph. As such there are many applications for film in radiotherapy. Two main types of film are used: radiographic film and radiochromic film. Both of these have a thin active layer (few μm thick) which undergoes some conversion after exposure to radiation. The end result is an increase in opacity which can be calibrated back to dose.

17.3.1 Radiographic Film

Radiographic film is perhaps most familiar, the film processed in darkrooms up until the digital revolution. Figure 17.3.1 shows the structure of the film. On the surface are small crystals with silver embedded in them, typically AgBr or AgI (Figure 17.3.1B). When the film is exposed to radiation (or more exactly to charged particles) a redox reaction happens which results in the silver being translocated to the surface of the crystal (Figure 17.3.1C). The unexposed crystals are removed through chemical reaction in a processor. In the areas where exposed silver remains, film has an increased opacity.

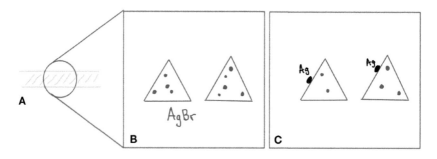

FIGURE 17.3.1
Structure of radiographic film. (A) Film. (B) Crystals containing silver (AgBr) on the surface of the film. (C) After exposure to electrons, the silver locates on the surface of the crystal and yields a darkening of the film.

17.3.2 Film Calibration

The increased opacity or optical density (OD) of a film is defined as $OD = \log_{10}\left(\dfrac{I_0}{I}\right)$, where I_0 is the input intensity of the light being used in the reader, and I is the intensity of that light after it passes through the film. The OD as a function of dose is shown in Figure 17.3.2. It starts at a baseline level (0.2 in this example), which is the background or "fog" of the film, i.e. the OD with no exposure. It then increases with dose and then saturates at some level (1.0 in this example). The saturation depends on the speed of the film. This characteristic curve is called the H&D curve after Herter and Driffeld who described this in the late 19th century.

Figure 17.3.3 from AAPM Task Group #69 (radiographic film for MV dosimetry) illustrates the energy-dependent response of radiographic film. This particular film formulation (Kodak XV film) has an optical density that is approximately twice as high at low energy (e.g.

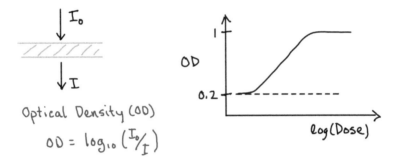

FIGURE 17.3.2
Dose calibration of film. After exposure the opacity increases. The optical density (OD) can be calibrated as a function of dose (right).

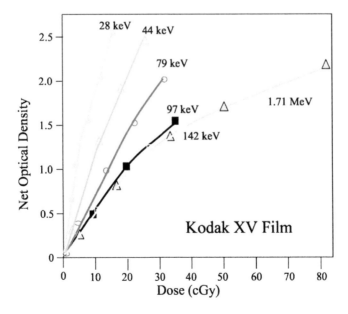

FIGURE 17.3.3
Energy-dependent dose response of radiographic film. (From AAPM Task Group 69, S. Pai, et al. 2007. TG-69: radiographic film for megavoltage beam dosimetry. *Med Phys* 34(6):2228–58.)

44 keV) vs. high energies (e.g. 1.7 MeV). This dependence manifests in several ways. If, for example, the film was calibrated in a high-energy beam (i.e. the H&D curve was established there) one would not want to use that same calibration for a low-energy beam. Also if the spectrum of a beam changes (e.g. by field size or depth) then the dose response may change.

Advantages of radiographic film include:

- **High resolution.** Photographic resolution.
- **Cost.** Relatively inexpensive.

Limitations include:

- **Processor required.** This a complex piece of equipment. A darkroom is typically needed and also chemical supplies.
- **Scanner required.** The optical scanner reads OD which can then be converted into dose.
- **Energy dependence.** Radiographic film shows a strong energy dependence in the dose response, with larger OD at lower energies.

17.3.3 Radiochromic Film

Radiochromic film (sometimes called by its trade names of GAFChromic or EBT®) operates by a completely different mechanism than radiographic film, though the end result is the same, i.e., a dose-dependent opacity that can be read and calibrated as described in Section 17.3.2. In radiochromic film there are small molecules, diacetylene monomers, embedded in a plastic substrate which is typically a few tens of mm thick. With exposure to radiation (or more exactly to electrons) there is a cross-linking of the monomers which results in an increased opacity to optical light. Note that there are some formulations of radiochromic film which have an opaque backing and cannot be read in "transmission mode" as shown in Figure 17.3.2. These are meant to be used for QA purposes like determining field shapes, where dose readings are not required.

There are several key advantages to radiochromic film:

- **High resolution.** As with radiographic film.
- **No processor.** This film "self-develops." The darkening happens without any chemical processing. The development darkening occurs quickly at first and then continues slowly over time until it reaches a saturation at approximately 24 hours post-exposure.
- **Energy independence.** Unlike radiographic film, radiochromic film has a response that shows very little dependence on energy.

Limitations:

- **Scanner required.** Like radiographic film, an optical scanner is required to read OD.
- **Complex calibration.** It is more complex to calibrate dose from radiochromic film. Protocols use the signal from all three color channels (red, green, and blue) to obtain a dose calibration.
- **Cost.** Relatively expensive.

Radiochromic films can be used in many ways, for example, embedding the film in a phantom and then irradiating it with a planned treatment. The film is converted into a dose map, and this dose map can be compared with the planned dose in the plane of the film. **The advantages and disadvantages of radiochromic film vs. radiographic film are a key concept.**

Further Reading

Dieterich, S., E. Ford, D. Pavord and J. Zeng. 2016. *Practical Radiation Oncology Physics*. Chapter 3. Philadelphia, PA: Elsevier.

Kahn, F.M. and J.P. Gibbons. 2014. *Kahn's The Physics of Radiation Therapy*. 5th Edition. Chapter 8. Philadelphia, PA: Wolters Kluwer.

McDermott, P.N. and C.G. Orton. 2010. *The Physics of Radiation Therapy*. Chapter 8. Madison, WI: Medical Physics Publishing.

Metcalfe, P., T. Kron and P. Hoban. 2007. *The Physics of Radiotherapy X-rays and Electrons*. Chapter 3. Madison, WI: Medical Physics Publishing.

Diodes, OSLDs and TLDs

IAEA. 2013. Development of procedures for in vivo dosimetry for external beam radiotherapy. IAEA Human Health Report Volume 8.

Van Dam, J. and G. Marinello. 2006. Methods for in vivo dosimetry in external beam radiotherapy. Brussels: ESTRO. Report.

Yorke, E., et al. 2005. Diode in vivo dosimetry for patients receiving external beam radiation therapy, report of AAPM Task Group 62. AAPM Report Number 87.

Films

Niroomand-Rad, A., et al. 1998. Radiochromic film dosimetry: Recommendations of AAPM Task Group 55. *Med Phys* 25(11):2093–2115.

Chapter 17 Problem Sets

*Note: * indicates harder problems.*

1. Which of the following devices is most appropriate for performing daily output verifications of a linear accelerator?

   a. OSLD

   b. Diode

   c. Radiographic film

   d. Farmer chamber

2. Which of the following devices is most appropriate for in vivo dosimetry measurements under a forced-air patient warming device used during anesthesia procedures (e.g. Bair Hugger™)?

   a. OSLD

   b. Diode

   c. Radiographic film

   d. Farmer chamber

3. What is the dose reading of an OSLD one minute after radiation? (See AAPM TG#191 for detailed information.)

   a. 40% low

   b. 5% low

   c. 5% high

   d. 40% high

4. Which of the following devices is most appropriate for establishing a primary dose calibration in a kV cone-beam CT?

   a. OSLD

   b. Diode

   c. Radiographic film

   d. Farmer chamber

5. A radiographic film calibrated in a 6 MV 10×10 cm beam is used to measure dose of 20 cGy in a 120 kVp diagnostic beam. What is the dose reading from the film? (See AAPM TG#69 for further details.)

   a. 200% low

   b. 10% low

   c. 10% high

   d. 200% high

6. An OSLD calibrated under standard conditions (6 MV 10×10 cm beam) is used to measure dose to a cardiac implanted electrical device. The OSLD is under buildup material on the surface of the patient over the device. The point is outside of the field. What is the dose reading from the OSLD? (See AAPM TG#191 for details.)

   a. 20% low

   b. 5% low

   c. 5% high

   d. 20% high

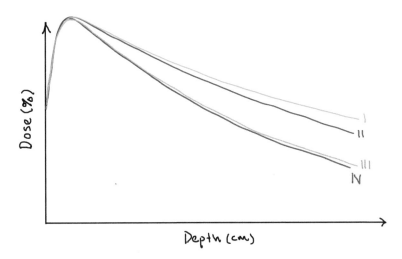

FIGURE PS17.1
Percent depth-dose curves.

7. Match the PDD plots in Figure PS17.1 with the field size and device used for measurement.

   a. 2×2 cm, 0.125 cm^3 ionization mini-chamber

   b. 2×2 cm, diode

   c. 30×30 cm, 0.125 cm^3 ionization mini-chamber

   d. 30×30 cm, diode

8. List two advantages and two disadvantages of radiochromic film compared to radiographic film.

9. * What is the effective Z value of an Al$_2$O$_3$:C OSLD? Describe how this affects the energy-dependent response of an OSLD.

10. * Compare the dose response of an Si diode placed at the beam entrance on a patient vs. the beam exit.

18

Quality Assurance (QA)

18.1 Principles of QA

18.1.1 Swiss Cheese Model of Accidents

The Swiss cheese model of accidents considers various error-prevention barriers to be like slices of Swiss cheese. The barriers (slices of cheese) are stacked in place and prevent errors, but none of these checks function perfectly, which is represented by the holes in the cheese. QA measures are like these Swiss cheese slices, and the goal of this chapter is to present best practices for QA. More information on this model can be found in the video.

18.1.2 Example: QA and Risk

Figure 18.1.1 shows an example error scenario of an output deviation on the linac. The cause might be an inadvertent change in the gain (signal) on the chamber. We analyze the risk of this through failure mode and effect analysis (FMEA). FMEA is presented in more detail in Section 27.3, but briefly three scores are assigned: severity, S, occurrence, O, and detectability, D, of the failure, each on a 10-point scale. These scores are multiplied together to give a final risk score (risk $= S \cdot O \cdot D$). In this example, we take the occurrence score as 10, i.e. we know that the output deviation has occurred in this case. For the severity we consider three possible scenarios: output deviation of 5% (taken as severity, $S = 8$), 2.5% ($S = 5$), and 1.5% ($S = 2$). These scores are not mathematically rigorous but are taken as an approximate indicator. One reference point is ICRU Report 24 which advocates for achieving a dose within 5% of that prescribed based on experiments and clinical outcomes, based on data from the 1970s. The final aspect is the detectability. With no QA in place the deviation is essentially impossible to detect, so we take the detectability score as 10 which corresponds to highly difficult to detect (see green in Figure 18.1.1). The final risk scores are shown at the right in blue.

Now we consider the effect of having a QA program in place. One key reference for this is AAPM TG-142 (Klein et al. 2009) which provides recommendations for linac QA. The frequency and tolerance for linac output measurements are as follows: daily (tolerance <3%), monthly (<2%), and annually (<1%). With these QA measures in place the detectability scores in Figure 18.1.1 change drastically, and there are much lower overall risk scores (gray in Figure 18.1.1). Note that the largest deviations would be detected by daily QA and the smallest deviations by annual QA. The example above illustrates the principles of QA and the functioning of different tolerance and frequency recommendations. This is considered in more detail in AAPM TG-100 (Huq et al. 2016) and will be discussed further in Chapter 30.

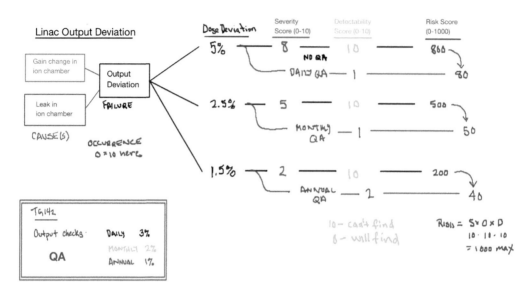

FIGURE 18.1.1
A risk-based view and the impact of QA.

18.2 QA of Linear Accelerators

18.2.1 Introduction and Reports

Key reports describing the QA of linacs and recommended tolerances and frequency of tests are the following: AAPM TG-142 from 2009, AAPM Medical Physics Practice Guideline (MPPG) 8a from 2017, and AAPM TG-198 due out soon. Table 18.2.1 summarizes the recommendations from these reports (note that this table lists the most important tests, not every test). The linac tests fall into two main categories: mechanical tests and dosimetry tests. Some tests are recommended to be performed daily, some monthly, and some only annually.

18.2.2 Dosimetry Tests

The first two dosimetry QA checks in Table 18.2.1 concern the linac output and the profile of the beam. Figures 18.2.1 and 18.2.2 show example devices that can be used to accomplish these tests. The example daily QA device shown (Figure 18.2.1) consists of eight ionization chambers, a central chamber to track output, and outer chambers to provide a measure of the beam profile. One of the inner chambers also has a filter to provide a check of the beam energy (quality) based on the expected relative signals.

On a monthly basis QA is performed with a Farmer chamber in plastic (Figure 18.2.2). On an annual basis a full measurement is performed in liquid water using the TG-51 protocol (see Chapter 16). The water tank shown in Figure 18.2.2B can be used to scan detectors across the field to provide an accurate measure of the beam profile. For an illustration of this see the video.

Beam quality measurements are required on a monthly and annual basis (Table 18.2.1). This can be accomplished by measuring the PDD or TMR of a beam at two depth points.

TABLE 18.2.1

Linac QA Test Recommendations from AAPM TG-42, MPPG 8a, and TG-198

Dosimetry Tests	Daily	Monthly	Annually
Output	3%	2%	1%
Profile	2%	**1%**	**1%**
Beam quality		1%	1%
Wedge output factor			2%
Output vs. dose rate			2%
Output vs. gantry angle			1%
Mechanical Tests	**Daily**	**Monthly**	**Annually**
Treatment couch position indicators		2 mm	
Gantry and collimator angles		1°	
Radiation isocentricity (coll, couch, gantry)			± 1 mm
Crosshair centering		1 mm	
Light-radiation field coincidence		2 mm	2 mm
Wedge placement		2 mm	
Field size	2 mm	2 mm	
Graticule		2mm	
Laser	**1 mm**	1 mm	
ODI	2 mm		
Leaf position		1mm	

The major QA checks are listed. Red: Checks that are only recommended in one of the reports. Bold: The three reports disagree on recommended tolerance.

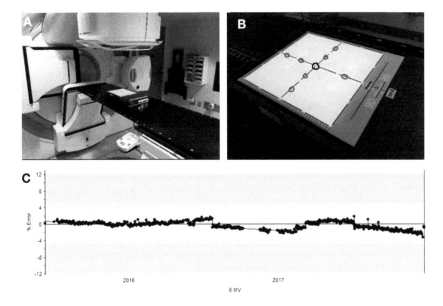

FIGURE 18.2.1
Daily check device. (A, B) Eight ionization chambers provide readings of output, beam profile, and energy. (C) Results over time with tolerance windows for low and high alerts.

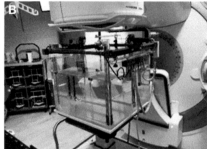

FIGURE 18.2.2
Devices for linac output measurements. (A) Farmer chamber in water-equivalent plastic for monthly QA. (B) Water tank with ion chamber for annual QA.

Recall that the falloff of the PDD or TMR depends on the energy spectrum of the beam (Section 10.1.3) so measuring PDD or TMR at two different depths provides an indicator of the beam spectrum. Typically a ratio is measured, e.g. TMR($d = 20$cm)/TMR($d = 10$cm), and the QA test is to confirm that this ratio is constant within the tolerance.

18.2.3 Mechanical Tests

For a detailed discussion of the mechanical QA tests listed in Table 18.2.1, see the video. Two QA tests are mentioned here: the laser alignment and the ODI.

Laser alignment. The laser system is an important component used in aligning the patient for treatment. Figure 18.2.3 shows the functionality of this system. Laser systems are installed on the linac and CT simulator, where the laser projects a thin line aligned to intersect at the isocenter. The two perpendicular lasers create a cross where they intersect the patient's skin

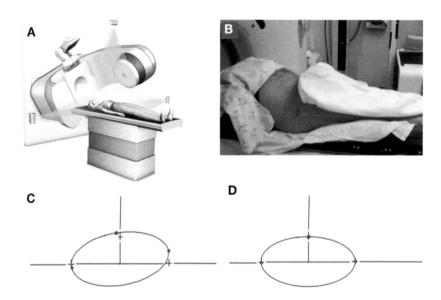

FIGURE 18.2.3
Laser systems for patient alignment. (A) On linac. Images courtesy of LAP Laser Inc. (B) Laser crosshair visualized on patient skin. (C) Three points marked on the patient (blue dots). (D) Patient position adjusted to align the skin marks with the laser.

(Figure 18.2.3B). This can be used to align the patient to isocenter as shown in Figure 18.2.3C. Three marks are on the patient (blue dots in figure). These might be tattoos or ink marks. The patient is moved and rotated until the marks line up with the lasers (Figure 18.2.3D). This aligns the isocenter to the intersection point of the three marks.

For the above system to work as intended the lasers must all be aligned with isocenter and must be colinear (i.e. directed along the same plane). The QA test for this can be accomplished by aligning graph paper with the known isocenter marked on the graph paper (Figure 18.2.4).

Optical distance indicator (ODI). The ODI provides a measurement of the SSD on the patient's skin as shown in Figure 18.2.5. A calibrated scale is projected from the linac. The SSD at the isocenter is read out by visualizing the intersection of the crosshair with the scale. The example shows an SSD of 95 cm. For a further illustration of this in action see the video. To QA the ODI the table is adjusted to an SSD as read on the ODI and then the table height is read out and compared to the expected height relative to the isocenter.

FIGURE 18.2.4
Laser alignment test. Graph paper is aligned with the isocenter using the crosshairs in the light field.

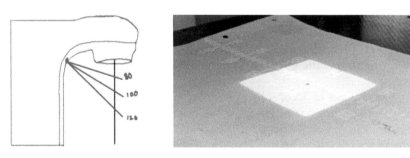

FIGURE 18.2.5
Optical distance indicator (ODI). The SSD is read out by visualizing the intersection of the crosshair with the projected scale.

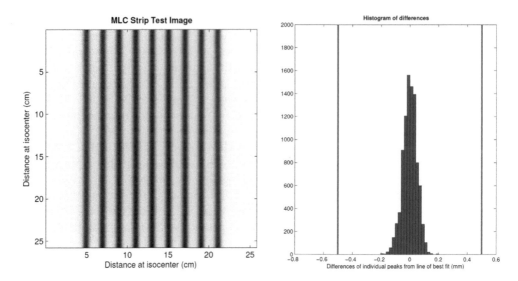

FIGURE 18.2.6
MLC leaf test. A thin strip is created with the MLC and moved across the field.

Leaf tests. The final QA test in Table 18.2.1 is for the MLC leaf positions. This is very important for a machine delivering IMRT or VMAT. MLC leaf tests can be varied and complex but for the purposes of presentation are simplified in the table as a single monthly test. Here we describe a few example MLC tests. A more complete description can be found in the AAPM reports.

One common test is the MLC strip test or "picket fence" (Figure 18.2.6). In this test a thin strip is created with the MLCs and is moved across the field to various locations and imaged using either film or an EPID. The positions of each MLC can be measured in these images and compared to the expected locations. Data in Figure 18.2.6 show that the leaves are well within 1 mm of the expected locations. It is also important to test the MLC leaf speed for VMAT delivery. Further details can be found in the video.

18.3 Patient-Specific QA

One valuable QA tool is a verification of the plan for a specific patient, and this is especially valuable for complex plans like IMRT or VMAT where simple calculations are challenging or impossible. **Patient-specific QA is an important part of medical physics practice and is a key concept.**

There are various ways to accomplish patient-specific QA. Figure 18.3.1 illustrates one method which relies on measuring the dose using a phantom. In this process the plan for a patient (Figure 18.3.1A) is transferred to a digital representation of the device on which the measurement will be made. In this example the ArcCheck (Sun Nuclear Inc.) is shown which consists of an array of diodes wrapped around the outside of a 21 cm diameter cylinder. The expected plan dose is calculated on the device in the treatment planning system (B). Then the device is aligned under the beam and the plan is delivered (C). The result is

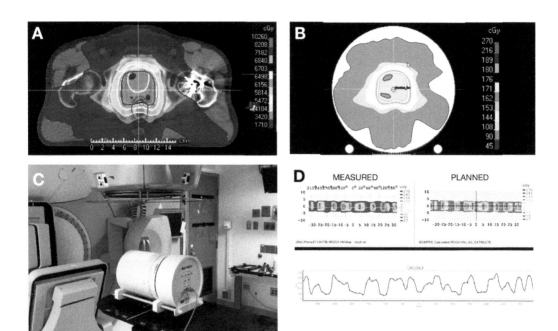

FIGURE 18.3.1

Patient-specific plan QA process. (A) Plan on a patient. (B) Same plan transferred to a model of the measurement device and recalculated. (C) Measurement device, an array of diodes around a cylinder. (D) Measured doses on the array vs. planned doses.

a collection of measured doses at each detector on the array vs. planned doses for those same points (D). For an illustration of this process see the video.

One issue which arises in comparing planned vs. measured doses is the steep dose gradients present in some plans, especially IMRT or VMAT plans. Figure 18.3.2 shows planned dose (blue line) vs. measured dose at various points (red dots). At the fifth point from the left, the dose difference between plan and measurement is very large. This is because it is in a region where is a large dose gradient, i.e. a large change in dose per unit distance. Measuring simply the dose difference, then, tends to exaggerate the discrepancies. To account for this, another metric can be used, the distance-to-agreement (DTA). DTA is the distance required in order to find the matching dose. This minimizes the effect in high-gradient regions but can exaggerate the discrepancies in low-gradient regions. One solution therefore is to combine the two methods into one metric. That is, dose difference and distance-to-agreement are both used together. This metric is the so-called "gamma index," γ. **Gamma analysis is important in the evaluation and the QA of IMRT and VMAT plans and is a key concept.**

The γ-index can be thought of visually as shown in Figure 18.3.2. For a particular measured point, one can find the closest point on the plan. γ is then the combination of dose difference for this point, ΔD, and DTA added in quadrature, $\gamma = \sqrt{\Delta D^2 + DTA^2}$. A scaling is set so that $\gamma = 1$ corresponds to some ΔD, and DTA. If one sets the γ criteria to 3%/3mm, for example, then every point lying within the 3%/3mm sphere will have $\gamma < 1$ and every point outside that sphere will have $\gamma > 1$. One can then further establish a "passing criterion" of $\gamma = 1$. That is, every point with $\gamma < 1$ passes and every point with $\gamma > 1$ fails. One then analyzes all the measured points and determines the percentage of points passing and the

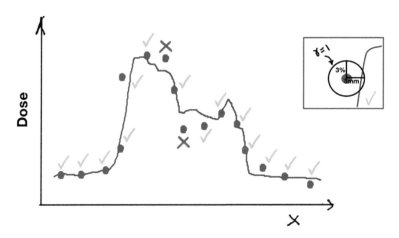

FIGURE 18.3.2
Dose and distance-to-agreement analysis. Planned dose (blue) vs. dose measured at various point (red).

percentage failing. In the example in Figure 18.3.2, of the 16 points 14 of them pass, or an 87.5% passing rate. In the actual measured plan shown in Figure 18.3.1D the passing rate is 98.6% with a 3%/3 mm criterion.

The gamma passing criterion can be changed. Although 3%/3 mm is commonly used in the community there is no "magic number." If the criterion is tightened the resulting passing rate will be lower. This can be visualized as shrinking the size of the sphere, which results in fewer points lying within the sphere. In the example in Figure 18.3.1D the passing rate is 98.6% with a 3%/3 mm criterion but is 92.2% with a criterion of 2%/2 mm. Note that the dose and DTA need not be the same number. One might choose, for example, a 2%/3 mm criterion.

18.3.1 Devices for IMRT QA

Measurement devices which use arrays of detectors are popular because they provide immediate readout and the dose-response of the detectors is well-understood. Numerous options are available (Tables 18.3.1 and 18.3.2). These include the following:

- Planar array device. The spacing between detectors determines the resolution with which the IMRT QA can be performed. Current devices commonly have detector spacings of 5 to 7 mm.
- Electronic portal imaging device (EPID) (QA StereoChecker, Standard Imaging Inc.). This has the advantage of very high resolution (0.2 mm), which is useful for small fields and/or high gradients which arise especially in the context of stereotactic radiosurgery and radiotherapy (Chapter 22).
- Non-planar arrays, including cylindrical arrays or biplanar arrays.

QA measurements can also be accomplished with film, most often radiochromic film (Section 17.3.3) embedded in a phantom. The advantage of film measurements is the extremely high resolution. The disadvantage is the lack of immediate readout and the need to establish careful calibration.

A final method of measuring delivered doses is to use the EPID, which is becoming increasingly common. For more information on the EPID see Chapter 19. EPID

TABLE 18.3.1

Planar Array Devices for Measurement-Based Patient-Specific QA

Device	Manufacturer	Detector Type	Resolution	Field Size
MapCheck2	Sun Nuclear	Diode	7 mm	20×20 cm
MatriXX	IBA Dosimetry	Ion chamber	7.6 mm	24.4×24.4 cm
Octavius 1500	PTW	Ion chamber	7.1 mm	27×27 cm
QA CrossChecker	Standard Imaging	Ion Chamber	5 mm	30×30 cm
Octavius 1000 SRS	PTW	Liquid-filled ion chamber	2.5 mm (center area)	5×5 cm
QA StereoChecker	Standard Imaging	EPID/scintillator	0.2 mm	20×20 cm

TABLE 18.3.2

Other Array Devices for Measurement-Based Patient-Specific QA

Device	Manufacturer	Detector Type	Arrangement	Resolution	Size (dia × length)
ArcCheck	Sun Nuclear	Diode	Cylinder	1 cm (on surface of 21 cm dia cylinder)	21×21 cm
Delta 4	ScandiDos AB	Diode	Biplanar array	5 mm (in center 6×6 cm), 1 cm (outer 20×20 cm)	22×40 cm

measurements have the advantage of very high resolution and ease of use. The EPID is available as standard equipment on most linacs. One of the challenges is the need to translate an EPID image readout into dose which is non-trivial due to detector response and the other issues. EPID dosimetry can be used to perform QA either before the treatment starts (as with the methods above) or even during treatment (see below).

18.3.2 Other Patient-Specific QA Approaches

The previous sections focus on measurements performed prior to QA treatment plan delivery. However, other approaches are possible for patient-specific QA including the following:

- *In vivo QA*. Measurement performed during treatment with the patient in the beam. Diodes or EPID may be used for this.
- *Transmission devices*. These are mounted on the head of the linac, and the beam is transmitted through them.
- *Calculation-based methods*. During delivery a computer log file is collected which records every MLC position, gantry angles, and the MU delivered. These data can be used to calculated fluence and dose in the patient which can be compared to the plan.

For further information on these techniques see the video.

18.3.3 IMRT QA References and Standards

There are several important references and resources related to patient-specific QA. AAPM TG-218, published in 2018, describes methods for IMRT QA and tolerance limits for tests. AAPM TG-119 form 2009 describes commissioning of IMRT systems and associated tests. It also includes downloadable data with standard test cases which can be planned and treated. AAPM TG-120 from 2001 discusses dosimetry tools and techniques for IMRT QA. In addition to these reports is the ASTRO Safety White Paper: Safety Consideration for IMRT published in 2011 which provides more general practice-level recommendations for IMRT programs.

18.3.4 QA: Review of Plans and Charts

Data suggest that plan and chart review by a physicist, therapist, or physician is one of the most potentially effective QA measures. Physics plan and chart review is the subject of the upcoming AAPM TG-275 Report. Physician peer review or "chart rounds" is also a key QA measure. An important reference here is the ASTRO Safety White Paper on Peer Review (Marks et al. 2013).

18.4 QA of Full Dosimetry System

This section considers, in brief, further QA measures which are designed to identify quality gaps in radiation dosimetry that may not be identified through the QA measures above. A more complete description of this topic, with illustrations, can be found in the video.

The first of these is an external audit, one form of which is a validation of the dose output of the delivery system under standard reference conditions. Mail order systems are available from the IROC-H Center and other organizations in which a TLD or OSLD is shipped to the clinic. This TLD or OSLD block is irradiated under standard conditions. The block is sent back to the center and the dose is analyzed and compared to the expected dose. In 2018 IROC-H made 9423 photon beam measurements with an average reading of 1.007 ± 0.015 (ratio of predicted to measured dose) and for electron beams 9155 measurements with an average reading of 1.003 ± 0.017 (D. Followill, private communication 2018).

A second class of QA measures of the overall system are those described in AAPM MPPG5a (Smilowitz et al. 2015) on commissioning and QA of the treatment planning system. The report recommends that validation checks be performed to compare planned vs. measured doses for the following fields/setups: MLC-shaped fields including small field and off-axis fields, different SSD, oblique angles, wedges, heterogeneous phantoms, and test cases from TG-119, MPPG5a, and clinical plans. In addition to this the report recommends end-to-end tests. **End-to-end tests are a key concept.** In this test, a phantom with an embedded detector (e.g. ion chamber or film, Figure 18.4.1) is scanned in the imagining system; a treatment plan is generated for this "patient"; the treatment is delivered, and the dose is read out and compared to the dose from the planning system. In this way the whole dosimetry system from end to end is tested.

Another validation test recommended by MPPG5a is a mail order phantom from an external agency. Figure 18.4.2 shows an example of the head-and-neck phantom from IROC-H. The phantom is mailed to the user and contains inserts with TLDs and film. The phantom is treated just like a patient, scanned, planned, and treated, and then sent back. The film and TLD results are then compared to the planned dose. A further detailed illustration of this is available in the video.

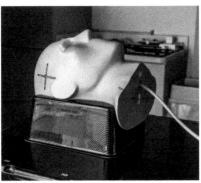
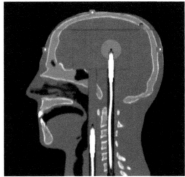

FIGURE 18.4.1
Phantom for an end-to-end test. The phantom is treated like a patient, and the delivered dose is measured. Here an ion chamber is shown, but film and other inserts are available. (Images courtesy of CIRS Inc.)

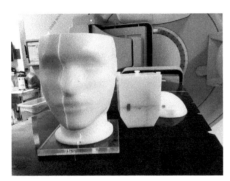
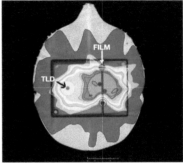

FIGURE 18.4.2
Head-and-neck phantom from IROC-H. The phantom includes inserts with embedded TLDs and film. A treatment plan is developed, and the phantom is irradiated. The measured dose is compared to the planned dose.

The value of this test is underscored by the relatively poor test results that are observed across institutions. With over 2400 irradiations performed on the head-and-neck phantom, IROC-H reports that only 87% of institutions pass the test with a 7%/4mm criterion (D. Followill, private communication 2018). Similar results are available for the other phantoms: liver, lung, prostate, and spine with pass rates of 71%, 85%, 85%, and 77% as of 2018. The poor pass rates underscore the necessity of this audit system.

Further Reading

Dieterich, S., E. Ford, D. Pavord and J. Zeng. 2016. *Practical Radiation Oncology Physics*. Chapters 5 and 12. Philadelphia, PA: Elsevier.

Ezzell, G., et al. 2009. IMRT commissioning: Multiple institution planning and dosimetry comparisons, a report from AAPM Task Group 119. *Med Phys* 36(11):5359–5373.

Hanley, J., et al. An implementation guide for TG-142: QA of medical linear accelerators, Report of AAPM Task Group 198. In preparation.

Huq, M.S., et al. 2016. The report of Task Group 100 of the AAPM: Application of risk analysis methods to radiation therapy quality management. *Med Phys* 43(7):4209–4262.

Kahn, F.M. and J.P. Gibbons. 2014. *Kahn's The Physics of Radiation Therapy*. 5th Edition. Chapter 17. Philadelphia, PA: Wolters Kluwer.

Klein, E.E., et al. 2009. Quality assurance of medical accelerators: Report of AAPM Task Group 142. *Med Phys* 36(9):4197–4212.

Low, D.A., et al. 2001. Dosimetry tools and techniques for IMRT, report of AAPM Task Group 120. *Med Phys* 38(3):1313–1338.

Marks, L.B., et al. 2013. Enhancing the role of case-oriented peer review to improve quality and safety in radiation oncology: Executive summary. *Pract Radiat Oncol* 3(3):149–156. doi:10.1016/j. prro.2012.11.010.

McDermott, P.N. and C.G. Orton. 2010. *The Physics of Radiation Therapy*. Chapter 18. Madison, WI: Medical Physics Publishing.

Metcalfe, P., T. Kron and P. Hoban. 2007. *The Physics of Radiotherapy X-rays and Electrons*. Chapter 11. Madison, WI: Medical Physics Publishing.

Miften, M., et al. 2018. Tolerance limits and methodologies for IMRT measurement-based verification QA: Recommendations of AAPM Task Group No. 218. *Med Phys* e53–e83. doi:10.1002/mp. 12810.

Smilowitz, J.B., et al. 2015. AAPM Medical Physics Practice Guideline 5.a.: Commissioning and QA of treatment planning dose calculations - megavoltage photon and electron beams. *J Appl Clin Med Phys* 16(5):14–34. doi:10.1120/jacmp.v16i5.5768.

Smith, K., et al. 2017. AAPM Medical Physics Practice Guideline 8.a.: Linear accelerator performance tests. *J Appl Clin Med Phys* 18(4):23–39. doi:10.1002/acm2.12080.

Chapter 18 Problem Sets

*Note: * indicates harder problems.*

1. Which dimension of risk is impacted by increasing the frequency of a QA test from annual to daily?
   a. Severity
   b. Occurrence
   c. Detectability
   d. Efficiency

2. What dimension(s) of risk was impacted when the technique for dose calibration was changed from the TG-21 protocol to TG-51 which was much more reliable to implement? (Check all that apply.)
   a. Severity
   b. Occurrence
   c. Detectability
   d. Efficiency

3. At a minimum how frequently should output checks in a water tank be performed?
   a. Daily
   b. Weekly
   c. Monthly
   d. Annually

4. Beam steering in a linac is the process of tuning magnetic fields in the waveguide in order to control where the electron beam collides with the target. Which QA check would directly identify a problem with beam steering in a photon beam?

   a. Output
   b. Profile
   c. Beam quality

5. Which QA check would be most sensitive to a miscalibration of the ODI?

   a. Output
   b. Field size
   c. Laser
   d. Couch translation

6. After an earthquake which linac QA test would be most likely to yield an error?

   a. Output
   b. Laser
   c. Light field-radiation field coincidence
   d. Graticule alignment

7. According to AAPM reports how often and at what tolerance should wedge output factors be checked? Describe how this check is performed.

   a. Daily
   b. Weekly
   c. Monthly
   d. Annually

8. According to AAPM reports how often and at what tolerance should MLC leaf positions be checked?

   a. Daily
   b. Weekly
   c. Monthly
   d. Annually

9. * What QA procedure directly tests the alignment of the field light with the collimator rotation axis? (Figure PS18.1)

   a. Output
   b. Profile
   c. Crosshair walkout
   d. Field size

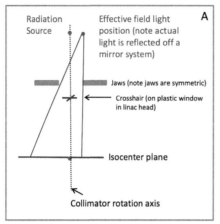
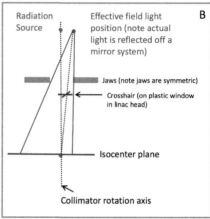

FIGURE PS18.1

10. * Describe how one might detect a misalignment of the radiation spot with respect to the collimator axis.

11. Which of the following are examples of patient-specific QA? (Check all that apply.)
    a. Pre-treatment phantom-based IMRT QA
    b. QA measurements with an IROC-H phantom
    c. In vivo diode measurements
    d. Dose output checks per the AAPM TG-51 protocol

12. Which QA measure could potentially detect a change in dose to a spine tumor as a result of ascites occurring after simulation?
    a. Pre-treatment phantom-based IMRT QA
    b. Pre-treatment secondary verification of MU calculations
    c. In vivo diode measurements
    d. In vivo EPID dosimetry measurements

13. Rank the following IMRT QA tests by expected number of points passing (lowest to highest).
    a. Diode array, 2%/2mm criteria
    b. Diode array, 3%/3mm criteria
    c. Film, 2%/2mm criteria
    d. Film, 3%/3mm criteria

14. Which device used in QA requires a correction for the response of the device as a function of the beam angle?
    a. OSLD
    b. Film
    c. Diode
    d. Ion chamber

15. Which device used in QA will be most sensitive to deviations in the high-gradient region of a head-and-neck plan?
    a. OSLD
    b. Film
    c. Diode array
    d. Ion chamber array

16. Which problem(s) might be identified by an end-to-end test? (Check all that apply.)
    a. Incorrect beam energy spectrum in TPS
    b. Inadequate model for beam profile in TPS
    c. Wrong scan on a patient used for treatment planning
    d. Miscalibration of image-guidance system

17. In the IROC-H prostate phantom which detector is used to measure gamma values TPS vs. measured?
    a. OSLD
    b. TLD
    c. Film
    d. Chamber

18. * Which IMRT QA method might detect a problem of an MLC leaf sagging under gravity? (Check all that apply.)
    a. Beam-by-beam QA on a flat phantom, gantry 0
    b. Beam-by-beam QA of fluence with a film on the gantry head
    c. Composite QA with an ion chamber
    d. Composite QA with film

19. * Which QA device might detect a wrong source size in TPS? (Check all that apply.)
    a. OSLD
    b. Film
    c. Diode array
    d. Chamber

19

Radiographic Imaging

19.1 Basic Principles of Radiography

Radiographic imaging is taken to mean anything that involves imaging with X-ray photons which are transmitted through the patient. This might include traditional "X-rays" (planar X-ray images) but also CT imaging (Section 19.2). Some of the topics here, such as data storage and standards, will also apply to other imaging modalities.

19.1.1 Contrast

One key goal in radiographic imaging is to distinguish different materials from each other (e.g. bone vs. muscle). That is, the contrast between materials in the image must be different. The process is illustrated in Figure 19.1.1, where photons emerge from a source (e.g. X-ray tube, Section 8.1) and pass through the patient. Some of these photons are transmitted through the patient and are registered on the detector which might be a film or an EPID (Section 19.1.3). To understand contrast, consider that the incoming photon fluence is I_0 and the transmitted photon fluence transmitted through 10 cm of muscle is I_1. If we calculate this for a monoenergtic X-ray beam of energy 30 keV using Equation 6.1, we find $I_1/I_0 = 0.0188$ (for further detail see the video). Similarly, we can calculate the transmission through 10 cm of muscle but with a 2 cm bone embedded. This yields $I_2/I_0 = 0.000251$. As expected the transmission through the bone is much lower due to the higher density and atomic number. The ratio of I_1 to I_2 is $I_1/I_2 = 74.8$. This is a measure of the contrast.

Contrast depends on the energy of the X-ray photons. If we perform the same calculation for a monoenergetic 25 keV beam we find $I_1/I_2 = 7928$. That is, the contrast between the regions with muscle and bone is much larger at lower energy. This is due to the increased effect of photoelectric absorption at lower energies (recall Section 5.1).

Images at megavoltage energies represent an even more extreme energy range and are relevant for imaging performed with a linear accelerator beam. Recall that in the megavoltage energy range there is very little Z-dependence of the mass attenuation coefficient. Therefore, any differences in contrast are due only to density and not composition. The impact of this is illustrated in Figure 19.1.2 where a megavoltage (MV) image from a linac is shown on the left vs. a kilovoltage (kV) image for the same patient on the right. High-density objects like bone are visible in the MV image, but the contrast in the kV image is much better because of the extra Z-dependent enhancement.

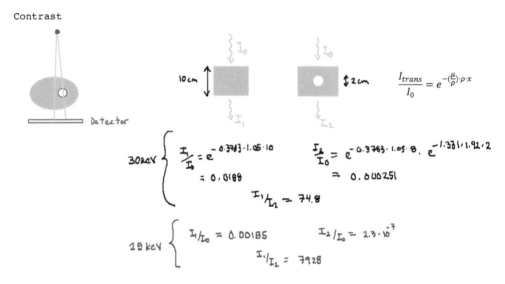

FIGURE 19.1.1
Contrast in radiographic images.

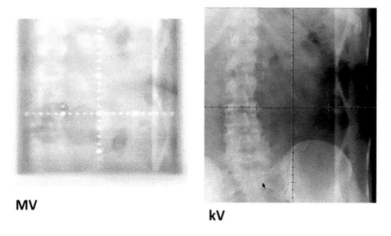

FIGURE 19.1.2
Megavoltage (MV) vs. kilovoltage (kV) imaging of a patient. The contrast is much better at kV energies due to the enhanced photoelectric effect.

19.1.2 Resolution

Another important aspect of image quality is whether objects are "sharp" or "blurry." The parameter describing this is resolution, or more exactly, spatial resolution. There are several aspects of the system which affect resolution. Figure 19.1.3 shows an object (e.g. a bone) placed at some distance a from the source. Below that is the detector, placed at some distance b from the object. The sharpness of the high-contrast object, like the edge of a bone, can be thought of as a penumbra. Recall from Section 9.1.6 that for a source with size, s, the penumbra is proportional to $s \cdot b/a$. The resolution increases as the object is moved closer to the detector (b/a decreases).

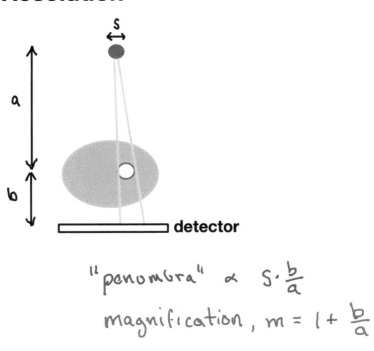

FIGURE 19.1.3
Spatial resolution and magnification, *m*.

 This is closely related to another parameter, which is magnification, *m*. To understand this, note that if the object has some actual physical size, *r*, it will appear larger on the detector due to the diverging rays from the source. If the size of the object on the detector is *R*, then by similar triangles we can see that $r/a = R/(a+b)$. This can be simplified to $R/r = 1 + b/a$. This is called the magnification, or "mag," represented by the symbol *m*.

$$m = 1 + \frac{b}{a} \tag{19.1}$$

Note that as *b/a* increases (object is moved farther from the detector) the mag increases, but the resolution decreases. There is a trade-off between these two parameters, mag and resolution.

19.1.3 EPID Detectors and Pixelization

Spatial resolution is also affected by the detector system used. This might be film (Section 17.3) or an electronic portal imaging device (EPID). Modern C-arm linacs have an EPID for the megavoltage (MV) beam as well as an EPID for the kV beam which is orthogonal to this. Within the EPID there are layers (Figure 19.1.4B). A photon entering the detector most commonly interacts with a conversion layer (copper plate) and Compton scatters, giving an electron. This electron interacts in the scintillator layer (green) which is typically a rare earth like gadolinium oxysulfide. In this interaction, light, i.e. optical photons, are created. These photons are registered in the detection layer (red) which is a photodiode

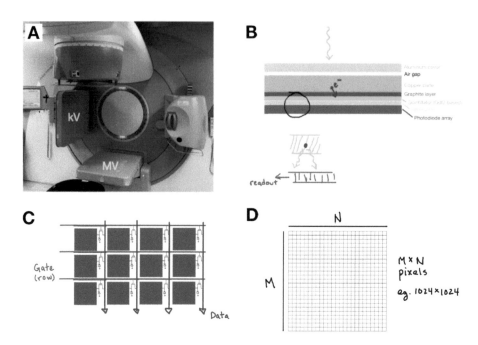

FIGURE 19.1.4
Electronic portal imaging device (EPID). (A) The EPID panels on a C-arm linac. (B) Layers in the EPID device. (C) Detector layer with electronics to read out each pixel. (D) Pixelization of the detector.

array, typically made of amorphous silicon, that provides some readout of the charge. This charge is approximately proportional to the number of photons.

The detector layer has electronics to read out each pixel (Figure 19.1.4C). This includes gate lines (the rows). The detector is read out row by row, and which row is being read out is controlled by the voltages on the transistors on each of the gate lines. In the vertical direction are the data lines, each one instrumented with electronics to allow readout of each pixel. The detector is read out row-by-row until a full two-dimensional image is formed.

The result is an image that is broken up into equal-sized squares or "pixels" (picture elements). In the example shown there are 1024 × 1024 pixels. **The pixelization of images is a key concept and influences the resolution of an imaging system.** To see an example, consider Figure 19.1.5 which shows the same image but with a different number of pixels. The image with more pixels (left) has a higher resolution than the one with fewer pixels (right). Note though that the number of pixels is not the only aspect that affects resolution. As

FIGURE 19.1.5
Pixelization and resolution. An image with more pixels (left) has a higher resolution than one with fewer pixels (right).

noted in the previous section, the mag and the source spot size also affect resolution, and no matter how many pixels are used it cannot compensate for a low mag or large spot size.

19.1.4 Noise and Exposure

Another important parameter that describes image quality is the noise in the image, i.e. how "grainy" the image looks (for example images see the video). In X-ray imaging a dominant source of noise is the random fluctuation of the number of photons registered in each pixel. That is, some pixels will randomly collect fewer photons and some will collect more. Mathematically this process of random counting is described by the Poisson distribution.

The Poisson distribution describes the probability of measuring k events, given that the average number of events is N, and is given by $P(k|N) = e^{-N} \dfrac{N^k}{k!}$. It is beyond the scope of this text to prove this mathematically, but the distribution is shown graphically in Figure 19.1.6. Note that as the average number of events becomes large (e.g. $N = 10$ in green) the distribution begins to look more like a Gaussian. It can be shown that the standard deviation of the Poisson distribution is $\sigma = \sqrt{N}$. The standard deviation represents the "spread" in potential measured values and is therefore a measure of the noise. In summary, then, the noise is given by $\sqrt{N}$ and the signal is given by N (i.e. the average value measured in the pixels). The signal-to-noise ratio (SNR), therefore, is given by:

$$\text{SNR} = \frac{N}{\sqrt{N}} = \sqrt{N} \qquad (19.2)$$

The signal-to-noise ratio (SNR) is an important parameter and represents a key concept. Equation 19.2 suggests that as more photons are registered in the detector, the signal-to-noise ratio gets larger. That is, the image becomes less noisy. The design of the detector influences the number of photons. One could, for example, make the conversion layer like the copper plate thicker (Figure 19.1.4B), resulting in more photons. One could also make the scintillator thicker or use more sensitive material. All of these steps would improve the detective quantum efficiency (DQE) of the detector, a measure of the number of optical photons produced for each X-ray photon that enters. There are trade-offs however. Making a thicker conversion layer or scintillator results in a "cloud" of photons or particles that is spread out over a larger area. This blurs out the image and makes the resolution lower.

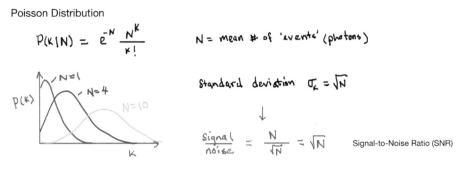

FIGURE 19.1.6

The Poisson probability distribution in a random counting process. The probability of measuring k events, given that the average number of events is N. The noise is quantified by the standard deviation of the distribution.

These trade-offs are quantified through a metric called the modulation transfer function (MTF). A description of the MTF is beyond the scope of this text, but it essentially describes the responsiveness of the detector as a function of the frequency of features in images, i.e. the spatial resolution.

There are other ways to increase the number of photons in the image which are under the control of the user. These include a longer exposure, a higher mA or a higher kVp. A thinner anatomical section will also result in more transmitted photons. The disadvantage of using higher mA or kVp is a higher exposure to the patient. Exposure is proportional to $mA \cdot kVp^2$ but note that this is at the skin of the patient. The transmitted exposure scales differently as approximately $mA \cdot kVp^5$ (see Bushberg et al., Chapter 6.5). Note that the exponent is approximate and some authors claim a scaling that is closer to $mA \cdot kVp^4$.

19.1.5 Scatter

A final parameter that affects image quality is scatter and this particularly influences SNR. Within a patient some photons are scattered to the detector (Figure 19.1.7A). That is, they do not travel in a straight line from the source to the detector. As such, they do not provide any information about what is attenuating the beam inside the patient. Scatter results in a signal that is larger (more photons = higher signal) but this is an unwanted signal. It does not provide any information about what is inside the patient.

Scatter is affected by the geometry of the imaging system. If the patient is farther away from the detector fewer scatter photons will reach the detector and vice versa (Figure 19.1.7B). It is also affected by the thickness of the patient (a thinner anatomical region will have less scatter) and by the field size (a larger field size will result in more photons scattering from a larger area).

There is a device which can be used to reduce scatter, the grid (Figure 19.1.7C). The grid consists of tiny fins of metal, arranged to point back towards the source. Any photon coming in along a line toward the source will get through. A scattered photon, however, does not come in along such a ray line so it will get stopped by the grid and will not make it to the detector. The grid, then, acts to reduce the scattered photons.

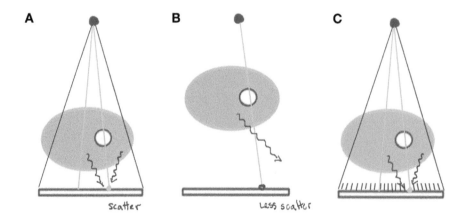

FIGURE 19.1.7
Scatter, geometry, and grids. (A) Scatter increases the signal at the detector but does not provide useful information about what is inside the patient. (B) Geometry affects scatter. (C) Anti-scatter grid.

19.1.6 DICOM

A final topic that is important to all of imaging is the representation and storage of image data. **This is a key concept.** The modern protocol that controls how images are stored, viewed, and transferred between systems is Digital Imaging and Communications in Medicine (DICOM). Prior to the early 1980s all medical images had their own format, vendor-specific formats, so there was no interconnectivity between systems. There was no common platform. In the mid-1980s the American College of Radiology and the National Electrical Manufacturers Association (NEMA) started a standardization initiative to organize this, and in 1985 the first standard was published for DICOM. Adoption of this standard took time, but by 1993 there was good buy-in and the system was further codified and published as a NEMA standard, DICOM 3.0, and that is in use today. For further examples of DICOM images including a list of freeware viewers for DICOM files see the video.

In the radiotherapy context an important development was "DICOM-RT." First published in 1996, this is an extension to DICOM for radiotherapy (RT) objects. It includes structure sets (contours), RT plan, RT dose, and RT images. It started to be used in cooperative group trials in 2003, and by 2013 the NCI required that the DICOM-RT standard be used in all trials. A new version of DICOM-RT is expected out soon which will include standards for region of interest templates, standard structures per disease site, and other changes.

Another term that one encounters is Picture Archiving and Communications System (PACS). This is not a communications standard like DICOM; rather it is a system for information storage, e.g. a hospital radiology PACS that serves the radiology images. PACS uses the DICOM standard to communicate between it and other systems.

19.2 Computed Tomography (CT)

The mathematics behind computed tomography (CT) dates back to the early 20th century (c.f. Radon transform and its uses). The modern incarnation of this for medical imaging was pioneered by Sir Godfrey Hounsfield at EMI research in the UK and, separately, Allan Cormack at Tufts. CT was first used in 1971 by Hounsfield to image a patient, and the first scanner was installed shortly after in the United States at the Mayo Clinic. Hounsfield and Cormack shared the 1979 Nobel Prize in medicine for this work.

19.2.1 Basics of CT Reconstruction

The basics of CT reconstruction can be understood from Figure 19.2.1. In this example the goal is to generate a reconstructed image of some object (the blue sphere) in a patient. An X-ray source (red) is used in combination with a detector, both of which rotate around the patient. At any given angle the detector provides a profile of photon fluences (Figure 19.2.1A). The process of CT reconstruction consists of "backprojecting" this signal through space. This essentially "smears out" the signal, resulting in the blue band in Figure 19.2.1B. That is, the backprojection indicates that some high-density object lies along the beam path, but the location along that path is not known. By rotating the source and detector around the object, however, different angles are obtained. Backprojections are performed at each angle, and this ultimately yields a reconstruction that is a faithful representation of the actual object. More properly, this algorithm is called "filtered backproject" because a frequency filter must first be applied to the image before backprojection

A   B

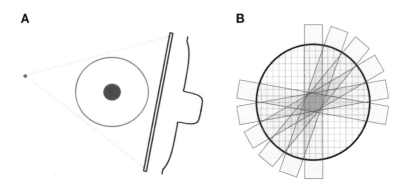

FIGURE 19.2.1
CT reconstruction. (A) Images are obtained at various angles. (B) Each image is backprojected through the reconstruction space resulting in CT reconstruction on a grid.

in order to get a mathematically exact reconstruction. For a further illustration of the CT acquisition and reconstruction process see the video.

19.2.2 Hounsfield Units

In CT imaging the transmission of the beam through the patient is measured. Recall from Equation 6.1 that $I = I_0 e^{-\mu x}$ where I_0 is the incident fluence intensity, I is the transmitted intensity, and the pathlength is x. All of these quantities are known or measured in CT, and therefore the linear attenuation coefficient, μ, can be derived. Each pixel in a CT image represents the value of μ in that pixel, with dark values corresponding to small values and bright values large values (Figure 19.2.2). The unit used for CT is the Hounsfield Unit (HU)

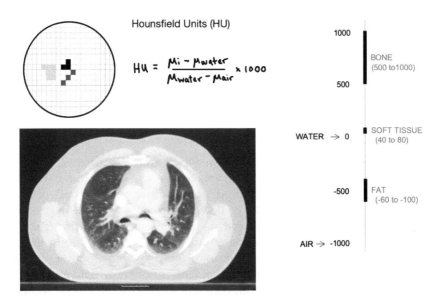

FIGURE 19.2.2
CT representation and Hounsfield Units (HU). Axial slice through the thorax of a patient showing (in order of brightness) lungs, fat, muscle, and bone.

defined as the linear attenuation coefficient relative to water. The Hounsfield Unit in some pixel i is:

$$\mathrm{HU} = \frac{\mu_i - \mu_{\mathrm{water}}}{\mu_{\mathrm{water}} - \mu_{\mathrm{air}}} \times 1000 \qquad (9.2)$$

The values for some key elements are water: 0 HU, air: –1000 HU and bone: 500–1000 HU. Note that higher density materials have large HU but material composition also plays a role, since μ depends on the material composition (recall the Z dependence of the photoelectric interaction in Section 5.1.3). There is also a dependence of μ on the spectrum of the X-ray beam. Softer beams for example have more photons at low energy where μ is larger (see Section 8.1.4). HU, therefore, depends on the spectrum of the X-ray beam. **The definition of HU and the dependencies on density, material composition, and the X-ray spectrum forms a key concept.** For radiation therapy purposes a conversion is required between HU and mass density which is used for dose calculations. The CT–density relationship is non-linear so a calibration must be established. For further detail and data see the video.

19.2.3 Fan-Beam Acquisition

CT data can be acquired in various ways. Figure 19.2.3 shows a fan-beam acquisition used in diagnostic scanners. Here the beam is collimated in the superior–inferior dimension to a relatively narrow extent, typically 1–5 mm. The detector (blue) can be a multislice detector consisting of many rows of detector lines in the superior–inferior dimension (e.g. 4, 8, 16, or up to 128 rows). A slice is reconstructed with some thickness, S, in the superior–inferior dimension. CT scans can be acquired in "serial" mode, that is slice-by-slice, but more commonly the acquisition is in helical mode (Figure 19.2.4). Here the patient slides

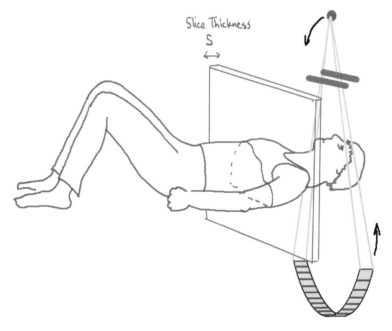

FIGURE 19.2.3
CT acquisition in a fan-beam geometry.

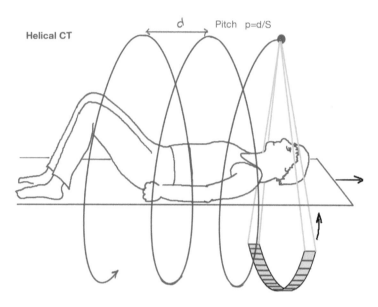

FIGURE 19.2.4
Helical CT scanning. Shown with an exaggerated pitch.

through the scanner bore as the tube and detectors rotate around the patient. The pitch, p, of the helical scan can be set and determines how "tight" or "loose" the spiral is. Pitch is defined as $p = d/S$, where d is the distance the table travels in one tube rotation. Larger pitch results in faster acquisition but a degraded slice sensitivity profile. After the helical CT is acquired slices are reconstructed in straight axial planes perpendicular to the table motion.

19.2.4 Image Quality in CT

The image quality parameters discussed in Section 19.1 also apply to CT: contrast, noise, and spatial resolution. We will leave aside contrast since the considerations are similar to those outlined in Section 19.1.1. Noise depends on collimation (or slice thickness), reconstruction algorithm, pitch, kVp, time (s), and mA. The last two should be obvious given the dependence of noise on the number of photons. CT reconstructions are limited by photon counting noise so it should be clear from Section 19.1.4 that signal-to-noise, $SNR \propto \sqrt{mA \cdot s}$.

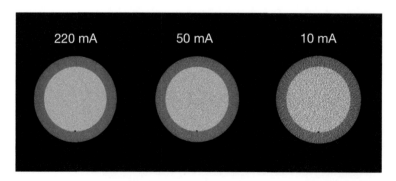

FIGURE 19.2.5
Noise in CT images and dependence on mA.

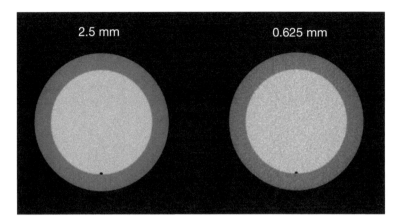

FIGURE 19.2.6
Noise in CT images and dependence on slice thickness.

Figure 19.2.5 illustrates this dependency. Similarly, the noise depends on slice thickness. Thicker slices have higher signal-to-noise ratio because more photons are collected. This can be seen in Figure 19.2.6.

Spatial resolution in CT images depends on the focal spot size, detector resolution, and pixel size, similar to the considerations for any radiographic image as outlined in Section 19.1.2. CT resolution also depends on the reconstruction filter used in backprojection and the slice thickness. In filter backprojection the size of the filter can be controlled to enhance higher or lower resolution features. Figure 19.2.7 shows an example with a commonly used phantom to test spatial resolution. When a bone filter is used for backprojection in

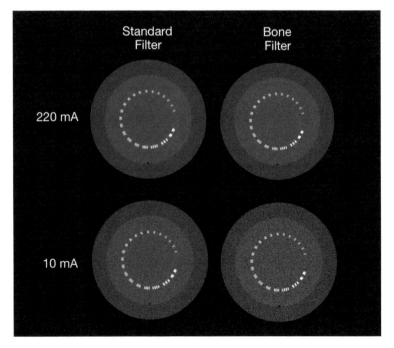

FIGURE 19.2.7
Spatial resolution in CT images.

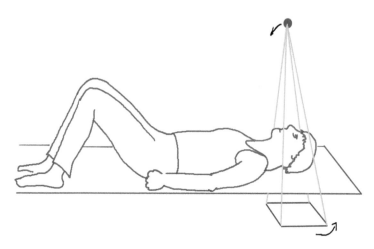

FIGURE 19.2.8
Cone-beam CT (CBCT) acquisition geometry.

reconstruction smaller features can be distinguished at the cost of a somewhat noisier image. Note that changing the mA, while it affects the noise, does not affect the spatial resolution.

19.2.5 Cone-Beam CT (CBCT)

In cone-beam CT (CBCT) the thin slice acquisition of a diagnostic scanner (Figure 19.2.3) is replaced with a large-size detector which irradiates a "cone" region through the patient, Figure 19.2.8. The main reason for this choice is that an entire volume can be acquired with a single revolution around the patient. This is an advantage especially on C-arm and other linac systems where it is not possible to rotate at the fast rotation speeds of diagnostic scanners (rotation periods of 1 sec or even 0.5 sec). In CBCT systems, roughly 700 images are acquired in a rotation while the patient remains in one position. This allows for imaging of the patient in treatment position just prior to treatment and allows for image-guided radiation therapy (IGRT) (see Chapter 21).

There are some distinct disadvantages of the cone-beam geometry. This includes degraded image quality due to possible motion of the patient (or of anatomy inside the patient) during the relatively long acquisition (note that the rotation period in diagnostic CT is <1 sec while in CBCT on C-arm linacs is 1 minute). Also scatter is larger in the cone-beam geometry which degrades image quality as discussed in Section 19.1.5. In addition, because of the larger scatter in CBCT, the HU are often not accurate representations of the linear attenuation coefficient. That is because the measured signal no longer depends just on attenuation but also includes scattered photons. This poses a challenge if CBCTs are to be used for therapy dose calculations.

19.2.6 CT Artifacts

CT images are subject to artifacts, some of which are shown in Figure 19.2.9. A: "cupping artifact." The density is apparently lower in the center of the image than in the periphery. This is due to the larger scatter contribution in the center of the image which is especially pronounced in CBCT. Also beam hardening plays a role, with rays through the center more attenuated and therefore a harder spectrum than rays through the periphery. B: "streaking artifact." This is due to beam hardening in the projection angles which pass laterally through the two high-density objects. This results in a harder spectrum which the

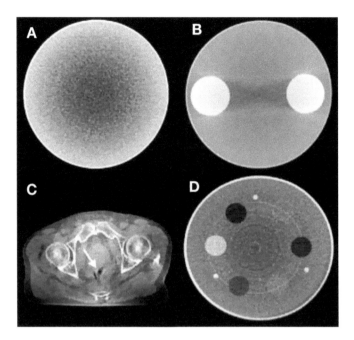

FIGURE 19.2.9
Artifacts in CT scans. (Reprinted with permission from *The Modern Technology of Radiation Oncology*, Volume 2, J. Van Dyk (Ed.) 2005.)

detector registers as fewer photons which results in a low-density streak between the two objects. C: motion during the scan. Here a gas bubble moves. D: "ring artifact." A faulty or miscalibrated pixel in the detector will reconstruct as a ring.

Further Reading

Bushberg, J.T., J.A. Seibert, E.M. Leidholdt Jr. and J.M. Boone. 2012. *The Essential Physics of Medical Imaging*. Chapters 4, 5, 7, and 10. Philadelphia, PA: Lippincott Williams & Wilkins.
McDermott, P.N. and C.G. Orton. 2010. *The Physics of Radiation Therapy*. Chapter 19. Madison, WI: Medical Physics Publishing.
RSNA Physics Education Modules: rsna.org/en/education/trainee-resources/physics-modules

Chapter 19 Problem Sets

Note: * indicates harder problems.

1. Rank the following sources by increasing resolution in the resulting radiographic image.
   a. Cobalt-60
   b. 6 MV linac
   c. 120 kVp X-ray tube for radiology, 1.0 mm focal spot
   d. 225 kVp X-ray tube for industrial parts inspection, 2.0 mm focal spot

2. How is the quality of an MV radiographic image impacted by adding a copper conversion plate? (Select all that apply.)
    a. Decreased signal
    b. Increased signal
    c. Decreased resolution
    d. Increased resolution

3. What is the storage size for an image that is 512 × 512 pixels with a 16-bit depth?
    a. 0.261 MB
    b. 0.524 MB
    c. 2.10 MB
    d. 4.19 MB

4. What is the pixel size of a picture printed out at 720 dots per inch (dpi)?
    a. 1.39 μm
    b. 3.53 μm
    c. 13.9 μm
    d. 35.3 μm

5. What is the pixel size of a 512 × 512 CT slice with a field of view of 30 cm?
    a. 0.059 mm
    b. 0.59 mm
    c. 1.17 mm
    d. 2.35 mm

6. True or False: The CT image in Problem 5 will have a higher resolution if reconstructed into a 1024 × 1024 image.

7. How long does it take to acquire a 50 cm scan for the following CT settings: tube rotation speed 0.5 sec, pitch 1.0, slice thickness 1.5 mm?
    a. 17 sec
    b. 27 sec
    c. 50 sec
    d. 167 sec

8. For each of the situations listed below indicate whether the signal-to-noise ratio in a CT slice will increase or decrease if all other scan parameters are unchanged.

Change	Signal-to-Noise (Increase or Decrease?)
mA changed from 5 to 10	
kVp changed from 80 to 120	
Region scanned is large adult pelvis instead of head	
Reconstruction filter changed from soft tissue to bone	
Slice thickness changed from 1 mm to 3 mm	

9. * Describe the artifacts in cone-beam CT if a full rotation is acquired but too few projection angles are used.

10. * Explain the physics of dual-energy CT and the potential clinical utility.

20

Non-Radiographic Imaging

20.1 Magnetic Resonance Imaging

This imaging modality relies on the properties of nuclei in matter (usually the hydrogen atom) to visualize soft tissue. By manipulating the magnetic spin of the nuclei one can get information about matter. This is the process of "nuclear magnetic resonance" (NMR). Starting in 1977, this technique began to be used to make images of living tissue. This technique, magnetic resonance imaging (MRI), uses a clever trick to turn nuclear spin information into an image. Because of the underlying physics, MRI is a sensitive probe of subtle soft tissue contrasts in the patient.

20.1.1 Nuclear Spin and Precession

In MRI a large magnetic field is applied to the patient (red arrow in Figure 20.1.1 denoted with the symbol for magnetic field "B"). This field is also called the "B_0 field" of the MRI unit, i.e. the main magnetic field. Field strengths of 1 to 3 Telsa (T) are typical (for comparison the earth's magnetic field is approximately 0.0001 T). Within this field are the nuclei in the atoms (blue sphere in Figure 20.1.1) each of which acts as its own magnet (the grey arrow). If the nuclear magnetic moment and the B_0 field are not aligned, there will be a torque on the nucleus and it will rotate or "precess" much like a gyroscope in a gravitational field. For an animation of this precession see Video 20-1.

The precession frequency known as the "Larmor frequency" and is given by $f(MHz) = \dfrac{\gamma B}{2\pi}$, where γ is a property of the nucleus called the gyromagnetic ratio. Note then that the larger the B field the faster the precession rate. For the hydrogen nucleus the Larmor frequency is 42.6 MHz/T. Therefore, in an MRI unit with a field strength of 1 T the hydrogen nucleus precesses at 42.6 MHz which is in the radiofrequency (RF) range.

The forces and torques described above are the result of magnetic fields interacting with each other. One of these fields is from the nucleus itself. The nucleus has a magnetic dipole moment denoted as μ. This can be thought of as analogous to a bar magnet which has a dipole moment ("north" and "south" poles). The gyromagnetic ratio and the magnetic dipole moment are related through the equation $\gamma = \dfrac{\mu}{I\hbar}$, where μ is the magnetic dipole moment, and I is a quantum mechanical property of the nucleus called spin. Note that some nuclei have a spin that is non-zero but in some others the spin is zero. The hydrogen nucleus, for example, has a spin of ½ but the helium nucleus has a spin of zero. MRI imaging is therefore not possible with helium because the magnetic dipole moment, $\gamma I\hbar$, is zero.

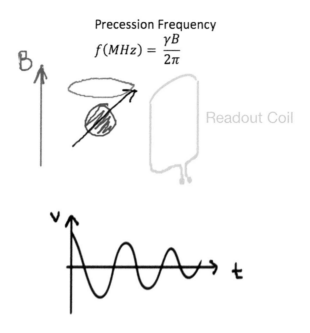

FIGURE 20.1.1
The nuclear magnetic moment (gray arrow) in the presence of an external magnetic field ("*B*") will precess or rotate with a frequency given by the formula. As magnetic moment rotates, the field in the readout coil changes (which results in a current) which can be read.

20.1.2 Signals and Spin Flips

The rotating nuclear magnets can generate a signal which can be measured. This key concept is illustrated in Figure 20.1.1 where the green wire loop represents a readout coil. As the nuclear magnetic moment precesses the amount of magnetic flux going through the coil will change. Recall from basic physics that a changing magnetic field results in a current (Faraday's Law). There is therefore a current through the coil can be read out (or equivalently a voltage). Note that a single precessing nucleus would produce a very small signal, but in a biological sample there are many billions of such nuclei all aligned and precessing.

The energy of a nuclear magnetic dipole depends on its orientation relative to the external magnetic field as shown in Figure 20.1.2. By pumping energy into the system the nuclear spins can be flipped from low energy (oriented with the field) to higher energy (oriented perpendicular to the field or anti-parallel). This energy is supplied in the form of RF radio waves at the Larmor frequency. If the spin is in a high-energy state (e.g. perpendicular to the *B* field) then over time it will naturally decay into the lowest energy state (aligned), precessing around as it does this. The process of excitation and decay is shown in an animation in Video 20-1.

20.1.3 Image Formation in MRI

In the example shown in Figure 20.1.3 the main B_0 field is oriented along the superior–inferior axis of the patient and has a value of, say, 1.0 T at the center of the patient. In addition to this main field, extra (much smaller) fields are applied using coils in the magnet to create a gradient along the axis of the patient. The field may be, say, 0.9985 T at the superior

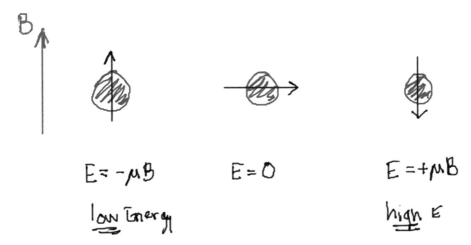

FIGURE 20.1.2
Orientation of the nuclear magnetic dipole and resulting energy.

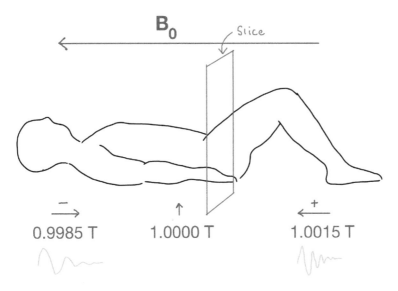

FIGURE 20.1.3
Slice selection in MRI. An extra magnetic field gradient is applied in the superior–inferior direction so that the field (and the Larmor frequency) varies slightly depending on the position in the superior–inferior direction.

aspect and 1.0015 at the inferior aspect. Because of this field gradient, the nuclei precess a little faster in the inferior region of the patient than in the superior regions. By selecting the frequency for excitation the slice of interest can be selected so that only that slice is being probed. Similarly, gradients in the other directions (e.g. left–right) can be used to determine the position from which signals emerge. This allows image formation in MRI.

20.1.4 Spin-Echo: TR and T_1-Weighting

To create signals in MRI the nuclear magnetic moments need to be manipulated and read out. To understand how that's accomplished, consider first the collection of all nuclear magnet moments in the sample. These have some overall net magnetization written as "M." M can be broken down into two vector components, the longitudinal component along the direction of the B_0 field written as M_z, and the transverse component in the x-y plan, written here as M_y (Figure 20.1.4).

The simplest component to understand is M_z. Consider what happens if one starts with a zero M_z component. This might happen after an RF pulse is applied to flip the spins into the x-y plane. At this point the nuclear magnet moments would start to decay to the ground state energy, i.e. to become aligned with the B_0 field. As this happens the M_z component grows (green curve in Figure 20.1.4). After some time, another RF pulse might be applied to orient the spins into the x-y plan, making $M_z=0$ again. This process can be repeated, and the repetition time is called TR in MRI. From Figure 20.1.4 one can see that the TR time determines how much the magnetization M_z is allowed to grow before the next spin flip happens.

By controlling TR one can control the contrast between different types of tissues as shown in Figure 20.1.5. This is because different tissues have a different relaxation rate for the M_z component. This rate is described by an exponential decay constant, T_1, which is the "spin-lattice" relaxation rate, i.e. the interaction of the nuclear spin is affected by the surrounding material, "the lattice." Consider what happens at a long TR value. Here most of the spins have decayed down and M_z is approximately the same (later times on the curve in Figure 20.1.5). Conversely at earlier times (short TR) the M_z components are still growing

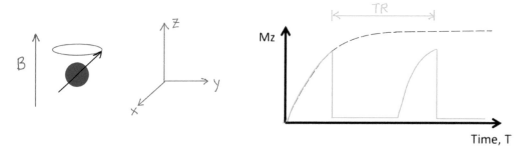

FIGURE 20.1.4
Longitudinal magnetization along the z-axis, M_z. As the nuclear magnetic moment decays to the ground state energy (aligned with the B_0 field) the M_z component grows. Pulse sequences repeated at some time TR can be used to probe M_z.

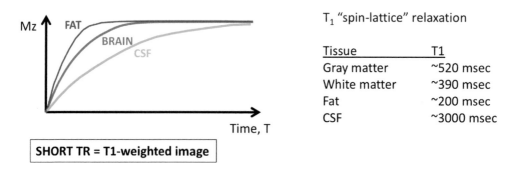

Tissue	T1
Gray matter	~520 msec
White matter	~390 msec
Fat	~200 msec
CSF	~3000 msec

FIGURE 20.1.5
Growth of the longitudinal magnetization, M_z, and T_1-weighted contrast. At shorter times there is a larger difference in M_z between the various tissues.

and there is a relatively large difference in M_z between the various tissues. Therefore, short TR is a T_1-weighted image. In such an image the signal from fat would be relatively large (bright) and the signal from liquids like cerebrospinal fluid (CSF) would be small (dark).

20.1.5 Spin-Echo: TE and T_2-Weighting

The other component of magnetization, M_y, is in the x-y plane (Figure 20.1.6). This can also be manipulated and read out to yield different information about the tissue. The M_y component decays due to "spin-spin" relaxation, i.e. the interaction of the nuclear spins with each other. The decay constant is T_2^*.

In order to read out the M_y component a series of pulses can applied as illustrated in Figure 20.1.7. The sequence begins with the spins all in the ground state, i.e. oriented along B_0. An RF pulse is applied which pumps energy into the system (see Figure 20.1.1) and begins to flip the spins. After a certain amount of time, the spins will be oriented in the x-y plane and the pulse is shut off (point "A" in Figure 20.1.7). This is called a 90-degree pulse because the duration of the pulse is such that the spins are flipped 90 degrees.

After the 90-degree pulse, the spins precess in the x-y plane as shown. Some spins have a slightly faster precession rate (those shown as green) and some are slightly slower (red) due

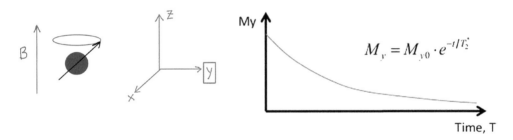

$$M_y = M_{y0} \cdot e^{-t/T_2^*}$$

FIGURE 20.1.6
Transverse magnetization in the x-y plane, written as M_y here. The M_y component decays with a decay time T_2^* due to "spin-spin" relaxation.

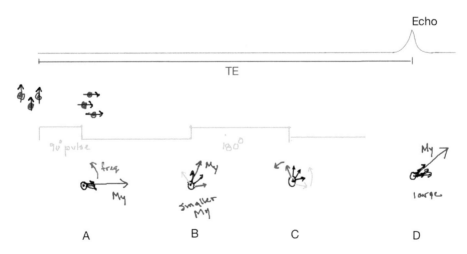

FIGURE 20.1.7
The spin-echo pulse sequence to recover the M_y component.

to local differences in the magnetic field. Over time the net magnetization M_y decreases because the spin are more spread out in angle and the vector sum is smaller.

At some point in time (B in the figure) another pulse is applied. This pulse is a 180-degree pulse which flips the spins to the opposite direction in the x-y plane. Now spins that are precessing faster (green) begin to catch up with spins that are slower (red). At some point in time (C in the figure) the spins have caught up to each other and the net M_y suddenly increases and there is an abrupt signal (or "echo") recorded from the large M_y. The time between the start of the 90-degree pulse and the echo is TE (time to echo), and it is set by the time delay between the 90-degree pulse and the 180-degree pulse. See the videos for further illustration of MRI sequences and the spin-echo process.

By using the spin-echo sequence above, M_y can be probed at different times as shown in Figure 20.1.8. The M_y measured in a spin-echo sequence decays with an exponential decay constant, T_2, which is the "spin-spin" relaxation rate, i.e. the interaction of the nuclear spins with each other. (Note that this is different than T_2^* above which includes effects of inhomogeneities in the magnetic field which are "reversed" by the spin-echo technique). As shown in Figure 20.1.8, the M_y measured in this way is different in different tissues and, in particular, there is more T_2 contrast for sequences that use longer TE times. In such an image the signal from fat or brain parenchyma would be relatively small (dark) and the signal from liquids like cerebrospinal fluid (CSF) would be large (bright).

Real MRI pulse sequences have somewhat more complexity than the basic features described above. For example, a slice selection pulse is applied in combination with the other pulses in order to select those nuclei in a particular slice, i.e. at the precession frequency determined by the gradients (see Figure 20.1.3). However, the basic principles apply, and images can be created like those in Figure 20.1.9 for a patient with a glioblastoma tumor. Note that the T_1-weighted image (left) has a short TR while the T_2-weighted images have a long TE; CSF is dark on the T_1 image and bright on T_2.

Of note, in the T_1 image shown in Figure 20.1.9 both the tumor and normal vasculature have bright signals. This is because the image was acquired after IV injection of the contrast agent gadolinium (Gd). In the presence of Gd the T_1 relaxation rate is shortened (due to local magnetic field effects of the metal). This results in a brighter signal (larger M_z) on T_1-weighted images. Thus, after IV injection the normal vasculature appears bright. The brain does not appear bright because the Gd chelate does not pass the blood–brain barrier. However, in the tumor where the blood–brain barrier is disrupted and there is leaky vasculature the Gd is present and the signal is enhanced.

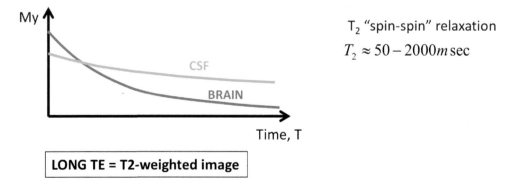

FIGURE 20.1.8
Decay of the M_y as measured with a spin-echo sequence. At long TE values, there is a larger difference between the M_y signals from various tissues.

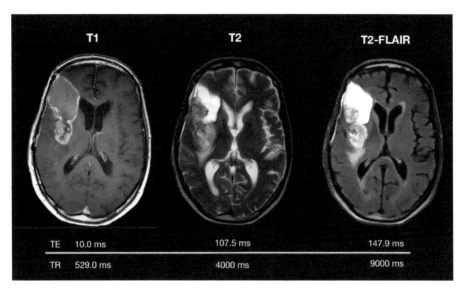

FIGURE 20.1.9
MRI images for a patient with a glioblastoma tumor using three different MRI sequences, T_1-weighted with short TR, and T_2-weighted with long TE.

20.1.6 Inversion Recovery (IR) Pulse Sequences

Figure 20.1.9 shows an example of an image formed with an inversion recovery pulse sequence, namely Fluid-Attenuated Inversion Recovery (FLAIR). This image is clearly T_2-weighted because of the long TE, but the CSF fluid appears dark. To understand how this is possible, consider Figure 20.1.10. The sequence begins with a 180-degree pulse which inverts the spins. The transverse magnetization M_z starts to grow as described above. At some time labeled "A" on the graph the M_z will be zero for a particular interest (here CSF). By starting the sequence at this time A after the inversion pulse, the signal from this select material will be nulled out.

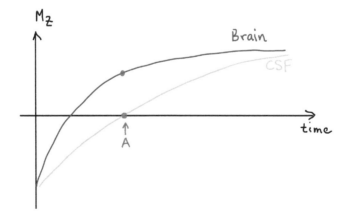

FIGURE 20.1.10
Inversion recovery sequence. The spins are first inverted 180 degrees and then allowed to decay. Starting the remaining sequence at time "A" nulls out the signal from the tissue of interest, here CSF.

20.1.7 Distortion and Artifacts in MRI

Distortion and artifacts in MRI are particularly important in the context of radiation oncology where accurate geometric definitions are essential. To understand artifacts, consider Figure 20.1.3 where we saw that changing the frequency corresponded to changing the position in the patient. This happens because the magnetic field is slightly at different positions. Equivalently, if the frequency were fixed but the magnetic field were changed this would result in changes in position. This, then, is a key concept underlying distortion and artifacts: changing the local magnetic field is equivalent to moving the pixel in space.

Figure 20.1.11 shows examples of how field distortions occur. Ideally the B_0 field would be completely uniform throughout the patient. Figure 20.1.11 (left) illustrates a case where there are non-uniformities at the edge of the bore opening where the field bows out. This results in distortions as illustrated in Figure 20.1.12 in a phantom test. This phantom consists of a grid of regularly spaced plastic rods which is scanned. Ideally the lines in the phantom image are regular and straight (Figure 20.1.12B), but in the presence of field inhomogeneities they are distorted (Figure 20.1.12C).

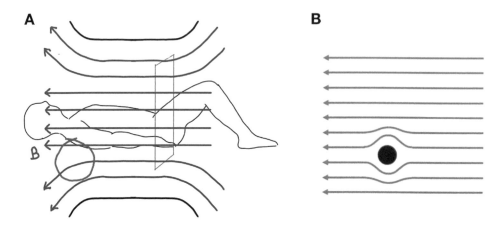

FIGURE 20.1.11
Magnetic field inhomogeneities that lead to distortion. (A) Inhomogeneities in the main magnetic field. The otherwise uniform magnetic field bulges out near the edge of the bore (red circle). (B) A metallic paramagnetic object distorts the field locally leading to a susceptibility artifact.

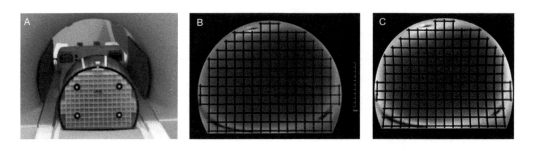

FIGURE 20.1.12
Phantom test object to measure the distortion in MRI. (A) The phantom being scanned in an MRI unit. (B) An image with little distortion. (C) Distortion near the edge due to inhomogeneities in the B_0 magnetic field. (Images courtesy of CIRS, Inc.)

Other classes of artifacts in MRI include the following:

- *Magnetic susceptibility artifact.* A metallic object in or near the field can distort the field (Figure 20.1.11B). This also results in an artifact in the image, most typically a signal void near the metallic object.
- *Gradient distortions.* This can be understood from Figure 20.1.3. The field gradient through the patient determines the position of each pixel. If the gradient is larger than expected the image would appear compressed in that dimension.
- *Chemical shift artifact.* This occurs at the interface of two very different materials. In this region the precession frequency of protons on one side is very slightly lower than the other due to the different chemical environments. This results in a physical shift of the signal. This is illustrated by an experiment which shows spin-echo MRI images of a chicken egg (see Weygand et al. *Int J Radiat Oncol Biol Phys*, 95(4) 1304–1316, 2016).

20.2 Nuclear Medicine and PET Imaging

20.2.1 Radioisotopes Used in Imaging

In nuclear medicine imaging, the patient is injected with small amounts of a radiopharmaceutical, i.e. radioactive isotope which is chemically conjugated to a compound that probes a biological process of interest. Various radioactive isotopes are in use as shown in Table 20.2.1. They undergo various modes of decay and have a range of half-lives.

The process of nuclear decay, the decay modes, and products are described in Section 2.2. To summarize briefly, important decay modes include beta-plus, beta-minus, electron capture, alpha decay, isomeric transition, and internal conversion. For a summary see Table 2.2.1. For beta decay specifically, the neutron-rich isotopes undergo beta-minus decay, have relative long half-lives, and are made in reactors, while the proton-rich isotopes undergo beta-plus decay, have relatively short half-lives, and are made in cyclotrons.

TABLE 20.2.1

A Selection of Radioisotopes Used in Nuclear Medicine Imaging, Including the Decay Mode (Isomeric Transition, Electron Capture, β^-, and β^+), the Half-Life, Dominant Energy of the Photons and the Method of Production

Isotope	Decay	Half-Life	Photon Energy	Production
^{99m}Tc	IT	6 hr	140 keV	Generator
^{123}I	EC	13 hr	159 keV	Cyclotron
^{131}I	β^-	8 dy	364 keV (average of many decay modes of daughter)	Reactor
^{201}Tl	EC	73 hr	69–83 keV	Cyclotron
^{111}In	EC	68 hr	171, 245 keV	Cyclotron
^{133}Xn	β^-	5.2 dy	81 keV	Reactor
^{18}F	β^+	110 min	511 keV (from annihilation of e+)	Cyclotron

The photons that emerge are often the result of a secondary processes (e.g. the electron-positron annihilation in ^{18}F or the decay of daughter products into a ground state).

20.2.2 Single Photon Emission Computed Tomography (SPECT)

An example of the clinical application of SPECT is the measurement of perfusion in the myocardium to evaluate coronary artery disease. This application employs ^{99m}Tc conjugated to 6-methoxyisobutylisonitrile (sestamibi) which is taken up in the vasculature of the healthy heart muscle. ^{99m}Tc is an excited nuclear state of Tc and is the daughter product of ^{99}Mo decay. ^{99m}Tc decays to its ground state, emitting a 140 keV photon which is used for imaging (note that other photon energies are present from higher energy excited states, though in smaller quantities).

SPECT is also used in oncologic imaging. Examples include ^{123}I for thyroid cancer evaluation and treatment or ^{99m}Tc-methylene diphosphonate (MDP) for imaging bone metastases. However, oncologic SPECT competes with PET which offers higher sensitivity, the ability to quantify uptake, and often higher spatial resolution.

The geometry of the SPECT camera system is illustrated in Figure 20.2.1. Here the radiopharmaceutical has accumulated in some region (red), undergoes decay, and emits photons. The photons interact with the scintillator in the head of the gamma camera which converts the gamma rays into optical photons. These are registered by the position-sensitive photomultiplier tubes. A collimator is placed in front of the scintillator. This provides some spatial localization in that a photon entering the camera at an angle will not be transmitted to the scintillator. The whole assembly rotates around the patient, and images are acquired (for an animation see Video 20-2). These data can be used to reconstruct the distribution of radioisotope within the patient. This is similar to the process with CT although this is an emission tomography instead of transmission. Note that many modern SPECT scanners have two gamma camera heads and are also coupled to a CT scanner, so that a CT scan is acquired just before or after the SPECT scan to allow co-localization.

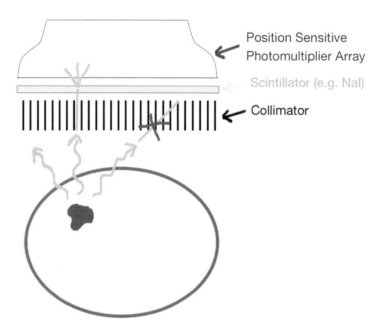

Position Sensitive
Photomultiplier Array

Scintillator (e.g. NaI)

Collimator

FIGURE 20.2.1
Geometry of the SPECT imaging system.

20.2.3 Positron Emission Tomography (PET): Isotopes and Uptake

PET imaging uses isotopes which undergo beta-plus decay and emit a positron. The classic example of this is ^{18}F which decays to ^{18}O+e$^+$ with a half-life of 109.8 minutes (see Section 2.2.2 and Figure 2.2.2 to review). To be useful in oncologic imaging, of course, the ^{18}F has to be attached to some compound of biological interest. The most widely used radiopharmaceutical for this is ^{18}F-fluorodeoxyglucose (^{18}F-FDG), a glucose analog that is taken up by cancer cells which are in anabolic metabolism (an altered cell metabolism that has been appreciated since the 1930s). Like glucose, this compound is taken up in tumor cells through an active process involving glucose transporter proteins. It is then phosphorylated and would go on through a glycolysis process. However, the FDG cannot go through the full glycolysis process because the –OH group which is normally present has been replaced by ^{18}F. FDG is therefore trapped in the cell. It is this avid uptake and trapping that makes ^{18}F-FDG useful for imaging cancer.

A typical ^{18}F-FDG scan protocol involves having the patient fast for at least six hours before the scan to get the glucose levels down. The patient is then injected IV with approximately 15–20 mCi and ^{18}F-FDG and then waits for typically one hour for uptake. Physical activity of the patient is kept to a minimum during this time to reduce uptake in the muscles. The scan then begins.

20.2.4 PET: Image Acquisition

PET image acquisition begins with the uptake of the radiopharmaceutical in the tumor (red) in Figure 20.2.2. The isotope decays, creating a positron (e+) which has some energy and therefore wanders some distance in the tissue (typically 1–2 mm). At some point it slows and undergoes an annihilation with an electron (e–) in the tissue. Recall that the positron is the anti-electron and that matter and anti-matter annihilate when they are in close proximity. This annihilation event produces photons, each with an energy of 511 keV. This energy is a result of the fact that energy must be conserved (i.e. the total energy after the annihilation should equal the total energy before which is two times the rest mass energy of the electron, mc^2 or 511 keV). Also, the photons emerge opposite to each other since momentum must also be conserved, i.e. zero overall momentum before and after the annihilation.

The photons are registered in the detector crystals in the PET ring (blue, Figure 20.2.2B) as events which are approximately coincident in time. Note also that since the energy

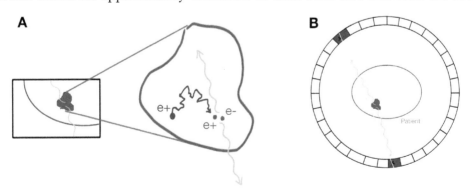

A

B

Patient

FIGURE 20.2.2
PET image acquisition. (A) The positron produced in the decay wanders in tissue and then annihilates with an electron producing two 511 keV photons emerging opposite each other. (B) These photons are registered as events coincident in time in the crystals of the PET detector ring and a line-of-response can be calculated (gray).

of the photons is known to be 511 keV the detector can be tuned to accept only photons within a small energy window in this band which allows for the rejection of other photons such as background events or scattered photons which have a lower energy. This makes PET detectors very sensitive.

Each coincident event, then, determines a line-of-response (LOR, shown in gray in Figure 20.2.2B). One knows then that the annihilation occurred somewhere along this LOR, though the exact location on the LOR for this one event is not known. As more decay events occur, more LORs are accumulated at different angles. This allows one to determine the underlying distribution of radioisotope through tomographic reconstruction.

Of note is the fact that the two annihilation photons do not emerge exactly anti-parallel to each other. There is a slight angle between them (see Figure 20.2.2B). They are non-colinear. This is because there is some small momentum imparted to the nucleus which must be balanced by a small momentum vector of the photons in the opposite direction. The end result of this is that the LOR does not intersect exactly with the position of the annihilation event. This is important because it is one factor that impacts the spatial resolution that can be achieved with PET.

Note that since roughly 2000 most PET scanners also have a CT unit attached (PET-CT). As with SPECT-CT, a CT scan is acquired just before or after the PET scan to allow co-localization. Also for PET imaging this CT scan can be used to apply an attenuation correction as discussed in Section 20.2.6.

20.2.5 PET: Resolution and Representation

The typical resolution of a PET images is approximately 4–5 mm. There are several factors affecting resolution:

- Non-colinearity of annihilation photons (511 keV photon do not emerge exactly anti-parallel as discussed in the previous section).
- Size of the crystals in the PET detector ring.
- Energy of the positrons that emerge from the decay (which determines how far they range in tissue before annihilating which blurs the spatial response).
- The algorithms and filters used in tomographic reconstruction.

Figure 20.2.3 shows an example ^{18}F-FDG PET image of a patient with lung cancer. FDG uptake can clearly be appreciated in the right posterior aspect of the lung. Higher activity is also present in the heart. For a more complete review of the images from this patient see

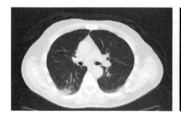 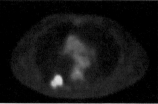 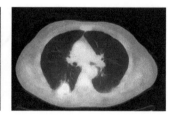

FIGURE 20.2.3
CT (left), ^{18}F-FDG PET (middle), and fused (right) images of a patient with a non-small cell lung cancer lesion in the right posterior aspect of the lung. (Images derived from the Cancer Imaging Archive, TCIA [www.cancerimagingarchive.net].)

Video 20-2 where activity can also be seen in the liver, bladder, and kidneys as it is cleared from the patient. Note that the resolution of the PET is substantially worse than CT.

Image values in PET scans are often reported as Standardized Uptake Value (SUV) which is defined on a voxel-wise basis as $SUV = \dfrac{\text{activity in the image}}{\text{injected activity/body mass}}$. There are many factors which can affect SUV so it is considered a semi-quantitative measure.

20.2.6 PET: Attenuation Correction

The annihilation photons are attenuated as they emerge from the patient, and this attenuation is not uniform. It depends on the tissue type and the location of the annihilation in the patient. The lung, for example, attenuates relatively less than other soft tissue (Figure 20.2.4A). If this effect is not accounted for the lung would appear to have activity in it. That is, more counts are observed emerging from the lung but this is not because the lung has activity in it, it is just because the attenuation is less (i.e. more photons emerge). Figure 20.2.4 C shows a PET image without attenuation correction. The lungs appear to have activity, and there are other artifacts as well: a hot "ring" around the skin and a bright signal at the lateral aspect of the liver. Attenuation correction can be applied by incorporating information from the CT during reconstruction (another reason to have a CT attached to the PET unit). Applying this correction yields a much more realistic image of the actual activity distribution (Figure 20.2.4B).

20.2.7 PET: Beyond FDG

Although FDG is the most widely used radiopharmaceutical for PET imaging, other compounds are in use as well. These probe various biological processes both in oncology and beyond. Table 20.2.2 shows a selection.

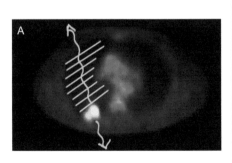
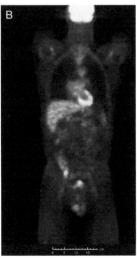
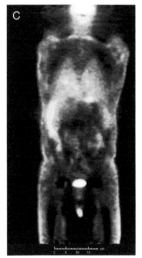

FIGURE 20.2.4

Attenuation correction in PET. The lung has lower attenuation (A, gray lines) which results in artifacts if not corrected for. (B) With attenuation correction. (C) Without attenuation correction. (Images derived from the Cancer Imaging Archive, TCIA [www.cancerimagingarchive.net].)

TABLE 20.2.2

Select Radiopharmaceuticals Used in PET Scanning and Their Applications

Isotope	Tracer Compound	Physiological Process or Function	Typical Application
^{11}C	Methionine	Protein synthesis	Oncology
^{11}C	Flumazenil	Benzodiazepine receptor antagonist	Epilepsy
^{11}C	Raclopride	D2 receptor agonist	Movement disorders
^{13}N	Ammonia	Blood perfusion	Myocardial perfusion
^{15}O	Carbon dioxide	Blood perfusion	Brain activation studies
^{15}O	Water	Blood perfusion	Brain activation studies
^{18}F	Fluoro-deoxy-glucose	Glucose metabolism	Oncology, neurology, cardiology
^{18}F	Fluoride ion	Bone metabolism	Oncology
^{18}F	Fluoro-mizonidazole	Hypoxia	Oncology—response to radiotherapy

TABLE 20.2.3

Radioisotopes Used in PET Imaging

Isotope	Half-Life (min)	Maximum Positron Energy (MeV)	Positron Range in Water (FWHM in mm)	Production Method
^{11}C	20.3	0.96	1.1	Cyclotron
^{13}N	9.97	1.19	1.4	Cyclotron
^{15}O	2.03	1.70	1.5	Cyclotron
^{18}F	109.8	0.64	1.0	Cyclotron
^{68}Ga	67.8	1.89	1.7	Generator
^{82}Rb	1.26	3.15	1.7	Generator

These compounds use different isotopes as summarized in Table 20.2.3, having different energies and half-lives. Some of the shorter half-life isotopes are quite challenging to work with, and the convenience of the 110-minute half-life of ^{18}F is one of the reasons for its widespread use.

20.3 Ultrasound

Ultrasound dates back to developments in underwater sonar during World War II that then were translated over into medicine like so many applications. Ultrasound has many applications in radiation therapy including image guidance for external beam radiation therapy and visualization during prostate brachytherapy. Further detailed information on ultrasound can be found in the video.

Further Reading

Bushberg, J.T., J.A. Seibert, E.M. Leidholdt Jr. and J.M. Boone. 2012. *The Essential Physics of Medical Imaging*. Chapters 12, 13, 18, and 19. Philadelphia, PA: Lippincott Williams & Wilkins.

Kahn, F.M. and J.P. Gibbons. 2014. *Kahn's The Physics of Radiation Therapy*. 5th Edition. Chapter 12. Philadelphia, PA: Wolters Kluwer.

McDermott, P.N. and C.G. Orton. 2010. *The Physics of Radiation Therapy*. Chapter 19. Madison, WI: Medical Physics Publishing.

Chapter 20 Problem Sets

*Note: * indicates harder problems.*

1. List three advantages and three disadvantages of a low-field MRI (e.g. <0.5 T).

2. Which of the following parameters controls the thickness of the slice in MRI?
   a. B_0 field strength
   b. Frequency bandwidth
   c. TE setting
   d. TR setting

3. Which MRI artifact would lead to a mislocalization of the target in stereotactic radiosurgery target definition?
   a. Inhomogeneity in the B_0 field of the magnet
   b. Beam hardening
   c. TE too long
   d. Phase wrapping

4. What is one advantage of placing an RF surface coil on the patient for MRI imaging?
   a. Lower B_0 field required
   b. Lower SAR
   c. Faster scan time
   d. Higher signal

5. What physical factor improves PET resolution in a small animal scanner for mice and rats vs. a human scanner?
   a. Range of the positron in tissue
   b. Lower injected activity
   c. Non-colinearity of annihilation photons
   d. Time of flight

6. Rank the following isotopes used in PET in order of improving spatial resolution in the resulting scan.
   a. ^{11}C E_{max}: 0.96 MeV
   b. ^{13}N E_{max}: 1.19 MeV
   c. ^{15}O E_{max}: 1.70 MeV
   d. ^{18}F E_{max}: 0.64 MeV

7. Discuss the factors underlying the more widespread clinical use of ^{18}F-FDG vs. other PET imaging tracers.

8. What is the depth resolution of a 3.5 MHz ultrasound unit? Assume the speed of sound in the tissue is 1540 m/s and that three cycles are used in each pulse. Discuss how this changes for a frequency of 5 MHz and what the disadvantages of that are.

    a. 0.66 mm

    b. 1.32 mm

    c. 1.75 mm

    d. 2.64 mm

9. * Describe what SAR is in the context of MRI, what affects it, and the relevant regulatory limits.

10. * Describe how the sensitivity of a PET scan depends on the injected activity. Is it monotonically increasing/decreasing? Why?

21

Image-Guided Radiation Therapy (IGRT) and Motion Management

21.1 Image-Guided Radiation Therapy (IGRT)

Image-guided radiation therapy (IGRT) is the use of imaging during or just prior to treatment with the goal of improving the accuracy of treatment delivery. IGRT saw expansive growth in the decade of 2000–2010 to the point where it is now available in virtually every clinic in North America and Europe.

21.1.1 CBCT on C-Arm Linac Systems

Though more general, the term "IGRT" is sometimes thought of as synonymous with cone-beam CT (CBCT) on a C-arm linac. In these systems an X-ray tube and flat-panel imager are mounted on the linac gantry (Figure 21.1.1), allowing the CT to be acquired in a cone-beam geometry with a single rotation of the gantry (see Section 19.2.5).

IGRT starts with a volumetric image acquired during CT simulation (Figure 21.1.2). This reference image is used during the treatment planning to define beams and calculate dose. During this process a treatment isocenter is defined on the reference image, which is the point where one intends to line up the patient during treatment. The second image set for IGRT is the "localization" image, i.e. the CBCT image acquired on the treatment unit with the patient in the treatment position. This image also has an associated isocenter which is the actual isocenter of the treatment unit and is the center of the CBCT image.

At the time of treatment a localization image is acquired with the patient in treatment position. This image is registered or "fused" with the localization image in the IGRT software. If the patient is not well-aligned with the intended isocenter in the localization image there will be an offset of the two images (see Figure 21.1.2 top right). Using the IGRT software the user can move the two images until they are well-aligned (Figure 21.1.2 bottom right). This alignment process produces a table correction, i.e. the shift of the patient in three-dimensional space that is required in order for the two isocenters to be aligned. These "shifts" are sent to the table on the treatment unit and the patient is moved. For a video animation of this process see the video.

Note that it is possible not only to translate the images relative to each other (i.e. move in the X, Y, and Z coordinate directions) but it is also possible to rotate around the three major axes (i.e. pitch, yaw, and roll). Some treatment units are equipped with patient support assemblies (i.e. "tables" or "couches") which enable this rotation, though the rotation is restricted to small ranges (<3 degrees). These are referred to as six degree-of-freedom (6-DOF) tables which refers to the fact that the table has three motions in X, Y, and Z plus three rotations in pitch, yaw, and roll.

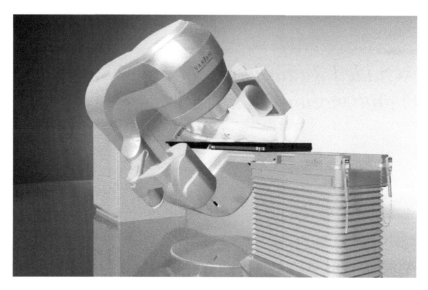

FIGURE 21.1.1
Cone-beam CT acquisition on a C-arm gantry. The X-ray tube (left) and flat-panel imager (right) are mounted perpendicular to the treatment beam. Images are acquired in one rotation of the gantry and reconstructed in the region traced out by the cone (green). (Image courtesy of Varian Inc.)

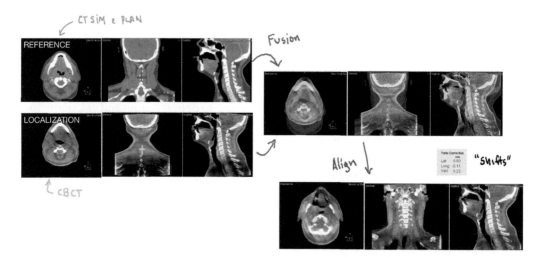

FIGURE 21.1.2
The IGRT process. Reference images acquired at CT simulation are registered or "fused" with the CBCT image acquired on the linac. The images are aligned which results in a coordinate registration or "shifts" that are transferred to the treatment unit. The patient support table is moved to these coordinates.

It is important to note that while IGRT enables an alignment of the patient to the intended isocenter this alignment is a "rigid registration," i.e. it assumes the anatomy of the patient is completely rigid and that a simple translation and rotation can be performed to make the patient align perfectly with the scan acquired at CT simulation. This is not always the case, though, of course. The neck of the patient might flex for example (as illustrated in the video) which means that it might be possible to align one region of spine but not the whole spine. Alternatively, the patient might gain or lose weight or the tumor might shrink through the course of treatment. All of these scenarios involve anatomical deformations that cannot be directly accounted for with IGRT. This is part of the motivation for adaptive radiation therapy (ART) which is an emerging radiation therapy approach in which the treatment plan is adapted during the course of therapy to adjust for changes in the tumor or the anatomy of the patient.

21.1.2 IGRT with Planar Images

IGRT is also possible with planar images (i.e. radiographs) taken either with a kV imager or an MV imager on the treatment unit. An example is shown in Figure 21.1.3 where the intent is to localize the center of the bone. Two images are acquired at angles perpendicular to each other (A and B). The center of the bone must lie along the green line in each image. From the composite of the two images then the center of the bone can be localized in three-dimensional space (C). This process is often called orthogonal imaging (since the angles are typically perpendicular to each other). Since there are two images the term "orthogonal pair" is often used, or "iso pair" since the images are often taken to verify that the patient is positioned correctly at isocenter.

To accomplish this alignment requires digitally reconstructed radiographs (DRRs) (Figure 21.1.4 top). This DRR is created by digitally tracing rays through the CT simulation scan and creating a planar X-ray image as it would look from the angle of interest. Figure 21.1.4 shows the process schematically where a DRR from the planning scan is registered with the orthogonal images acquired at the treatment machine. Note that is possible to align not only bone but also implanted metallic fiducials or hardware in the patient. For a further illustration of this process see the video.

21.1.3 Other IGRT Technology

The above sections focus on IGRT with a C-arm linac, but other technologies are available as well. These are described in some detail in Section 9.2 about linear accelerators. One of the earliest of the modern IGRT-capable systems is the TomoTherapy device (Figure 21.1.5, left, Accuray Inc.). From the outside this device appears similar to a CT scanner, but inside there is a compact linear accelerator along with an MV imaging detector. The system is therefore capable of generating MV CT scans. Newer models ("RadiaExact" units) also have

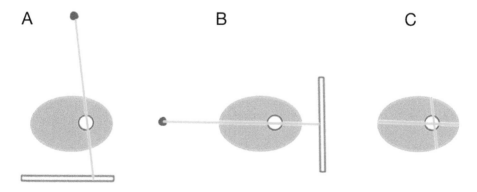

FIGURE 21.1.3
Alignment of the patient with orthogonal imaging. The position of an object (e.g. center of the bone here) must be at the intersection of the lines from images A and B.

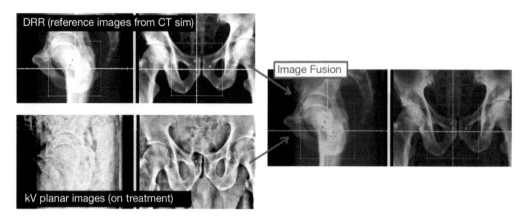

FIGURE 21.1.4
IGRT with planar imaging. Digitally reconstructed radiographies (DRRs) are created from the reference scan from CT simulation at the isocenter defined in treatment planning. These images are registered or fused with the orthogonal images acquired on the treatment unit, here kV planar images.

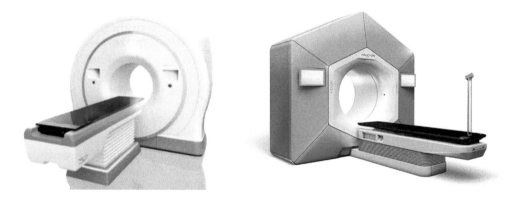

FIGURE 21.1.5
Other IGRT-capable radiation therapy devices. The TomoTherapy system (left). (Image courtesy of Accuray Inc.) The Halcyon system (right). (Image courtesy of Varian Inc.)

a kV tube and imager which allows for kV-based IGRT scans. Further information about this system can be found in AAPM Task Group 148 (Langen et al. 2010). Another IGRT-enabled system to emerge recently is the Halcyon system from Varian Inc. (Figure 21.1.5 right) and the associated Ethos® system which adds software to this system to allow for adaptive radiotherapy (ART). The Halcyon platform is a ring gantry system with a linear accelerator operating at 6 MV. A cone-beam CT scan can be acquired in 17 seconds.

Other systems supporting IGRT include the following:

- CyberKnife. Stereoscopic imaging (at intervals of approximately 1 second). See Section 9.2 and AAPM Task Group 135 (Dieterich et al. 2011).
- ExactTrac (Varian Inc./Brainlab Inc.). Stereoscopic X-ray imaging system coupled with a standard C-arm gantry linac. See AAPM Task Group #104 (Yin et al. 2009).
- Ultrasound guidance. Ultrasound probe is registered to the linac isocenter and used to guide treatment. This has found use in various forms over the years especially for prostate cancer treatments.

- Implanted electromagnetic (EM) transponder beacons (e.g. Calypso®, Varian Inc.). The transponders are miniature antenna coils in a glass envelope which can be implanted via catheter in tissue. An RF antenna array panel is placed just above the patient during treatment which allows for the real-time tracking of the position of the beacons during treatment.
- Surface imaging. Examples include AlignRT (Vision RT, London, UK) and C-RAD (C-Rad AB, Uppsala, Sweden). For further information see AAPM Task Group 147 (Willoughby et al. 2012).

21.1.4 MR-Guided Radiation Therapy (MRgRT)

MR-guided radiation therapy provides soft tissue visualization not only just before treatment but even during treatment. The first patient treatments with a commercial MRgRT system occurred in 2014 with the MRIdian system (ViewRay Inc.) which uses a low-field strength magnet at 0.35 T. In the original unit three ^{60}Cobalt treatment sources were used in order to avoid the electromechanical challenges of operating an MRI unit and a linear accelerator together. However, more recent versions of the system employ a 6 MV linear accelerator. The other MR-guided system commercially available circa 2019 is the Unity system (Elekta Inc.) which employs a 1.5 T magnet. Further information about these technologies can be found in Section 9.2.

There are some unique physical phenomena that occur with radiation therapy beams in the presence of a strong magnetic field (which is always present in these units) and that can affect the dosimetry. Most important is the fact that electrons do not travel in straight lines in the presence of the magnetic field. This can be especially pronounced in air cavities or low-density regions like the lung where electrons can bend back toward the source ("electron return effect") and create regions of enhanced dose at the chest wall. In the air cavity of an ion chamber the magnetic fields can disturb the equilibrium, and corrections for dosimetry may be required. Even in tissue-equivalent homogeneous regions the magnetic field can alter the shape of otherwise flat and symmetric beam profiles in some cases.

21.1.5 IGRT Use Scenarios

One useful way to categorize the types of IGRT is by the manner in which they are used, namely:

- Online IGRT: Imaging immediately prior to the treatment.
- Offline IGRT: Imaging at time of treatment but corrections applied prior to the next treatment session.
- Real-time IGRT: Imaging continuously throughout treatment. Intervene to stop or adjust the treatment as needed.

Table 21.1.1 illustrates the various IGRT technologies enabling these options. Further information and background can be found in the video.

21.1.6 Availability of IGRT, Practice Patterns, and Evidence of Effectiveness

Data are emerging about the availability of IGRT and practice patterns. In their survey of radiation therapy clinics in the United States, Nabivizadeh et al. (2016) found that 92%

TABLE 21.1.1

Capabilities of Some Major IGRT Technologies

	kV/MV Planar Imaging	CBCT/MV-CT	CyberKnife	MR-Guided RT	Ultrasound	Radiofrequency (EM) Beacon Transponders
Soft tissue visualization	✗	✓	✗	✓	✓	✗
Fiducials	✓	✓	✓	✓	✗	✓
Online IGRT	✓	✓	✓	✓	✓	✓
Offline IGRT	✗	✓	✗	✓	✗	✗
Real-time IGRT	✓	✗	✓	✓	✓	✓

of respondents reported having some form of volumetric IGRT capability, typically cone-beam CT. (Note that this 92% was in addition to portal imaging capabilities, which were available in essentially every clinic.)

Figure 21.1.6 shows that the use of IGRT varies greatly by disease site. For example, for brain cancer treatments, CBCT/MV-CT is most often used and performed on a weekly basis. CBCT/MV-CT is also the most often used form for head-and-neck cancer treatments but in that site it is used on a daily basis. Contrast this with breast cancer treatments where the IGRT is most often portal imaging performed on a weekly basis.

IGRT technology has evolved rapidly since the year 2000. Figure 21.1.7 shows survey data from Simpson et al. (2010) demonstrating the growth of various types of IGRT over time. Prior to 2000, IGRT of any form was not widely available, but then starting in the late 1990s the availability grew quickly, first with MV planar images via EPID devices and then with kV planar and volumetric imaging (mostly CBCT) starting especially after 2004. As noted above, volumetric IGRT is available in over 90% of clinics in the United States, circa 2019.

What evidence is there of the clinical benefits of IGRT? The review article of Bujold et al. (2012) examined this question and concluded that although there was little direct evidence, there is a wealth of data to support the benefits of IGRT in terms of improved control rates, lower toxicities, and increased treatment options for patients. However, data continue to emerge regarding the clinical value of IGRT in various disease sites. For more information see the video.

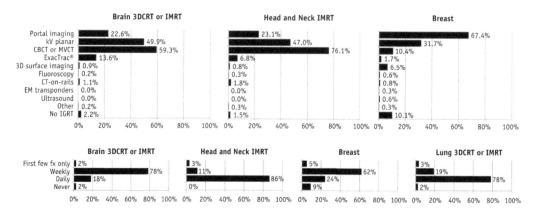

FIGURE 21.1.6

Types of IGRT used (top) and frequency of use (bottom) by disease site from 2014 survey of clinics in the United States. (From Nabivizadeh et al., 2016.)

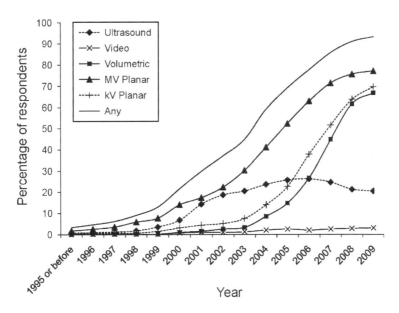

FIGURE 21.1.7
Availability of IGRT technologies over time from surveys of US clinics. (From Simpson et al., 2010.)

21.1.7 Quality Assurance (QA) of IGRT and Imaging Systems

There is a wealth of information on QA of imaging systems (Table 21.1.2). Table 21.1.3 lists some of the key QA requirements in the context of IGRT.

A further description of these QA tests can be found in the video. Some important example QA tests are shown in the figure below. These include:

- Planar imaging QA (either MV or kV): Figure 21.1.8.
- Cone-beam CT QA using the "CATPhan" (i.e. "Customer Acceptance Test Phantom," so named by the vendor since it is the test device used during acceptance testing of the IGRT system): Figure 21.1.9.
- Measurement of image-based positioning and registration, performed daily: Figure 21.1.10.

21.2 Motion Management

21.2.1 Inter- and Intra-Fraction Motion

Patient motion and variability can be classified on two time scales:

- *Inter-fraction motion.* Variability between treatment fractions. Example: Shift of patient relative to isocenter, roll of patient, neck flex, bladder filling.
- *Intra-fraction motion.* Motion during a treatment fraction. Examples: Respiratory motion, peristalsis, coughing, muscle flex.

TABLE 21.1.2

Quality Assurance Guideline Documents for IGRT Systems

Year	Report	QA Topic
2014	MPPG 2 (version a)	X-ray-based IGRT systems
2012	TG179	CT-based IGRT systems
2009	TG142	Linac QA
2011	TG135	CyberKnife
2010	TG148	TomoTherapy
2017	TG132	Image registration
2003	TG66	CT simulator QA
Other Documents		
2014	ACR-AAPM	ACR-AAPM technical standards for medical physics: IGRT (updated regularly)
2014	ACR-ASTRO	ACR-ASTRO practice parameter for IGRT (updated regularly)
2013	ASTRO	Safety considerations for IGRT
2018	TG180	Dose from IGRT (update of TG75)
2009	TG104	kV IGRT
2005	CAPCA	Canadian Quality Control Standards for EPID
2012	TG147	Non-radiographic localization systems
2011	TG154	Ultrasound for external beam treatment of prostate

TABLE 21.1.3

Key Quality Assurance Tests for IGRT Systems: Not Included Here Are Recommendations for the CyberKnife (TG135), TomoTherapy (TG148), and Other Specialized Systems

Test	kV/MV Planar			CBCT		
	MPPG 2.a	TG179	TG142	MPPG 2.a	TG179	TG142
Positioning and registration	D		D	D	D	D
Imaging vs. treatment isocenter	D		D	D	D	D
Geometric dimensions	A		M	A	A	M
Spatial resolution	A		M	A	A	M
Contrast	A		M	A	A	M
Noise			M	A	A	M
Uniformity			M	A	A	M
Dose	A		A	A	A	A

One way of reducing inter- and also intra-fraction motion is patient "immobilization." Figure 21.2.1 demonstrates some of the immobilization devices in common use, including a thermoplastic mask which is heated and then molded to the patient (A), a vacuum-sealed plastic bag which forms to the body contour of the patient and moldable head holder which controls the position and flexion of the head (B), and leg rolls and hand holders to control the position of the arms, shoulders, and legs (C). The goal is not to keep the patient completely immobile but to make a setup that is comfortable for the patient and allows a reproducible positioning from day to day. Regardless of whether IGRT is used or not, it is important to have good patient immobilization. Even with the use of IGRT not all variability in patient positioning can be controlled or eliminated. IGRT may be used to align the C spine for example, but if the neck flexes then the cranium and the T spine may not be well-aligned even if the C spine is aligned. For an illustration of this see the video.

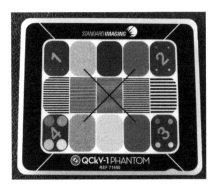

FIGURE 21.1.8
Planar imaging showing the phantom (left) and the resulting image (right) in this case of kV imaging. A similar (metallic) phantom is available for MV imaging. This tests both resolution (number of visible line pairs per millimeter) and contrast (blocks on outer edges).

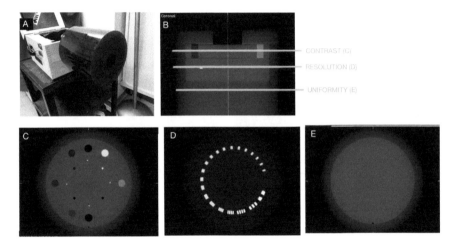

FIGURE 21.1.9
(A) Phantom for CBCT image QA. Different sections are present in the phantom as shown on a coronal image (B), including a section for measuring contrast (C), spatial resolution (D), and uniformity across the image (E).

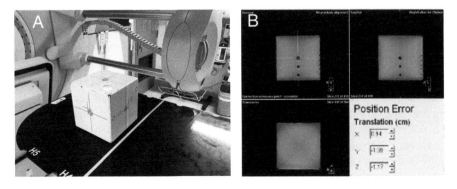

FIGURE 21.1.10
Phantom for CBCT positioning and registration QA tests. (A) A phantom with embedded objects (air holes) is aligned with a known offset from isocenter. (B) Cone-beam CT registration is performed, and the translational shifts are compared to the known shifts.

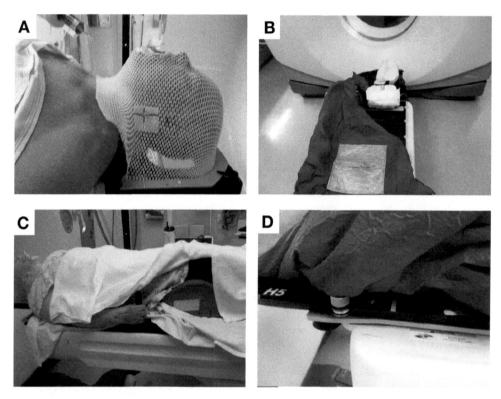

FIGURE 21.2.1
Immobilization devices for patient treatment. (A) Thermoplastic mask, (B) vacuum-sealed plastic bag and mold-able head holder, (C) leg roll and hand holders, and (D) table indexing system.

It is important to note that immobilization devices also have an effect on the dosimetry of the beam (e.g. attenuation and bolusing). This is considered in more depth in AAPM TG176 (Olch et al. 2014).

Immobilization devices often support a table indexing system as shown in Figure 21.2.1D which allows the immobilization device to be placed on the table in exactly the same way for each treatment. The table coordinates can then be programmed into the delivery control system to ensure that the table is positioned at the same location for each treatment, which serves as a safety feature to ensure the treatment of the correct location.

21.2.2 Respiratory Motion

Respiratory motion is an important component of intra-fraction motion and affects the lung, liver, and other organs in the abdomen. It must be accounted for in the process of planning and delivery of radiation therapy in order to ensure local control. A key reference for this is AAPM Task Group 76 published by Keall et al. (2006) with an update pending.

Key goals for respiratory motion in radiation therapy are to: (1) measure the motion and (2) manage it during treatment, either through some active system or by accounting for it in treatment planning. Respiratory motion can be measured with fluoroscopy. Though this may provide some visualization of a tumor in the lung, the soft tissue visualization is limited and, unless there is a dual-camera system in use, the images are acquired only from a single angle.

21.2.3 4DCT

In the earlier 2000s a technique was developed which allowed for dynamic visualization of motion using CT scans, called "4DCT." This technique was first described by Ford et al. (2003). For the first 4DCT movie see the video.

Historically, 4DCT requires some form of input signal of respiration. One such system is an infrared marker block placed on the abdomen of the patient which is viewed by a camera attached to the end of the table (Figure 21.2.2). As this block moves up and down a breathing trace is generated (Figure 21.2.2 C). Other systems are available for measuring the respiratory signal such as a belt with an air bellows and a pressure transducer to measure motion, and more recently "deviceless" 4DCT systems are available which track respiration via signatures in the images themselves.

4DCT reconstruction is accomplished by correlating the breathing signal with the acquisition of the CT. In this way the CT slices can be assigned to the breathing phase to which they belong. A further description of this is found in the video.

4DCT is subject to possible artifacts, the main one being irregular breathing of the patient. Note that at any given slice location only one respiratory cycle is acquired. (Even though 4DCTs are often presented as continuous movie loops they are acquired over only one breath of the patient, not multiple breaths.) Therefore, if the patient breathes irregularly this irregularity will show up in the reconstructed scan. An example is a patient taking a deep breath at one point in the scan which results in an offset of the chest wall at

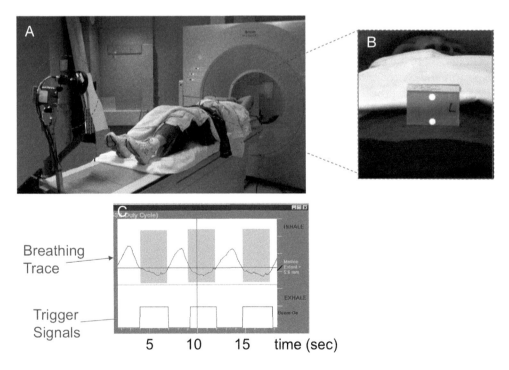

FIGURE 21.2.2
System for measuring respiratory motion consisting of a camera (A) which views and a block with infrared markers placed on the abdomen of the patient (B). Respiratory motion creates a breathing trace (C). This signal is used to reconstruct 4DCTs or to provide trigger signals to the linac beam if a gated treatment is being used.

one particular slice and a "ghost" diaphragm, i.e. a small section of diaphragm appearing superior to and disconnected from the main diaphragm. For an example see the video.

4DCTs have been acquired from many thousands of patients and patterns have emerged. Typically, the motion of tumors in the lung is largest in the superior–inferior direction, followed by the anterior–posterior direction, and smallest in the right-left direction. Motion can be up to 2 cm or even more in some patients. The motion is not in one plane but rather traces out a circular pattern. These are generalizations, however, and some patients do not fit this pattern. For an example see the video. It is important therefore to measure the motion in each patient.

Finally, substantial respiratory motion has been observed not only in the lung but also in the abdomen and even the pelvis. The liver moves substantially with respiration as do the kidneys and other organs as well. Therefore, although 4DCT is most useful for lung tumors, it also has applications beyond the lung.

21.2.4 Respiration and Margins

The simplest way of managing respiration is to define appropriate margins to ensure that the tumor will be adequate during treatment. Recall the definitions of volumes from ICRU62 and other reports (see Section 13.1). Based on the above discussion, we know that the GTV may move with respiration. Therefore, one simple way to account for this motion is to input all the phases of the 4DCT scan into the contouring or treatment planning software and define an internal GTV (iGTV) which includes all of the motion from the 4DCT. This is a conservative and simple approach to ensure that the entire GTV incorporates all the motion.

One important aspect to note about respiration is that the patient typically spends more time at an end-expiration phase vs. end-inspiration (see the breathing trace in Figure 21.2.2). On average, then, the position of the tumor is typically at a slightly more superior location than simply the center of all breathing phases.

There are other methods to handle GTV definition aside from inputting all the phases of the 4DCT scan into the contouring or treatment planning software. One can define a Maximum Intensity Projection (MIP) image (largest Hounsfield unit observed in each voxel over all respiratory phases) or average scan (average HU in each voxel). Another approach, when 4DCT is not available, is a "slow scan" whereby a normal CT is acquired but with a tube rotation speed that is slow enough to capture an entire respiratory phase. The main disadvantage of the averaging scans is that the border of a tumor is blurred relative to faster scans or 4DCT.

21.2.5 Respiratory Gating

Another technique to manage and reduce respiratory motion during treatment is respiratory gating. In this technique the linac beam is turned on and off in cycle with the breathing trace as shown in Figure 21.2.2C. The breathing trace is monitoring during treatment with a device such as the marker block. When the patient respiration enters some predefined window a trigger signal is sent to the linac to turn it on. If gating is employed it is important to use a CT scan in treatment planning that represents the position in the gating window during treatment.

An advantage of respiratory gating vs. simple margin expansion is that smaller margins might be used since the tumor moves less within the gating window. Challenges include loss in efficiency and longer treatment time (the beam is not always on); potential

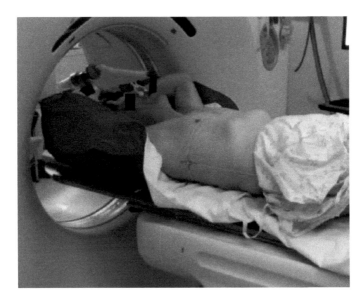

FIGURE 21.2.3
Breath-hold using the Active Breathing Coordinator system (ABC, Elekta Inc.). Airflow is measured with a spirometry system through a hose. When a preset volume of air is achieved, the beam or imaging system is turned on.

irregularities in the breathing can result in a mispositioning of the tumor. Note also that gating relies on a respiratory signal that is not a direct measurement of the tumor position but rather a surrogate for that position which can introduce uncertainties.

21.2.6 Breath-Hold Treatment

A final way to reduce motion during treatment is to have the patient hold his or her breath during treatment. One challenge with this approach is that the breath-hold air volume must be the same on every cycle and the same from CT simulation to treatment. Devices are available to assist with this including the Active Breathing Coordinator (ABC) shown in Figure 21.2.3. Here, the airflow is measured with a spirometer system connected to a flexible tube in the patient's mouth. When the volume reaches a certain pre-defined value a valve closes and the patient's breathing is arrested. A signal is sent to the treatment device to turn the beam on. This system can be used to treat an inhale (or potentially deep-inspiration breath hold if the patient is so coached) or can be used at exhale.

There are challenges with this system. Some patients may have compromised lung function and find it difficult to hold their breath for long periods. There is also a loss in efficiency. The beam remains off as the patient is coached through breathing maneuvers and into a breath-hold and only then is the beam or imaging system turned on. Efficiencies can be well below 50% and even less than 10%.

21.2.7 Compression

A final system for controlled respiratory motion is compression, typically applied to the abdomen of a patient as shown in Figure 21.2.4. This compression can limit the respiratory motion of a tumor and other anatomy or at least reminds the patient to breath in a shallow and regular manner. Other compression systems are available as well.

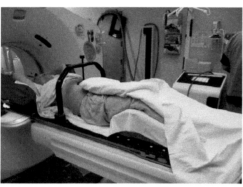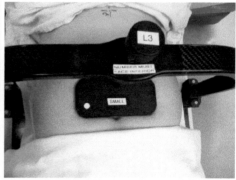

FIGURE 21.2.4
Abdominal compression. A plate is pressed against the patient's abdomen.

Further Reading

Bujold, A., T. Craig, D. Jaffray, and L.A. Dawson. 2012. Image-guided radiotherapy: Has it influenced patient outcomes? *Semin Radiat Oncol* 22(1):50–61. doi:10.1016/j.semradonc.2011.09.001.

Dieterich, S., et al. 2011. Report of AAPM TG 135: Quality assurance for robotic radiosurgery. *Med Phys* 38(6):2914–2936.

Ford, E.C., G.S. Mageras, E. Yorke and C.C. Ling. 2003. Respiration-correlated spiral CT: A method of measuring respiratory-induced anatomic motion for radiation treatment planning. *Med Phys* 30(1):88–97.

Keall, P., et al. 2006. The management of respiratory motion in radiation oncology report of AAPM Task Group 76. *Med Phys* 33(10):3874–3900.

Langen, K.M., et al. 2010. Quality assurance for helical tomotherapy: Report of the AAPM Task Group 148. *Med Phys* 37(9):4817–4853.

Nabavizadeh, N., et al. 2016. Image guided radiation therapy (IGRT) Practice patterns and IGRT's impact on workflow and treatment planning: Results from a National Survey of American Society for Radiation Oncology members. *Int J Radiat Oncol Biol Phys* 94(4):850–857. doi:10.1016/j.ijrobp.2015.09.035.

Olch, A.J., et al. 2014. Dosimetric effects caused by couch tops and immobilization devices: Report of AAPM Task Group 176. *Med Phys* 41(6):061501. doi:10.1118/1.4876299.

Simpson, D.R., et al. 2010. A survey of image-guided radiation therapy use in the United States. *Cancer* 116(16):3953–3960. doi:10.1002/cncr.25129.

Willoughby, T., et al. 2012. Quality assurance for nonradiographic radiotherapy localization and positioning systems: Report of Task Group 147. *Med Phys* 39(4):1728–1747.

Yin, F.F., et al. 2009. The role of in-room kV X-ray imaging for patient setup and target localization. AAPM Task Group 104.

Chapter 21 Problem Sets

*Note: * indicates harder problems.*

1. Which of the following are advantages of kV planar imaging over CBCT for use in IGRT? (Select all that apply.)
   a. Improved visualization of soft tissue
   b. Lower dose
   c. Potential for real-time tracking
   d. Ability to support adaptive replanning

2. How far does the edge of a target region 10 cm from isocenter move for a rotation of 3 degrees?
   a. 3.0 mm
   b. 5.2 mm
   c. 7.5 mm
   d. 14.2 mm

3. What is the typical dose from a CBCT scan of the pelvis?
   a. 0.02 cGy
   b. 0.2 cGy
   c. 2.0 cGy
   d. 20.0 cGy

4. List two advantages and two challenges of MR-guided radiation therapy compared to CBCT for lung cancer treatment.

5. How does the CT number uniformity compare for a 20 cm diameter phantom vs. a 40 cm diameter phantom in a QA test of CBCT QA?
   a. More uniform for 40 cm
   b. Less uniform for 40 cm
   c. Same. CT number is independent of size

6. How does noise vary with mA in a QA test of CBCT?
   a. Noise increases with mA
   b. Noise decreases with mA
   c. Same. Noise is independent of mA in CBCT

7. Consider a GTV modeled as a cylinder with a base of 3 cm diameter and a height of 3 cm. What is the volume of the GTV without motion (no sup-inf expansion) and with motion (expand height by a total of 0.5 cm)?

	No Motion	0.5 cm Motion
A	6.8	7.9
B	21.2	24.7
C	27.0	31.5
D	84.8	98.9

8. List three advantages and three challenges of breath-hold treatment for radiation therapy of left-sided lung cancer.

9. In 4DCT with spiral acquisition of a patient with a 6 sec breathing period how much does the table move in one breathing cycle if the following parameters are used: pitch $p = 0.5$, tube rotation speed $T_{rot} = 1.0$ sec, slice thickness $S = 3$ mm? How many independent slices can be reconstructed?

   a. 3 mm

   b. 4.5 mm

   c. 9 mm

   d. 18 mm

10. How much does the diaphragm move if the lungs are inflated from 2 L to 4 L? Assume the lungs are cylinders with a height twice that of the diameters.

   a. 3.5 cm

   b. 4.4 cm

   c. 5.6 cm

   d. 6.8 cm

22

Stereotactic Treatments

22.1 Stereotactic Radiosurgery (SRS)

Stereotactic Radiosurgery (SRS) is typically a single treatment of a high dose to a target in the brain. It is used to treat a variety of conditions including metastatic cancer lesions, acoustic neuromas, meningiomas, arterial venous malformations (AVMs), trigeminal neuralgia, and (less commonly) functional disorders such as obsessive-compulsive disorder or essential tremors. It is beyond the scope of this text to discuss the clinical aspects of these in any detail, but they span the range from highly aggressive cancers to slow-growing tumors and benign conditions.

Some technical characteristics of SRS are the following:

- Single fraction treatment
- >5 Gy per fraction: Often in the range of 12–20 Gy, but sometimes higher
- Target diameter <3.5 cm (brain)
- Nominal delivery accuracy <1 mm
- No PTV margins (CTV in some cases)

The word "stereotactic" derives from the Greek "stereo" (having to do with space) and "taxis" (arrangement or placement) so it can be translated as "3D placement." Originally this meant the placement of a biopsy needle or a probe at a location in the brain. The stereotactic system allows for highly accurate placement. The stereotactic coordinate system is defined by the frame which is affixed to the cranial bones of the patient. These frames and corresponding coordinate systems were developed in the 1970s for radiosurgery applications, an early one being the Brown–Roberts–Wells (BRW) system and later the Cosman–Roberts–Wells (CRW) system. These are still in use today.

Stereotactic systems can also be used to delivery radiosurgery, i.e. to place radiation beams at a precise location in the brain instead of a needle or probe. The example system considered in detail here is the GammaKnife® (Elekta Inc.) developed by the Swedish neurosurgeon Lars Leksell in the 1960s, but the general principles of radiosurgery apply across other vendor systems from standard C-arm linear accelerators to other devices.

The SRS frame is shown in Figure 22.1.1. It affixes to the patient's head with pins. The patient receives a local anesthetic at the pin site, and the pins do not push through the bone but rather clamp into it to keep the frame rigidly in place for the duration of the procedures which can be hours long. Although seemingly very invasive, this is usually well-tolerated by patients.

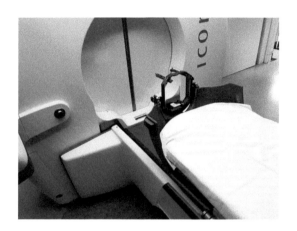
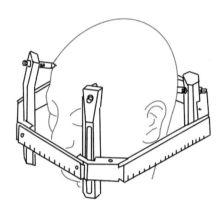

FIGURE 22.1.1
The SRS Leksell G-frame used in the GammaKnife unit. The frame is affixed to the patient's head with pins. The patient's head is locked into the holder on the tabletop (left panel).

22.1.1 SRS Treatment Delivery

The GammaKnife device employs 192 high-activity ^{60}Co sources arranged in rings concentrically around the patient (Figure 22.1.2A). Each source is collimated by long tungsten tubes all oriented toward a common isocenter (red circle inside the brain). Three collimation sizes are available: 4, 8, and 16 mm (red, blue, and green in the figure). The patient couch position determines the location of the isocenter. The SRS coordinate system is determined by the frame, and the frame, in turn, is locked into the table. Therefore, the table position can be used to target the intended location in SRS coordinate space.

Similar systems operate in linac-based SRS except that the table position cannot be controlled with the necessary accuracy so typically the table is fixed and the SRS frame itself is moved with specialized devices. In this way, sub-millimeter accuracy is possible. Also in

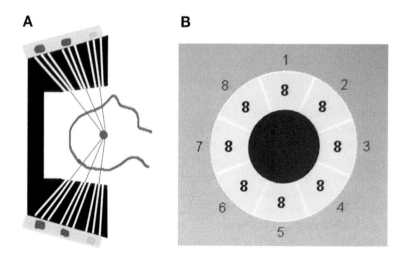

FIGURE 22.1.2
Treatment geometry for the GammaKnife unit. (A) 192 ^{60}Co sources directed to a common isocenter (red circle). (B) Each ring is divided into eight sectors.

linac-based SRS a tertiary collimator system is usually affixed to the head of the linac. This is often a "cone," a high-density metal with a conical hole through the center. This places the collimation aperture close to the patient and thereby improves penumbra. The gantry can be rotated to trace out arcs around the patient in various planes.

22.1.2 SRS Planning and Dose Distributions

Treatment plans with SRS differ from conventional radiation therapy, mainly in the fact that they are allowed to be much more heterogeneous. This can be appreciated in Figure 22.1.3A where the dose is not even throughout the target but is very hot in the center and cooler at the periphery. This is seen in the dose profile (Figure 22.1.3B) and can also be appreciated in the DVH for this plan (Figure 22.1.3B) which shows a high-dose hot spot of 35 Gy within the target.

Heterogeneous dose distributions are a common feature of both SRS and SBRT, and the reason for this is that they provide a sharp dose gradient at the periphery of the target (note how quickly the dose falls off at the periphery of the target in Figure 22.1.4B). This sharp dose falloff means that nearby critical structures can be spared. It also means there are extra demands on localization since if the target moves even a small amount it will be in a low-dose region.

One variable that affects the dose falloff is the isodose line for which the prescription is written. In Figure 22.1.4A the plan is designed to deliver 20 Gy to the 57% isodose line at the periphery of the target (prescription isodose lines of approximately 50% are common in planning with GammaKnife). Other choices are possible, however. For example, the plan could be made to deliver 20 Gy to the 70% isodose line at the periphery of the target. This would result in a more homogeneous dose distribution within the target (a hot spot of 20 Gy/0.70 = 28.6 Gy instead of 20 Gy/0.57 = 35.1 Gy). However, such a plan would not have

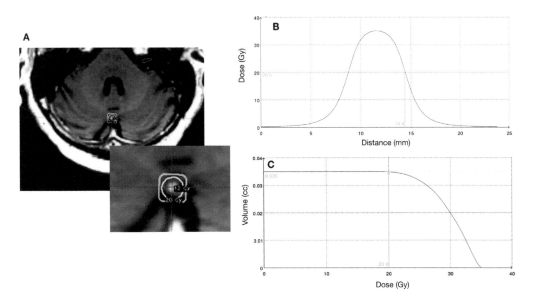

FIGURE 22.1.3

SRS plan for a single metastatic lesion in the mid-cerebellum. (A) Isodose lines for the plan prescribed at 20 Gy to the 57% isodose line using a single isocenter treatment. (B) Dose line profile through the lesion. (C) Dose volume histogram.

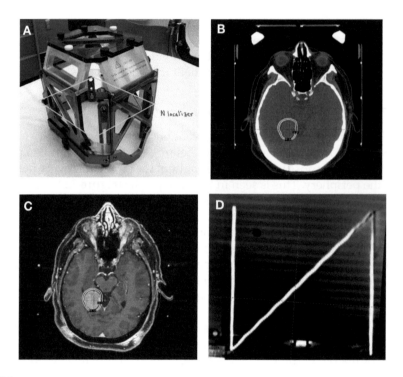

FIGURE 22.1.4
"N"-localizer system. (A) Fiducial box affixed to the frame. (B) CT image showing markers. (C) MR image also showing the fiducials. (D) The N localizer shown in a sagittal cut through the CT scan.

as steep a dose gradient and would be less sparing of nearby normal tissue. These same patterns of inhomogeneous dose distributions and prescription dose dependency are also present in SBRT plans (Section 22.2).

Figure 22.1.3A showed a plan with a single isocenter treatment (i.e. a single "shot" in GammaKnife parlance). However, because brain lesions are not typically perfect spheres more sophisticated planning approaches are often required. In the system considered here this is accomplished by using multiple treatment isocenters each with a different weighting to achieve the desired dose distribution.

A typical workflow for a patient treatment proceeds as follows:

- Patient arrives early, and the frame is placed.
- Images are acquired with the frame on (MR and/or CT and/or angiography).
- Treatment planning is performed to create a plan with the desired dose distribution.
- The plan is sent to the delivery control system. This includes the locations of the isocenters and dose/time for each.
- Quality checks are performed.
- Treatment begins.

The actual time for treatment on GammaKnife ranges from roughly 15 min to 4 hours or more, depending on the complexity of the plan, the prescribed dose, and the age of the ^{60}Co sources. Linac-based SRS treatment times are typically much shorter. Overall the whole process can range from approximately 2 to 12 or more hours.

22.1.3 Stereotactic "N"-Localizer System

With this basic understanding of SRS treatment delivery and planning, we now turn to the stereotactic coordinate system to better understand it. In order to procced with treatment planning the SRS coordinate system needs to be defined relative to anatomical locations in the patient. This means identifying the stereotactic coordinates on the images themselves (CT, MR, or sometimes radiographs from angiography).

The key to accomplishing this is the "N"-localizer box method (Figure 22.1.4), developed in 1978 by the neurosurgeon Russell Brown. This is an important innovation which allows for the use of images in stereotactic cases. The box attaches rigidly to the frame which is on the patient's head. The box has fiducial markers embedded in the shape of the letter "N" (Figure 22.1.4A, red). There is one N-fiducial on each side of the patient. A CT or MRI image acquired with the frame on will yield a slice through the fiducial box (shown in green). Within that slice the N-fiducials are visible as three bright spots (Figures 22.1.4B and C). By measuring the distance between the fiducials on the image, the location and orientation of the image slice plane can be determined in stereotactic space. For a further illustration of this see the video.

One special consideration is localization for AVM cases where angiography images are used. These are not volumetric so the N-localizer method is not possible since there are no image slices. Instead a different localizer is affixed to the frame with four markers in it. Projection images are acquired, and the location of the markers on the panels can be visualized on these images.

The above discussion assumes that a frame is used for the SRS treatment. However, "frameless" SRS is becoming more common. In frameless SRS, the patient is localized during treatment not with a frame but with an imaging system. This system can be either orthogonal planar images (e.g. CyberKnife or ExacTrac sytems) or a cone-beam CT-based system. The key requirement here is that the patient movement during treatment be minimized and/or some near-real-time tracking system be employed to compensate for movement.

22.1.4 Fractionated Treatment

Treatments of targets can be fractionated, that is, delivered over more than one fraction. The treatments are still planned and delivered at a high dose per fraction over a short course (e.g. 8 Gy $\times$ 3 fractions) but are not delivered in a single fraction. As such this would not be called "SRS" (which is reserved for single-fraction treatments) but rather Fractionated Stereotactic Radiation Therapy (FSRT) or an equivalent term.

Because of the need for repeat treatments it is typically not practical to employ a frame each time. These treatments are therefore usually "frameless" and rely on an IGRT system to accurately localize the lesion for treatment (see Chapter 21). Some systems include cone-beam CT systems on a C-arm linear accelerator and other devices, orthogonal planar imaging systems (e.g. CyberKnife or ExacTrac systems), and the GammaKnife ICON® which uses CBCT and a mask system. For a more detailed explanation and illustrations see the video.

22.1.5 Quality Assurance (QA) for SRS

A valuable resource for the QA of SRS techniques and technologies is AAPM Medical Physics Practice Guideline 9a (Halvorsen et al. 2017). This document outlines key QA tests, which include the following:

- Verification of radiation isocenter: Daily
- Measurement of imaging isocenter: Daily
- End-to-end tests: Annually

There are various ways to accomplish these QA tests, and there is a good deal of overlap with the QA tests described in previous chapters. However, the requirements for SRS are more stringent due to the very high dose per fraction and the fact that small or no margins are used.

An example of the isocenter and imaging verification tests is shown in Figure 22.1.5 for the GammaKnife. The QA device employs a diode mounted in the frame holder (Figure 22.1.5B) which is scanned across the beam to determine the location of the radiation isocenter. The device also has metal BB fiducials embedded in it (Figure 22.1.5B). A cone-beam CT scan is performed on the device, and the location of the fiducials is used to determine the location of the imaging system in SRS coordinate space.

Linear accelerator systems use different methods to accomplish these localization tests. One important QA procedure is the Winston–Lutz test, pioneered by Wendell Lutz and colleagues at the Joint Center in the 1980s. In this method a metallic BB is placed at the center of the SRS coordinate space, traditionally through the use of lasers but more recently with the use of the imaging system. Portal films are then acquired of this BB using a small field with the treatment beam. The offset of the BB with respect to the field edges then determines the location of the radiation isocenter (center of the beam) with respect to the intended treatment location (BB position).

As with any radiation device, output verifications must also be performed on a regular basis (daily, monthly, and annually). These tests are described in more detail in Section 18.2. There are extra requirements for SRS treatments related to the fact that the fields used are very small; one of the main requirements is the use of appropriately sized small detectors (Figure 22.1.6) and the application of the relevant correction factors. See Section 16.1.

Finally, the accuracy of SRS treatments relies on the geometric integrity of the imaging system. In other words, any distortion in the imaging system will translate into

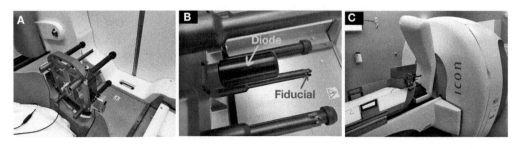

FIGURE 22.1.5
Daily test device used for SRS QA in the GammaKnife system. (A) The test device mounted in the frame holder. (B) A diode in the device (inside the black plastic housing). (C) Cone-beam CT determines the imaging isocenter via embedded fiducial markers.

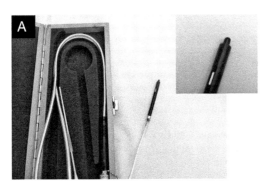
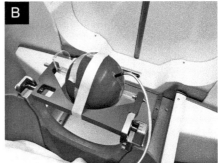

FIGURE 22.1.6
Output verification in SRS. (A) Small-volume detectors are a key requirement. Here the A16 microchamber is shown (volume 0.007 cm^3). (B) Chamber in the phantom. Note that the phantom is plastic and not water equivalent.

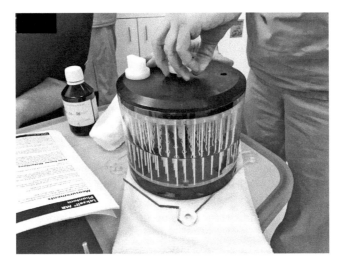

FIGURE 22.1.7
Device for QA of the geometry of the MR imaging system. The location of embedded plastic rods can be visualized on MR images.

a mislocalization of the target. QA of imaging systems, therefore, is a key requirement in SRS services. Figure 22.1.7 shows an example of one phantom device used for QA of the MRI imaging system. The device is filled with liquid (e.g. a CuSO$_4$ solution) and has plastic inserts which have a long T1 relaxation rate and therefore can be visualized as voids in the MR images. This device is mounted into a head frame with the fiducial N-localization box attached, and images are acquired as for patient treatment. Similar devices are available for CT QA.

22.2 Stereotactic Body Radiation Therapy (SBRT)

Stereotactic Body Radiation Therapy (SBRT) refers to high dose-per-fraction treatment outside the cranium. Although the dose fractionation schemas vary widely, a common feature

TABLE 22.2.1

Sites Where SBRT Is Commonly Used and Example Trials, Dose Fractionation Schemas, and Margin Criteria

Site	Example Trial	Criteria	Example Dose Fractionation	Margins
Lung	RTOG-0236 RTOG-0915	Inoperable Stage I/II Non-small cell lung cancer; restricted to lesions in the peripheral lung	20 Gy × 3 34 Gy × 1 12 Gy × 4	ITV + 5 mm (if 4DCT)
Liver	RTOG-1112	Hepatocellular carcinoma (HCC)	Up to 10 Gy × 5	4–20 mm permitted; goal of <10 mm
Spine	RTOG-0631	Metastatic lesions in the spine		

is >5 Gy per fraction (and often >10 Gy/fraction) delivered in very few fractions, sometimes only one. SBRT is also sometimes referred to in the literature as Stereotactic Ablative Body Radiotherapy (SABR).

22.2.1 SBRT Sites and Dose Protocols

SBRT is currently used most commonly in lung and liver cancers as well as metastases in the spine. According to a 2017 survey of North American practices collected by AAPM Task Group 275, SBRT is offered by 81% of practices. The clinical rational for the use of SBRT varies by site. The spine is one of the most common sites of metastases, and disease in the spine can cause severe pain and compromise neurological function if it extends into the epidural space. SBRT is effective at alleviating pain and preventing or reversing neurological deficits. SBRT is also used in other sites and disease conditions including prostate (ultra-hypofractionation), breast, head and neck, and oligometastases in various sites.

Table 22.2.1 illustrates key clinical features of SBRT with reference to cooperative group trials. SBRT typically aims to treat well-circumscribed tumors that are small in volume (usually <5–7 cm diameter in the thorax or abdomen). SBRT also employs relatively small margins for the PTV which allows for more OAR sparing.

22.2.2 SBRT Treatment Planning

Compared to plans with standard fractionation, SBRT plans are, by design, much more heterogeneous. This is illustrated in the example lung SBRT plan in Figure 22.2.1. The dose profile along the line through the center of the PTV shows a distribution which is peaked at the center (Figure 22.2.1B). Here, the prescription is set to 18 Gy × 3 (i.e. 54 Gy total) to the isodose line which is 63% of the maximum dose. That means there is a hot spot inside the tumor of 85.7 Gy (i.e. 54 Gy/0.63 = 85.7 Gy). This dose heterogeneity can be observed in the DVH as well (Figure 22.2.1C). The DVH for the planning target volume (PTV) is not a sharp falloff step-function as would be expected for a target that had a relatively uniform dose in it. Rather it shows a gradual falloff with increasing dose, meaning that some regions of the PTV receive low dose while others receive high dose.

The dose heterogeneity in SBRT plans is purposeful and allows for a steep dose gradient at the periphery of the tumor and an improved sparing of normal tissue as with SRS (Section 22.1.2). One of the key properties of SBRT plans is that the prescription dose is set to a relatively low isodose value (Figure 22.2.1). Compared to a standard fractionation plan

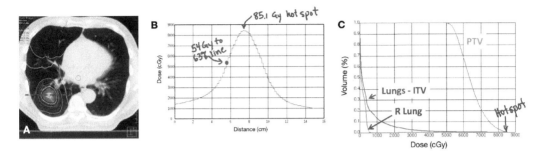

FIGURE 22.2.1
Example lung SBRT plan. (A) Isodose distribution. (B) Dose profile along the line shown in (A). (C) DVHs for various regions of interest.

where the dose prescription might be to the 98% or 99% isodose line, an SBRT plan might have a prescription set at 70% for example. RTOG-0813, for example, specifies that the dose should be prescribed to the 60–90% isodose line. If the periphery of the tumor is covered by a low isodose line, then, by definition, there will be some spot at a higher dose. This hot spot should lie within the PTV.

There are additional dosimetric goals for SBRT plans. One is to limit cold spots within the PTV, and there are dose metrics that can be specified to ensure this. RTOG-0813, for example, specifies PTV V100% > 95% and PTV V90% > 99%. This ensures adequate dosimetric coverage of the PTV.

Second, one has to consider not only the PTV but surrounding normal tissue as well. A key reference for this is the series of HyTEC papers from AAPM (see reading list below) which discuss the dose tolerances for normal tissues under hypofractionated radiation delivery schemas like SBRT. In SBRT planning it would be possible to generate a plan that "overcovers" the PTV, and although this might achieve the dosimetric goals for the PTV, it would deliver a large dose to surrounding normal tissue that is not acceptable.

Several dose metrics are in use which can ensure low dose to normal tissues. One is the conformity index (CI) defined as

$$\frac{\text{Volume of the prescription isodose line}}{\text{Volume of the PTV}}$$

There are also metrics for low-dose spillage such as R50% which is defined as

$$\frac{\text{Volume of 50\% isodose line}}{\text{Volume of the PTV}}$$

and $D_{2cm}(Gy)$ which is the maximum dose at 2 cm from the PTV. Finally, a metric commonly used in lung SBRT planning is V_{20Gy}, which is a meaningful dosimetric endpoint for predicting lung toxicity (see HyTEC Report for lung SBRT).

Though the limits for these dose metrics vary with the protocol, some commonly used limits are CI < 1.2 and lung $V_{20\%}$ < 10 Gy, though some protocols allow for CI up to 1.5 and lung $V_{20\%}$ up to 15 Gy before they are considered a major violation. The appropriate limits for $R_{50\%}$ and $D_{2cm}(Gy)$ are somewhat harder to define since what is achievable in a plan depends on the PTV size. For smaller PTV volumes it is possible to achieve smaller values for $R_{50\%}$ and $D_{2cm}(Gy)$, and so the limits are typically dependent on the PTV size. More details can be found in protocols (e.g. RTOG-0236) and in the video.

22.2.3 Recommendations for Safe and Effective SBRT

Several reports provide technical and practice recommendations to ensure safe and effective SBRT treatments. These include AAPM Task Group #101 (Benedict et al. 2010), and the Safety White Paper on SBRT from ASTRO (Solberg et al. 2012). Key recommendations include the following:

- Commissioning
  - Special care must be taken in the detectors used in commissioning and the associated correction factors.
  - 6 MV beams should be used.
  - MLC leaf widths of ≤5 mm should be used. There is little advantage to leaves narrower than 5 mm.
  - End-to-end tests should be conducted.
- Treatment planning
  - Motion assessment should be used for thoracic and abdominal sites.
  - Dose calculation grid size <2 mm should be used.
- Imaging and delivery
  - IGRT should be used for alignment.
  - Do not use a body frame alone.
  - A radiation oncologist should approve alignment images prior to each treatment.
  - A qualified medical physicist should be present at the first fraction and available for the remaining fractions.

Further Reading

AAPM 2014 Summer School on SRS/SBRT. See virtual library on AAPM and meeting program. https://www.aapm.org/meetings/2014SS/default.asp.

Benedict, S.H., et al. 2010. Stereotactic body radiation therapy: The report of AAPM Task Group 101. *Med Phys* 37(8):4078–4101.

Dieterich, S., E. Ford, D. Pavord and J. Zeng. 2016. *Practical Radiation Oncology Physics*. Chapter 17. Philadelphia, PA: Elsevier.

Halvorsen, P.H., et al. 2017. AAPM-RSS medical physics practice guideline 9.a. for SRS-SBRT. *J Appl Clin Med Phys* 18(5):10–21.

Solberg, T., et al. 2012. Quality and safety considerations in stereotactic radiosurgery and stereotactic body radiation therapy: Executive summary. *Pract Radiat Oncol* 2(1):2–9. doi:10.1016/j.prro.2011.06.014.

HyTEC Organ-Specific Papers

Optic NTCP: https://www.redjournal.org/article/S0360-3016(18)30125-1/pdf
H&N TCP: https://www.redjournal.org/article/S0360-3016(18)30107-X/pdf

Lung NTCP: https://www.redjournal.org/article/S0360-3016(18)34014-8/pdf
Liver TCP: https://www.redjournal.org/article/S0360-3016(17)34525-X/pdf
Liver NTCP: https://www.redjournal.org/article/S0360-3016(17)34527-3/pdf

Chapter 22 Problem Sets

*Note: * indicates harder problems.*

1. List three advantages and three disadvantages of frame-based treatment for cranial SRS on a linear accelerator.

2. What is an approximate dose gradient in a linac-based SBRT plan at its steepest point?
   a. 10 Gy/mm
   b. 1 Gy/mm
   c. 0.1 Gy/mm
   d. 0.01 Gy/mm

3. How does the value for D_{2cm} (dose at 2 cm from the PTV) change if a lung SBRT prescription is modified from 18 Gy × 3 to an isodose line at 75% of the maximum dose vs. 10 Gy × 5 prescribed to an isodose line at 80% of the maximum dose?
   a. Decreased to 87% of original
   b. Decreased to 99% of original
   c. Increased to 103% of original
   d. Increased to 123% of original

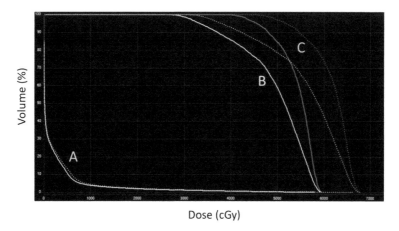

Dose (cGy)

FIGURE PS22.1
Example DVHs for an SBRT thoracic treatment plan (Problems 4–6).

4. Match the three DVH curves in Figure PS22.1 (solid lines) with their corresponding structure.

   iGTV

   PTV

   Lungs

5. Match the two DVH curves in Figure PS22.1 (solid vs. dotted lines) with the corresponding prescription.
   a. 12 Gy × 4 to the 82% isodose line
   b. 12 Gy × 4 to the 72% isodose line

6. What are the approximate values for the following dosimetric parameters from the DVHs shown in dotted lines in Figure PS22.1? Comment on the acceptability of this plan.
   a. PTV V100%
   b. PTV V90% (= 4320 cGy)
   c. Lung V_{20Gy}

7. According to the HyTEC Report and also QUANTRC and cooperative group protocols, what is the recommended method for defining the lung structure in SBRT plans?
   a. Lung_R + Lung_L
   b. (Lung_R + Lung_L) – iGTV
   c. (Lung_R + Lung_L) – PTV
   d. Each lung separately

8. How does the lung V_{20Gy} [%] for the structure "Lungs" compare to that for "Lungs-GTV" for an SBRT treatment of 18 Gy × 3?
   a. Higher
   b. Lower

9. Which parameters used in treatment plan optimization are associated with a more conformal lung SBRT plan? (Select all that apply.)
   a. Nine beams used instead of three
   b. Overall prescription dose
   c. Hot spot in PTV allowed up to 130% of prescription dose vs. 120%
   d. Larger R50% allowed

10. Select an SBRT treatment plan for a lung tumor in your clinical TPS and reoptimize this plan as if it were conventionally fractionated IMRT/VMAT. Compare the mean lung dose and the PTV V100% and D2%.

23

Total Body Irradiation (TBI) and Total Skin Electron Therapy (TSET)

23.1 Total Body Irradiation

23.1.1 TBI: Background and Dosimetric Goals

Total body irradiation (TBI) is one key step in preparation for bone marrow transplants in the treatment of lymphoma or leukemia. This technique was a major advancement in the treatment of this disease, and E. Donnall Thomas of the Fred Hutchinson Cancer Research Center was awarded the Nobel Prize in Medicine or Physiology in 1990 for its development. Prior to receiving a bone marrow transplant from either a donor or themselves, the patient receives a conditioning regimen that includes TBI. TBI can be delivered with a myeloablative intent, i.e. to eradicate stem cells in the bone marrow, in which case a typical prescription would be 12–15.5 Gy delivered in six to ten fractions BID (twice per day) sometimes with a boost to the testes in male patients. Alternatively, TBI can be delivered in a non-myeloablative fashion for the purpose of suppressing the immune system prior to transplant. In this case the dose is much lower (e.g. 2 Gy in one fraction; note that circulating T lymphocytes are very radiation-sensitive.

Key dosimetric goals of TBI treatments include the following: a uniform dose throughout the body (many protocols call for less than ±10% variation), limiting the lung dose (using blocks or other techniques), and controlling the dose rate in order to prevent variable dose rate-dependent radiobiological responses (typical dose rates are 5–15 cGy/min at midplane). Typically the dose is prescribed to midplane at the level of the thickest part of the patient, namely the umbilicus in adults and the head in children.

23.1.2 Key Features of TBI

Figure 23.1.1 shows one setup approach for TBI treatment of patients. Here the patient is standing, the linac photon beam is directed horizontally, and the patient is treated AP/PA (only the AP beam is shown here). Some of the key features include the following:

- A large SSD (typically >400 cm). This ensures that tall patients will fit in the field and ensures dose homogeneity in the beam direction (see Section 23.1.3 below).
- A "spoiler" which is a plastic sheet (approximately 1/4" thick) in front of the patient. When photons interact with this plastic, Compton-scattered electrons are generated and reach the patient. This creates extra superficial dose in the patient which is desirable in this treatment where the bone marrow in the ribs for example is at fairly superficial depths. This device is called a spoiler because it "spoils" the skin-sparing effect that is normally present in a high-energy photon beam.

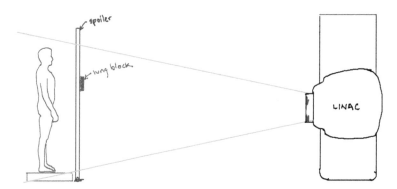

FIGURE 23.1.1
Example setup for a TBI treatment.

- Lung blocks when needed (e.g. for myeloablative regimens). For an AP/PA treatment the lungs are easily blocked. A typical protocol calls for 2 HVL thickness blocks applied for half of the treatment fractions. Note that the dose under the block is relatively high even for thick blocks (e.g. 2 HVL) because they are placed relatively far from the patient surface and there is scatter of the photons in the air and in the patient.

- A compensator (shown here as a red structure in the beam near the gantry head). This compensator corrects for the fact that some parts of the patient are thin (neck) while some are thicker (e.g. abdomen). The compensator is custom-designed for each patient and is thicker in the regions where the patient is thin in order to deliver uniform dose.

23.1.3 Dose Homogeneity

Figure 23.1.2 summarizes the dosimetric effects of using two geometrically opposed beams, as is almost always used in TBI treatments. This concept is also addressed in Section 12.2. Here two beams are used to treat the patient (red and blue as shown in Figure 23.1.2A). The final dose in the patient is the sum of the dose from these two beams (black). In order

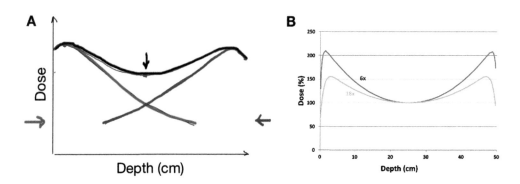

FIGURE 23.1.2
Dose homogeneity as a result of treating with two parallel-opposed beams. (A) The final dose in the patient is the sum of the dose from the two beams used for treatment (red and blue). (B) Dose is more homogeneous for higher energy beams due to less rapid falloff of the depth dose. Similarly, using a larger SSD will result in a more homogeneous dose.

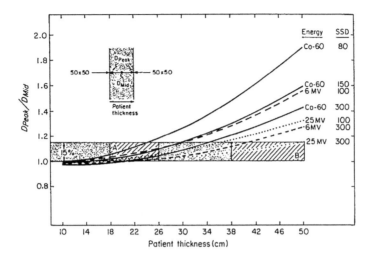

FIGURE 23.1.3
Dose homogeneity index for various TBI scenarios. (From AAPM Task Group 29, Van Dyk et al. 1986.)

to deliver some desired dose to the midplane point, a higher dose has to be delivered to tissue near the surface. There is, in other words, an inhomogeneity in the depth direction.

The inhomogeneity can be reduced by using beams that have depth doses which fall off less steeply with depth. One way to accomplish this is to use higher energy beams as shown in Figure 23.1.2B. Here, 18 MV beams result in a much more homogeneous dose than the 6 MV beams. Another way to improve homogeneity is to employ a larger SSD. At larger SSD the depth dose falls off less rapidly due to a relatively smaller impact of the inverse square falloff. Therefore, the two beams will add together to form a more homogeneous dose at larger SSD. To understand this SSD-dependent effect consider a very small SSD (e.g. 40 cm). In this situation each extra cm of depth in tissue would have a large reduction of dose due to inverse square fall off. Conversely at large SSD (e.g. 400 cm) each extra cm of depth in tissue would have only reduction of dose due to inverse square falloff. Finally, dose homogeneity is also improved with smaller separations in the patient. Again this is due to less dose falloff in the patient at the depth of midplane when the anatomy is thinner. **The factors controlling dose homogeneity are a key concept, namely energy, SSD, and separation.**

Dose homogeneity can be quantified by the ratio of dose at the peak (D_{peak}) to dose at midplane (D_{mid}). Values of D_{peak}/D_{mid} that are closer to 1 represent more homogeneous treatments. Figure 23.1.3 shows values of D_{peak}/D_{mid} for TBI treatments and the dependencies. To summarize the above discussion of the trends in Figure 23.1.3, dose homogeneity is improved for: (1) higher energy beams, (2) larger SSDs, and (3) smaller thicknesses of the patient.

23.1.4 TBI Setup Techniques and Devices

There are numerous possible variations for patient setup and considerations include the following:

- *AP/PA patient standing.* One challenge is that sick patients may be unable to stand for long periods. Some centers use harness restraints for patients for safety.

- *AP/PA patient lying down decubitus (on side).* One challenge with this setup is the accurate application of lung blocks; patients may move, and the lung on the downward side is compressed. Some centers use alternating right/left decubitus on alternating treatment fractions.
- *Lateral beams.* Patient standing or (more often) sitting or lying down. This is tolerated more easily by patients, but it is not possible to apply lung blocks. Also, the dose is less uniform in the depth direction vs. an AP/PA treatment because of the larger separation.

Pediatric patients receiving TBI require special considerations for setup. One technique is to place the patient on a low platform on the floor of the room and direct the gantry downward. This provides an extended SSD and allows for easy access of anesthesia. All but the oldest patients will fit in a single field. A spoiler is required here also, and may be placed on a tabletop holder over the patient. Plastic compensator sheets and blocks may be placed on top of this spoiler.

23.1.5 Dose Verification with In Vivo Measurements

Various detectors may be used for verification of dose on the patient including diodes, OSLDs, TLDs, and MOSFET. Diodes are very often used because of convenience and immediate readout. Note that whatever detector is used it must be calibrated for the TBI geometry. Using the calibration obtained under standard conditions (100 SSD, 10×10 cm field) would not accurately represent the dose in the TBI setup where the beam spectrum and scatter conditions are very different. Note also that detectors such as a diode often have a buildup cap integrated on them, and this needs to be taken into account in the calibration.

In vivo detectors can be employed at various points on the patient, e.g. forehead, neck, chest, umbilicus, and/or extremities. The detector is typically left on during treatment and read out when the treatment is complete. In this way the detector registers both the entrance dose (e.g. the AP beam) and the exit dose through the patient (e.g. PA beam). As such it provides a verification of the calculation, the patient separation, and the construction of the compensator at the point(s) measured.

23.2 Total Skin Electron Therapy (TSET)

23.2.1 Background and Dosimetric Goals of TSET

Total skin electron therapy (TSET) is used in the treatment of cutaneous T-cell lymphoma, a type of non-Hodgkin lymphoma, the most common form of which is mycosis fungoides. A typical dose schema is 36 Gy in 36 fractions delivered four days per week. A key dosimetric goal of this treatment is to deliver a high superficial dose but a low dose to internal organs. As such, electron beams are an optimal choice. Low-energy electron beams are used, typically 6 MeV, in order to limit the X-ray bremsstrahlung dose deep in the patient. Recall from Section 15.1.5 that these X-ray bremsstrahlung photons are produced when electrons interact with components in the head of the machine and the patient themselves and the production efficiency of these photons is lower at lower energies.

23.2.2 TSET Beam Delivery and Dose Uniformity

A two-beam technique is used to deliver TSET as shown in Figure 23.2.1. With this technique the beam is not aimed directly at the patient but rather at an angle. This reduces the X-ray component which is highest along the central axis of the beam. The delivery is performed at extended distance (typically 2 to 4 meters) in order to encompass the whole patient in the field(s). At extended distance there is substantial scatter and attenuation of electrons in the air. Therefore, it is often desirable to operate the linac in a special high-dose-rate mode.

A key dosimetric goal of TSET is to deliver a dose that is uniform over the surface of the patient (typically to within 10–15%). This is achieved by using multiple fields from different angles around the patient. The patient rotates and executes a series of poses at different angles relative to the beam in order to deliver these different angles. This is illustrated schematically in Figure 23.2.2 which shows a six-field technique: three beams from three different angles on one day (red arrow) followed by three beams from three different angles on the following day (blue arrows). It is also possible to rotate the patient during treatment.

Uniformity is also achieved by using a scatterer (Figure 23.2.1). Here the scatterer is placed near the exit window of the linac instead of near the patient as in TBI (c.f. Figure 23.1.1). Placing the scatterer near the exit window of the linac (vs. near the patient) results in a narrower angle for electrons that are scattered which results in a deeper depth-dose curve.

Some regions of the patient do not see a direct electron beam, for example the scalp, chin, perineum, and soles of the feet. These regions can be treated with an electron boost field using conventional techniques. One also has to be aware of skin folds which can create regions of low dose. The skin folds can be minimized and/or treated with electron boost fields using conventional techniques.

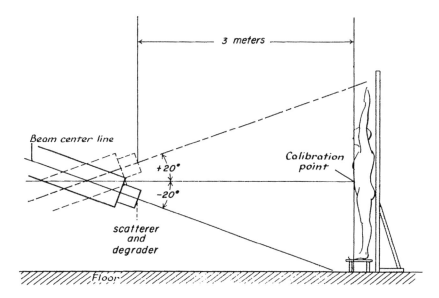

FIGURE 23.2.1
TSET treatment geometry using a two-field technique to avoid the X-ray component along the central axis of the beam. (From AAPM Task Group 30, Karzmark et al. 1987.)

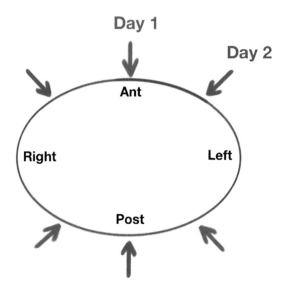

FIGURE 23.2.2
Schematic of beam delivery for TSET. Three beams are delivered each day from different angles around the patient.

Further Reading

ACR-ASTRO Guidelines for the Performance of Total Body Irradiation. 2017. https://www.acr.org/-/media/ACR/Files/Practice-Parameters/TBI.pdf.

Dieterich, S., E. Ford, D. Pavord and J. Zeng. 2016. *Practical Radiation Oncology Physics.* Chapter X. Philadelphia, PA: Elsevier.

Dieterich, S., E. Ford, D. Pavord and J. Zeng. 2016. *Practical Radiation Oncology Physics.* Chapter 24. Philadelphia, PA: Elsevier.

Kahn, F.M. and J.P. Gibbons. 2014. *Kahn's The Physics of Radiation Therapy.* 5th Edition. Chapters 14 and 18. Philadelphia, PA: Wolters Kluwer.

Karzmark, C.J., et al. 1987. Total skin electron therapy: Technique and dosimetry, report of AAPM Task Group 30. AAPM Report No. 23.

Metcalfe, P., T. Kron and P. Hoban. 2007. *The Physics of Radiotherapy X-rays and Electrons.* Chapter 5. Madison, WI: Medical Physics Publishing.

Van Dyk, J., J.M. Galvin, G.P. Glasgow and E.B. Podgorsak. 1986. The physical aspects of total and half-body photon irradiation, report of AAPM Task Group 29. AAPM Report No. 17.

Chapter 23 Problem Sets

*Note: * indicates harder problems.*

1. What is the impact of accidentally delivering a TBI treatment at 400 cm SSD if the standard setup is meant to be at 450 cm SSD? (Select all that apply.)
   a. Lower skin dose
   b. Higher dose at midplane
   c. More homogeneous dose
   d. Lower dose to head and feet

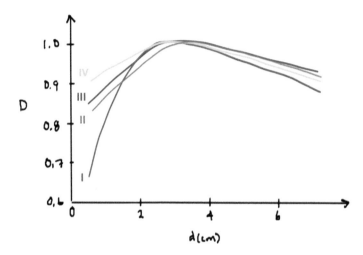

FIGURE PS23.1
Depth-dose curves.

2. What is the approximate dose at midplane for TBI treatment at 410 cm SSD if the dose for a standard setup at 400 cm SSD is 200 cGy?
   a. 190 cGy
   b. 195 cGy
   c. 205 cGy
   d. 210 cGy

3. Match the photon PDDs in Figure PS23.1 with the corresponding scenario.
   a. TBI treatment at 18 MV
   b. TBI treatment at 6 MV
   c. 6 MV TBI treatment with spoiler left off
   d. 18 MV 10×10 cm field at 100 SAD

4. In commissioning a linear accelerator for TBI treatment what steps may be taken to increase the homogeneity throughout the patient? (Select all that apply.) Discuss the trade-offs for each.

a. Employ as low an energy beam as possible
b. Design a patient setup with as large an SSD as possible
c. Use a very thin beam spoiler
d. Design for treatment with AP/PA beams

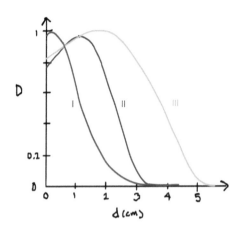

FIGURE PS23.2
Depth-dose curves for electron beams.

5. What shielding considerations must be taken into account if TBI treatments are to be started in a particular vault that was previously used for a low-energy (6 MV) linac?

6. How does the relative dose under a lung block for TBI treatment of a pediatric patient compare to a TBI treatment of an adult?
   a. Decrease
   b. Increase
   c. Stay approximately the same

7. Match the electron PDDs in Figure PS23.2 with the corresponding scenario.
   a. 6 MeV 10×10 field at 100 SSD
   b. 9 MeV 10×10 field at 100 SSD
   c. 9 MeV TSET field

8. What is the impact of delivering a TSET treatment with the gantry directed straight at the patient vs. at an angle?

9. * What are the possible causes of a low reading of in vivo dose using a diode during TBI treatment?

10. * Describe how Cherenkov radiation can be used as an in vivo dose measurement tool in TSET treatments.

24

Particle Therapy

In Chapter 15 we examined the behavior of one type of charged particle for therapy, the electron. Electron beams provide a dose that falls off quickly with depth, but electrons also undergo substantial scattering in tissue and execute a "random walk" as they penetrate through tissue which affects the dose distribution characteristics. In this chapter we examine heavier charged particles, the proton and other particles. The first section focuses specifically on proton radiotherapy beams.

The proton has a rest mass approximately 2000 times larger than the electron and therefore it undergoes many fewer large-angle scatter events as it penetrates through tissue. If the electron is a pea, the proton is a bowling ball. This bowling ball "barrels right through" tissue. This, as we will see, has profound implications for therapy with these beams.

24.1 Basic Physics of Proton Therapy Beam Production

24.1.1 Overview and Indications of Use

The main advantages of therapy with proton beams can be appreciated in Figure 24.1.1. The dose near the surface is somewhat lower than in a photon beam, but, even more importantly, the proton beam has a very sharp falloff at a certain depth. There is essentially no dose beyond this depth. Proton beams, therefore, can provide much lower doses to normal tissues past this depth.

This manifests in treatment plans. Figure 24.1.2 shows an example plan for a whole CNS treatment of a pediatric patient. Here posterior proton beams are used, and because of the sharp falloff there is very little dose anterior to the vertebral bodies. This provides substantial dose sparing of the heart, bowel, and other organs.

An example proton plan for treatment in the brain is shown in Figure 24.1.3. Of note is the fact that proton therapy plans often use many fewer beams than photon therapy, e.g. a single beam in Figure 24.1.3A or two beams in Figure 24.1.3B. Because of the sharp falloff in dose, low doses to normal tissue can be achieved with relatively few beams.

The question of which disease sites are appropriate for proton therapy is complex and is driven by a variety of factors including the relatively higher cost of treatment and lack of level-one evidence. For a recent discussion of evidence-indicators for proton therapy see Misra et al. (2017). This and other documents outline disease sites where proton therapy plays a role. These include the following (with the caveat that this is not an exhaustive list):

- Pediatric cancers: Due to the long life-expectancy of the patient, protons have a potential advantage of reduced secondary malignancies due to the lower doses to normal tissues

- Tumors of the central nervous system (CNS)
- Eye cancers
- Sarcomas
- Advanced or unresectable head-and-neck cancer
- Hepatocellular carcinomas
- Re-irradiation cases: To achieve reduced doses to normal tissues

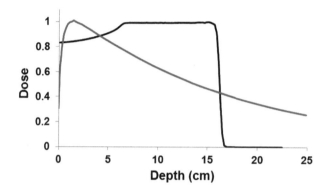

FIGURE 24.1.1
Depth-dose curve for a 6 MV photon beam (red) compared to a proton beam (black).

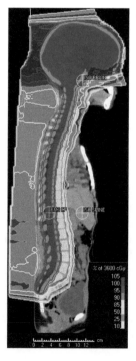

FIGURE 24.1.2
Example proton treatment plan for a whole CNS treatment of a pediatric patient.

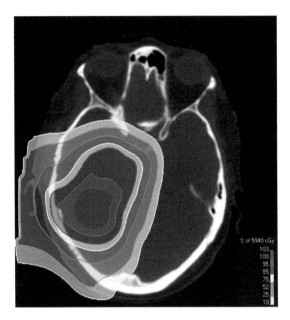

FIGURE 24.1.3
Example proton treatment plans.

24.1.2 Historical Overview, Expansion, and Costs

The first use of proton therapy on humans was in 1954 at the University of California, Berkeley labs as pioneered by the physicist Robert Wilson and the physician Cornelius Tobias. The mid-2000s saw the beginning of a rapid expansion of proton therapy centers in the United States and globally. Part of what has driven the relatively slow growth of proton therapy over the decades is the dimension of cost and complexity. For more information see the video.

24.1.3 Physics of Pristine Bragg Peak

To understand the physics of proton beams we begin with the depth-dose curve of a roughly monoenergetic proton beam as shown in Figure 24.1.4. At a shallow depth the

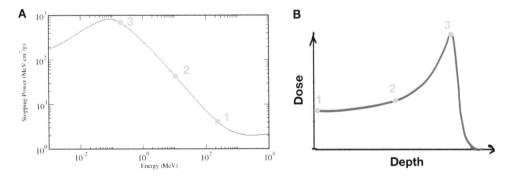

FIGURE 24.1.4
Depth-dose curve and stopping power of a proton beam. (A) The stopping power of the proton (energy loss per unit length) increases as the energy decreases. (B) The dose increases with depth. As more energy is lost the stopping power increases. The dose reaches a maximum at the Bragg peak. (Data from NIST pstar.)

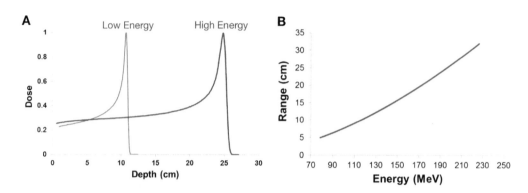

FIGURE 24.1.5
Depth–energy relationship for proton beams. (A) The Bragg peak forms at a deeper depth for higher energy beams. (B) The relationship between the range of the proton in water and the incident energy of the proton beam. (Data from NIST pstar.)

proton has a high energy and the dose deposited is relatively low. This is a reflection of the relatively small stopping power at high energies (Figure 24.1.4A). As the proton traverses through tissue, more energy is lost and the stopping power increases (e.g. "point 2" in Figure 24.1.4). This is a runaway process. The more energy the proton loses, the more its stopping power increases and so the more energy is lost and so on. Eventually the proton loses almost all its energy and the resulting dose is very high ("point 3" in Figure 24.1.4). This is referred to as the Bragg peak. After the Bragg peak, the proton has lost essentially all of its energy and it goes no further. Note that the peak in Figure 24.1.4 is often referred to as the "pristine" Bragg peak, that is, the Bragg peak that would be generated if there were no modifiers in the beam as will be discussed in the next section(s).

The range of the proton (and the depth at which the Bragg peak forms) depends on the energy of the proton as it enters the patient. Higher energy protons have more energy to lose and penetrate deeper. This is illustrated in Figure 24.1.5 which shows the dependence of the depth of the Bragg peak on the incident energy of the protons. Note that range of energies for therapy proton beams is approximately 90 to 230 MeV, although there are some specialized beams at lower energies (e.g. 60 MeV) for treating lesions at more superficial depths as in ocular melanoma.

24.1.4 Spread-Out Bragg Peak (SOBP)

While the pristine Bragg peaks shown in the last section illustrate the physical principles of proton beams, they are not useful for treating tumors since the dose is highly localized in the depth dimension. It would not provide for a uniform dose to a macroscopically large tumor. This is illustrated in Figure 24.1.6 where it can be appreciated that the single high-energy beam (red) does not provide an even dose throughout the target. However, it is possible to add other lower energy beams which provide pristine Bragg peaks at shallower depths (e.g. blue, green in Figure 24.1.6). The various energy beams add with each other, and by delivering the appropriate dose for each beam a spread-out Bragg peak (SOBP) can be made. The SOBP (black curve in Figure 24.1.6) has a uniform dose over some predetermined range in depth which covers the target.

The highest energy beam provides the dose at the distal end of the target (red in Figure 24.1.6). To generate the beam at lower energy (e.g. green in the figure) a "modulator" is introduced into the beam. This is a low-Z plastic device which reduces the energy of

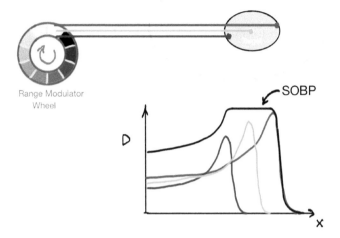

FIGURE 24.1.6
Creation of the spread-out Bragg peak (SOBP). Beams of different energy are superimposed to create an SOBP with a uniform dose over some range in depth (black). This is accomplished with a range modulator wheel with different thicknesses of plastic.

the beam and therefore the range. A range modulator wheel is one system for introducing these range modulators into the beam in a controlled fashion. The modulator wheel has different thicknesses of plastic in the various sectors tuned to provide an SOBP when the wheel is spun. There are other systems to accomplish this such as ridge filters.

Proton therapy beams are characterized by a range (determined by the highest energy of beam) and a modulation (determined by the modulator wheel or equivalent system). An example is shown in Figure 24.1.7. The range and modulation can be selected as needed in the treatment planning process. Figure 24.1.7B shows two beams with the same range but different modulation. The beam with the larger modulation has a higher surface dose due to the addition of many pristine Bragg peaks, each of which contributes dose at the

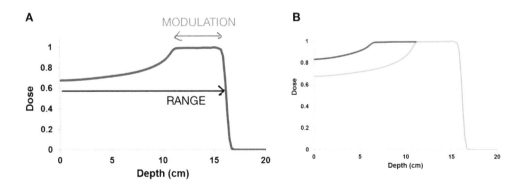

FIGURE 24.1.7
Range and modulation effects in proton beams. (A) An example beam R16MOD5 (i.e. range 16 cm, modulation 5 cm). (B) Two beams with the same range but different modulation. The beam with the larger modulation has a higher surface dose.

entrance surface. This effect is important to note because skin dose in proton beams can sometimes be a challenge.

24.1.5 Beam Shaping with Compensators

Now we turn to the problem of how to create a three-dimensional dose distribution that conforms to the target. As shown in Figure 24.1.8, with an open beam, the SOBP forms an approximately rectangular region and the dose does not conform well to the tumor target. One way to address this problem is to introduce a compensator (Figure 24.1.8B). The compensator thickness is adjusted in order to modulate the beam at different points and make it conform to the tumor shape at the distal edge. If the patient surface is sloped or irregular (Figure 24.1.8C) then there will be a variable depth to each point at the distal edge of the tumor. However, this can also be taken into account with the compensator in such a way that the distal aspect conforms to the tumor. While this approach can be used to achieve good dose conformality on the distal edge of the tumor, the proximal edge is not conformal because it is set back by the distance of the SOBP which may not conform to the complex shape of the tumor. This is one disadvantage of this approach.

The effects of inhomogeneities in proton beams are different than in photon beams. Figure 24.1.9, for example, shows the effect of a 3 cm thick bone in the beam. In the photon beam the bone causes a decrease of the dose in the region distal to the bone by 11% in

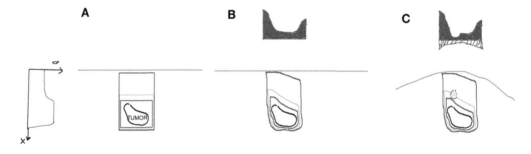

FIGURE 24.1.8
Dose distributions in proton beams (A) with an open beam, (B) with a compensator, and (C) with a compensator that also accounts for the sloped surface of the patient and for an inhomogeneity.

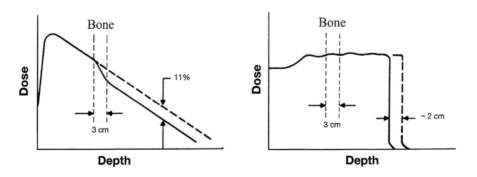

FIGURE 24.1.9
Effect of an inhomogeneity in a photon beam (left) vs. a proton beam (right). (From D.T.L. Jones et al. ICRU Report No. 78, Prescribing, Recording, and Reporting Proton-Beam Therapy, 2007. 7(2):1–210. Reprinted by permission of SAGE Publications.)

this example. In the proton beam, however, the effect is more dramatic. The distal edge of the SOBP is pulled back by approximately 2 cm which may result in the tumor being underdosed. If the inhomogeneity were a low-density region (e.g. the lung with a density of approximately 0.3 g/cm^3) the distal edge of the SOBP would travel farther then it would in higher density tissue.

Figure 24.1.8C illustrates this effect and how the compensator can be designed to account for it. Here the bone is shown in gray. The compensator is made thinner in the region where the beam passes through the bone (see Figure 24.1.8C) which compensates for the bone and produces a conformal dose distribution. However, an additional problem arises that the inhomogeneous region may move due, for example, to variability in the patient setup. It is also possible to accommodate this in the construction of the compensator by manipulating the shape, making the edges near the homogeneity smoother and therefore less impacted by motion.

24.1.6 Beam-Shaping Systems

Figure 24.1.10 shows an example proton beam delivery system. The beam first enters two scatterers which act to broaden the beam. The two-scatterer system uses dose more efficiently than a single scatterer. It then enters a range shifter, the thickness of which is chosen to achieve the desired range for the distal edge of the SOBP. Then the beam enters the range modulator wheel which will create the SOBP, then through a collimator, and finally through a compensator. The applicator and compensator are specific to the patient and the beam being treated.

Typically a cutout of the required shape for the patient treatment is constructed from brass. Multi-leaf collimators are not widely used in proton therapy centers and the collimator aperture is typically milled on site. The compensator is also usually made on site and is constructed of a low-Z material like wax or plastic to reduce scattering of the beam.

Data on the penumbra of proton beams are shown in Figure 24.1.11. Note that the penumbra of the proton beam increases with depth. This is due to the multiple Coulomb

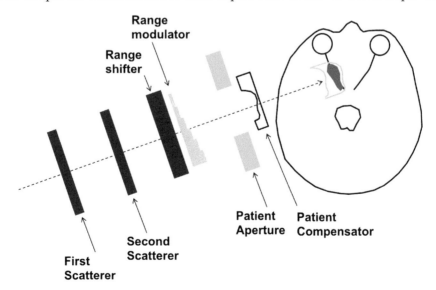

FIGURE 24.1.10
Proton beam delivery system showing the major components.

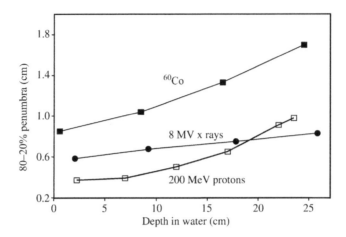

FIGURE 24.1.11
Penumbra of proton beam vs. photon beam. (From D.T.L. Jones et al. ICRU Report No. 78, Prescribing, Recording, and Reporting Proton-Beam Therapy, 2007, 7(2):1–210. Reprinted by permission of SAGE Publications, Inc.)

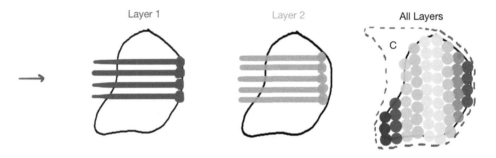

FIGURE 24.1.12
Spot scanning or pencil beam scanning (PBS). A magnetic steering system scans the beam in one layer. The proton beam energy is reduced and then the next layer is scanned. This process is repeated until the entire tumor is irradiated. The resulting dose distribution is more conformal than is possible with a scattering or uniform scanning system (dashed line, right).

scattering events as the proton passes through tissue. The penumbra is approximately similar to a modest energy photon beam. This is an important point, namely that the penumbra of a proton beam is not substantially better than a photon beam.

An alternative delivery technique to the one described above is spot scanning or pencil beam scanning (PBS). The approach is illustrated in Figure 24.1.12. Here a magnetic steering system scans the beam back and forth across the tumor. First one layer is scanned, then the energy of the beam is changed which moves the depth of the Bragg peak and another layer is scanned (Figure 24.1.12). This process is repeated until the entire tumor is irradiated. This allows for a more conformal dose distribution. Also there are fewer devices in the beam path which has a potential advantage of reducing the neutron dose. Many proton centers have moved to this technique in recent years.

24.1.7 Cyclotrons and Synchrotrons

Because the rest mass of the proton is approximately 2000 times heavier than the electron, small linear accelerators are typically not sufficient to accelerate a proton to the energies needed for therapy purposes. Instead there are two main technologies: the cyclotron and the synchrotron. These are used for accelerating both protons and other heavier ions.

The cyclotron, the older of the two, uses a high-frequency oscillator and a magnetic field to accelerate charged particles (Figure 24.1.13). The cyclotron has two half circles to which an electric field is applied. The particle, a proton say, moves due to the force of this field. The magnetic field is oriented perpendicular to the path of the proton. Therefore, as the proton moves it experiences a force and is bent to travel in a circle (recall from Chapter 1 that magnetic fields exert forces on charged particles). As the proton travels through the system its energy increases due to the applied electric field. Therefore, it travels in a spiral with increasing radii. Recall that particles with higher energy travel in circles with larger radii. By picking off the proton at a specific radius, the energy can be selected. Modern cyclotron design employs superconducting magnets with magnetic fields of 5–10 T, making them very compact.

The other main technology for accelerating charged particles is the synchrotron. In this system the particle beam travels around in a vacuum tube. The beam is bent around into a circle by bending magnets located at various points along the beamline. Each time the beam goes around it travels through an accelerating cavity which applies an electric field (radiofrequency) which accelerates the particle. At each pass through the system, the particle is accelerated more. When it reaches the desired energy the beam is extracted from the ring. In such a system, particles can reach extremely high energies, and this is the technology used at major national and international particle accelerator laboratories like Fermilab or CERN.

There are some key differences between cyclotrons and synchrotrons. Some cyclotron designs provide a continuous beam of particles whereas synchrotrons provide repeated very short bursts (or "spills") of the beam as bunches of charged particles are extracted. The average beam current from a cyclotron is much higher than a synchrotron. Another difference between the two systems is the energy switching. In a synchrotron the energy of the beam can be switched very quickly while it is difficult to switch the energy of a

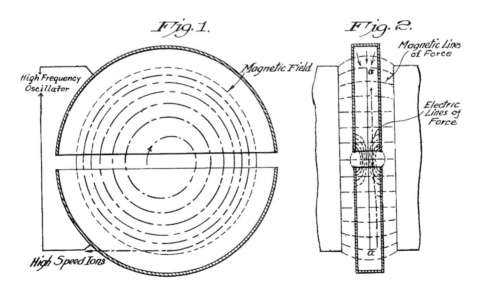

FIGURE 24.1.13
Cyclotron system for accelerating charged particles. (From U.S. Patent 1948384 Ernest O. Lawrence, Method and Apparatus for the Acceleration of Ions, 1934.)

cyclotron quickly and it often operates at a fixed energy. Finally, cyclotrons are much more compact than synchrotrons. Gantry-mounted cyclotrons are now being manufactured for proton therapy (Mevion Inc.).

Finally, it is worth noting that although cyclotrons and synchrotrons dominate the market for proton therapy devices, other technologies are being pursued as well. For example, researchers at CERN are developing a proton beamline using four coupled linacs operating at 3 GHz. This is a 20-m-long system that provides a 70 MeV beam with fast energy switching.

24.2 Proton Planning, Quality Assurance, and Ion Beams

24.2.1 Proton Dose and Relative Biological Effect (RBE)

As a first step to understanding proton planning we consider how dose is conceptualized and reported in proton therapy. As with all radiation therapy the units of dose are still energy per unit mass, J/kg (i.e. Gy). However, in therapy with proton beams and heavier charged particles it is important to consider the relative biological effect (RBE) of the particle.

In proton beams there are data to suggest that the RBE may be higher than in a photon or electron beam, and there are active debates and discussions around this. However, ICRU Report No. 78 (Jones et al. 2007) recommends applying a value of RBE = 1.1 for proton beams. Data indicate that the RBE increases at the end of the Bragg peak where the particle energy gets low and the linear energy transfer (LET) goes up. Here the RBE may be substantially increased. Nevertheless, ICRU Report No. 78 recommends applying a constant RBE value of 1.1. There is also the question of what nomenclature to use when reporting dose from proton or heavy ion beams. ICRU Report No. 78 recommends stating dose as "D_{RBE}", i.e. dose (in Gy) where an RBE weighting factor has been applied. However, other older units are in use still, e.g. "Cobalt Gy Equivalent" (CGE) or "Gray equivalent" (Gy or Gy(E)).

24.2.2 Proton Treatment Planning

There are some considerations in proton therapy that are different than in photon therapy. One is the way that margins are handled. In photon therapy there is often a uniform expansion for margins (e.g. GTV-to-PTV, Figure 24.2.1). However, in proton therapy the direction of the beam must be taken into account. More margin is needed at the distal edge of the target because uncertainties have a bigger impact in this region. Recall that the dose falls off abruptly distal to the SOBP, so a small uncertainty in range can have a large impact in this region.

There are uncertainties in the range of the proton beam which result from uncertainties in the CT numbers that are used in treatment plans and the stopping power of materials. Recall that CT numbers do not provide information about the composition of the material but are some combination of material and density. This translates into an uncertainty in the range. This is accommodated in proton therapy by adding a margin on the distal aspect of the PTV. One example of this is from Moyers et al. (2001) at Loma Linda, namely *Distal margin* = 0.035 × (depth CTV_{distal}) + 3 mm, where depth CTV_{distal} is the depth of the distal edge of

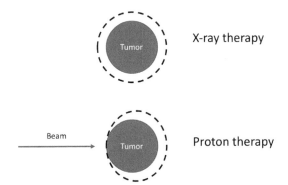

FIGURE 24.2.1
Margins in proton therapy. The dashed line shows a schematic representation of the margin for photon (top) vs. proton therapy (bottom).

the CTV. There is a 3.5% extra margin per unit length due to uncertainties in CT numbers and the conversion to stopping power. Also 3 mm is added to account for uncertainties such as the beam energy from the accelerator, devices in the beam path, and compensator construction. Though the numbers vary, formalisms like this are used at many centers.

Proton therapy is different than photon therapy in other ways as well. These include: (1) the effects of motion are more complex and potentially impactful. A tumor moving into the distal end of the proton beam range for example may be substantially underdosed. (2) Potential RBE effects especially near the end-of-range (see Section 24.2.1). (3) Anatomical changes can have a large effect on proton beams. If, for example, a patient were to gain weight, the radiological depth of a target may change. This might have a minor impact on the dosimetry from an MV X-ray beam, but may have a major impact on proton dosimetry (e.g. target out of range). Similarly, breathing and the subsequent change in lung density can result in the proton beam ranging much farther into tissue than planned. (4) IGRT technologies. In many proton therapy centers constructed to date there have been limited IGRT capabilities and typically CBCT is not available. This situation is rapidly changing, however, as new centers come online and are upgraded.

24.2.3 Proton Therapy Quality Assurance (QA)

QA considerations in proton beam therapy are somewhat different than for photon therapy. QA recommendations for proton therapy machines. QA must be performed for the spot, including the location, size, and depth of each spot. Some challenges include the high instantaneous dose rate in each spot (>1000 Gy/min), and the dynamic creation of the SOBP which means that it is not possible to use a QA system with a scanning detector. Further information on QA techniques and devices for proton therapy can be found in the video.

24.2.4 Heavy Ion Therapy

In heavy ion beam therapy different nuclei can be employed including He, C, Ne, Si, and Ar. Here we will focus on carbon ion therapy, because this is the most frequently discussed species for heavy ion therapy with the most facilities in operation. Heavy ion nuclei are much more massive than the proton and so require much more energy input to accelerate to the appropriate energy for therapy applications. Heavy ion beams require larger accelerators and larger bending magnets, resulting in much larger gantries. Such facilities are more costly and complex to construct and maintain.

Some key features of heavy ion vs. proton therapy beams are the following:

- Many of the properties are broadly similar to a proton beam. There is still a Bragg peak and the LET still increases with depth (increasing stopping power as the energy of the particle decreases).
- For the same energy, the range in tissue is much less, e.g. 25 cm for a 62 MeV proton beam but 8 mm for a carbon beam at 62 MeV/u (MeV per nucleon).
- LET is much higher than the proton beam. LET can be well in excess of 100 keV/μm.
- Dose is deposited beyond the Bragg peak. This is due to nuclear fragments created through inelastic scattering of the carbon ion. These fragments are relatively high LET, and the contribution is larger for higher energy beams and heavier nuclei.
- Penumbra of a carbon beam is smaller than a proton beam (see Figure 24.1.11) due to less multiple Coulomb scattering of the heavier particle.

For further description see the video.

24.2.5 Relative Biological Effect (RBE) of Heavy Ion Beams

Tumors and tissue have a different biological response to heavy ion beams vs. a photon or electron beam, and the basis for this is the different LET in these beams. To understand this consider the diagram in Figure 24.2.2. In a photon beam (left), dose is deposited at interaction sites that are spread out over a wide region of space, a region that is much larger than the cell. This is because the electrons that are created by Compton-scattered photons travel long distances (typical cm range in tissue). The locations where they deposit dose are randomly distributed over a large region. By contrast a high-LET particle delivers

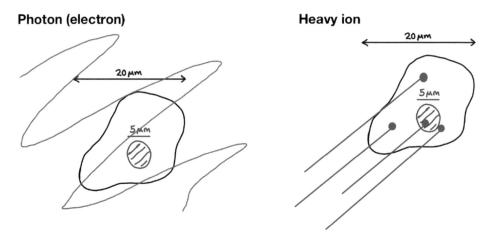

FIGURE 24.2.2
Dose deposition in a low-LET beam (left, photon/electron) vs. a high-LET beam (right, heavy ion). A typical cell and nucleus (blue) are shown for scale.

TABLE 24.1

Example LET and RBE for Various Beam Types

Particle Type	LET	RBE
[60]Co photon	0.3 keV/μm	1
Proton	2 keV/μm (higher at end of Bragg peak)	1.1
Carbon ions	100 keV/μm (varies strongly with depth)	3 and larger

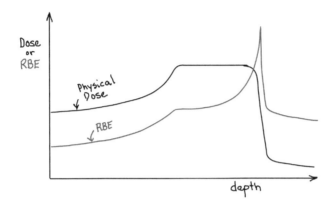

FIGURE 24.2.3
Physical dose and RBE in a carbon ion therapy beam. LET increases with depth which drives and increasing RBE. Although the physical dose is uniform in the SOBP, the biologically weighted dose is very non-uniform.

energy along a well-localized track in space. At the highest LET the region of dose deposition can be small on the cellular scale (Figure 24.2.2, right).

A primary response of cells to radiation is the DNA double strand breaks (DSB) and their subsequent repair (or lack of repair). Experiments have shown that the number of DSB per unit length of chromatin is independent of LET so that is not an explanation for the enhanced LET-dependent effect. However, the spacing of DSB is closer in high LET vs. low LET, and also high-LET beams produce clustered lesions on the DNA which are more difficult for the cell to repair. Because of these effects, the RBE in a high-LET heavy ion beam is larger than in the low-LET photon or electron beams. Table 24.1 shows example values of LET and RBE for different beam types.

Figure 24.2.3 illustrates the impact of this in a carbon ion treatment beam. Here the physical dose is constructed to be uniform in the SOBP, but the LET (and RBE) increase dramatically with depth in this beam. Therefore, the biologically weighted dose will be very non-uniform in this beam. This underscores the importance of including LET and RBE effects into treatment planning with heavy ion beam therapy.

Further Reading

Dieterich, S., E. Ford, D. Pavord and J. Zeng. 2016. *Practical Radiation Oncology Physics*. Chapter 8. Philadelphia, PA: Elsevier.

Jones, D.T.L., et al. 2007. Prescribing, recording, and reporting proton-beam therapy. *ICRU Report 78* 7(2):1–210.

McDermott, P.N. and C.G. Orton. 2010. *The Physics of Radiation Therapy.* Chapter 20. Madison, WI: Medical Physics Publishing.

Mishra, M.V., et al. 2017. Establishing evidence-based indications for proton therapy: An overview of current clinical trials. *Int J Radiat Oncol Biol Phys* 97(2):228–235.

Moyers, M.F., D.W. Miller, D.A. Bush and J.D. Slater. 2001. Methodologies and tools for proton beam design for lung tumors. *Int J Radiat Oncol Biol Phys* 49(5):1429–1438.

Chapter 24 Problem Sets

*Note: * indicates harder problems.*

1. What physical process accounts for the increase in penumbra of the proton beam with depth (cf. Figure 1.7 from ICRU Report 78)?

    a. Inelastic scattering

    b. Multiple Coulomb scattering

    c. Pair production

    d. Increased LET

2. List three advantages of a cyclotron for proton therapy vs. a synchrotron.

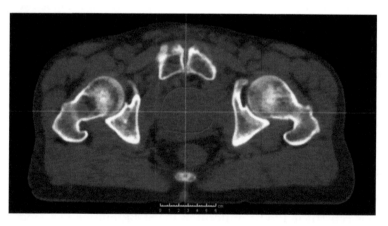

FIGURE PS24.1
CT cross-section at the level of the prostate.

3. What is one advantage of treating with lateral beams vs. AP/PA for prostate cancer in the image shown in Figure PS24.1?

    a. Less uncertainty in stopping power

    b. More confidence in rectal sparing

    c. Reduced skin dose

    d. Smaller margins

4. What margin is required on the distal aspect of the PTV for a left lateral beam in the image in Figure PS24.1 given the margin formalism of Moyers et al. (2001)?
   a. 0.83 mm
   b. 3.7 mm
   c. 6.8 mm
   d. 10.0 mm

5. What is the range of a 200 MeV proton in muscle using the continuous slowing down approximation? (See NIST PSTAR website for data https://physics.nist.gov/PhysRefData/Star/Text/PSTAR.html.)
   a. 4.3 mm
   b. 1.4 cm
   c. 12.4 cm
   d. 25.2 cm

6. What are some disadvantages of carbon ion beam therapy vs. proton beam therapy? (Check all that apply.)
   a. Wider penumbra
   b. Larger dose beyond Bragg peak
   c. Higher cost
   d. Lower RBE

7. What is a possible cause of a wrong spot location on daily QA tests?
   a. Incorrect magnet field strength
   b. Aperture not correct
   c. Range shifter too thin
   d. Snout length too long

8. Sketch the physical dose distribution required from a carbon beam in order to make a constant biological dose in the spread-out Bragg peak.

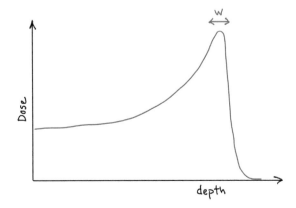

FIGURE PS24.2
Bragg peak and width.

9. * What proton beam parameter affects the width, W, of the pristine Bragg peak shown in Figure PS24.2?

10. * In a cyclotron system how does the width, W, change when a smaller range beam is used?

    a. Decreases
    b. Increases
    c. Stays the same

25

Radiation Protection

25.1 Dose Equivalent and Effective Dose

25.1.1 Dose Equivalent

The first term to define is dose equivalent, H.

$$H = D \cdot W_R \tag{25.1}$$

where
   D is the dose in Gy and
   W_R is a radiation weighting factor.

The units of H are the Sievert, Sv. An older unit is rem (for "Roentgen-equivalent man"). 1 Sv = 100 rem. The radiation weighting factor accounts of the relative biological effect (RBE) of the radiation type in question. Recall from Section 7.3.2 that RBE depends on the type of particle (e.g. photon, electron, proton) and the LET of the particle. The definition of W_R is put forth by the International Commission on Radiation Protection, ICRP. Note that this factor used to be called the "Q" factor, or radiation quality factor, and this term is still in use by the Nuclear Regulatory Commission in the United States. Values of W_R are shown in Table 25.1. There is agreement among the different reports and agencies as to the appropriate values with the exception of neutrons.

25.1.2 Effective Dose

Dose equivalent itself is a useful quantity but it will not fully account for the effect of radiation exposure on an organism. To account for this a different quantity is introduced which includes a factor for the differing radiosensitivity of each organ. This is the effective dose, E, defined as:

$$E = \sum_T W_T \cdot H_T \tag{25.2}$$

The sum is over all organs, T. H_T is the dose equivalent in that organ, and W_T is a tissue weighting factor which accounts for the radiosensitivity of that organ. The units of effective dose are also Sieverts, Sv. Values for organ weighing factors are given in Table 25.2. Note that the values in the two different relevant reports do not completely agree with each other.

TABLE 25.1

Radiation Weighting Factor, W_R

Radiation Type	W_R (from NCRP 116)	W_R (from ICRP 103)	US NRC (Quality Factor)
Photons and electrons	1	1	1
Neutrons (energy dependent)	5–20 (peak at 20 at ~1 MeV)	2.5–5	2–11 (peak at 11 at ~1 MeV)
Protons	2	2	Not listed
Alpha particles	20	20	20

TABLE 25.2

Organ Weighting Factors from NCRP Report #116 (Meinhold et al. 1993) and ICRP Report #103

Organ	W_T (from NCRP 116)	W_T (from ICRP 103)
Gonads	0.2	0.08
Red bone marrow	0.12	0.12
Colon	0.12	0.12
Lung	0.12	0.12
Stomach	0.12	0.12
Bladder	0.05	0.04
Breast	0.05	0.12
Liver	0.05	0.04
Esophagus	0.05	0.04
Thyroid	0.05	0.04
Skin	0.01	0.01
Brain	Included in "remaining organs"	0.01
Salivary glands	Not included	0.01
Bone surfaces	0.01	0.01
Remaining organs	0.05	0.05

25.2 Risk Models, Dose Limits, and Monitoring

25.2.1 BEIR VII Report: Deterministic and Stochastic Effects

A key reference which examines the impact of radiation exposure and presents consensus findings is the BEIR VII Report, Biological Effect of Ionizing Radiation (2006). At the highest doses the effects are deterministic, i.e. the impact can be predicted. Relevant threshold dose levels and effects are 5 Sv bone marrow depletion, 10 Sv GI syndrome, and 20 Sv CNS syndrome. At lower doses the effects are stochastic, i.e. random in nature. The main effect is a radiation-induced cancer. The model that that BEIR VII report advocates is a linear no-threshold (LNT) model for excess risk of cancer (Figure 25.1.1). Above about 100 mSv there are data on risk from the survivors of the nuclear bombing of Hiroshima and Nagasaki during World War II. Below 100 mSv, however, there are few data and so various models are possible, including even a radiation hormesis model where low levels of radiation are beneficial. The BEIR VII report concludes that there is no evidence for a low dose threshold and advocates the LNT model as a conservative estimate of risk.

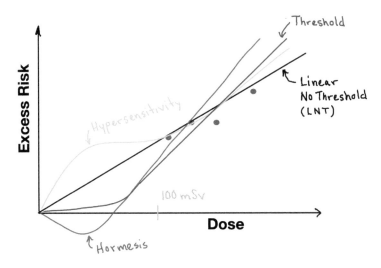

FIGURE 25.1.1
Models for excess risk from radiation exposure.

TABLE 25.3

Maximum Permissible Doses from NCRP Report #116 (Meinhold et al. 1993)

	Basis	Maximum Permissible Dose (Annual Limit)
Occupational limits	Stochastic effects	50 mSv (5 rem) And cumulative limit of 10 mSv × age
	Deterministic effects	Lens: 150 mSv (15 rem) Skin, hands, and feet: 500 mSv (50 rem)
Public limits	Stochastic effects	Continuous exposure: 1 mSv (0.1 rem) Infrequent exposure: 5 mSv (0.5 rem)
	Deterministic effects	Lens and extremities: 50 mSv (5 rem)
	Embryo or fetus	0.5 mSv (0.05 rem) in a month once pregnancy is known

25.2.2 Exposure Limits

These risk models provide some rationale for determining what the allowable limits of radiation exposure might be. Table 25.3 shows maximum permissible doses from NCRP Report #116 (Meinhold et al. 1993) which is what is followed in the United States. There are two categories of limits, one for occupational workers and one for the general public. **The LNT model and the maximum permissible dose limits are key concepts that are useful to remember.**

25.2.3 Background Exposure

To put the dose limits in context it is useful to know the rates of background exposure radiation due to natural sources such as cosmic rays. In the United States, the average background rate is approximately 3 mSv/year, i.e. three times higher than the public dose limit from the NCRP. In addition there is exposure from medical testing, which in the general public on average results in an additional 3 mSv/year/person. This is largely from CT scanning.

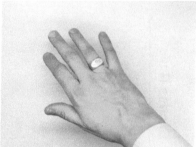

FIGURE 25.2.1
Personnel exposure monitoring badges.

TABLE 25.4

Release Criteria for Patients Undergoing Brachytherapy
Implants from NUREG-1556 (2019)

Radioisotope	Column 1 Activity Limit	Column 2 Dose Rate Limit
^{125}I implant	9 mCi (0.33 GBq)	1 mR/hr (0.01 mSv/hr)
^{103}Pd implant	40 mCi (1.5 GBq)	3 mR/hr (0.03 mSv/hr)

25.2.4 Exposure Monitoring

Exposure can be monitored using badges worn on the body (Figure 25.2.1) or on the finger for brachytherapy procedures. These devices have embedded OSLDs to record dose and are typically mailed in to the monitoring company for readout once per month. Recommendations for monitoring can be found in NCRP Report #102 (Gregg et al. 1989). The report recommends that monitoring should be in place for people whose exposure might exceed 10% of the allowable limits, the device should be visible, worn on the trunk above the waist, and should not be worn if one undergoes a procedure as a patient. The person should be notified in writing if their exposure exceeds 10% of the dose limit.

25.2.5 Criteria for Releasing Patients

The final topic for this section is the criteria for releasing patients from the hospital after they have undergone a procedure such as a brachytherapy implant. In the United States, the release criteria are determined by the NRC and regulations can be found in NUREG-1556 (2019). The criteria depend on the isotope being used. Examples are shown in Table 25.4. The patient can be released without specific instructions if the activity is less than that listed in Column 1 or the dose rate measured in air at 1 m from the patient is less than that shown in Column 2. There are other alternative criteria for releasing the patient in these regulations which are more complicated and include a specific calculation of dose to the patient and the public. More details can be found in the NUREG.

25.3 Shielding and Survey Meters

In radiation therapy and diagnostic rooms shielding is required in order to keep the dose to the surrounding areas within acceptable limits. A key reference in this context is NRCP

Report #151 (Deye et al. 2005). This report provides a clear and readable presentation of issues related to shielding as well as a wealth of data to enable calculations. We first consider the case of calculating the required shielding for a linac.

25.3.1 Shielding Calculation Formalism

There are three key sources of radiation for which shielding is required: primary, scatter, and leakage (Figure 25.3.1). This is a key concept in shielding calculations. Primary is the radiation in the direct beam itself, scatter is the radiation scattered from the patient, and leakage is the radiation escaping through the head and waveguide assembly of the linac itself. First, we consider an example of shielding for a primary barrier (Figure 25.3.2).

primary $B = Pd^2/WUT$

Leakage $B = \dfrac{Pd^2}{WXT}$ Leakage $\leq$ 0.1% $\cdot$ N

$B = \dfrac{Pd^2}{0.001 \, WT}$

Scatter

FIGURE 25.3.1
Sources of radiation for shielding calculations.

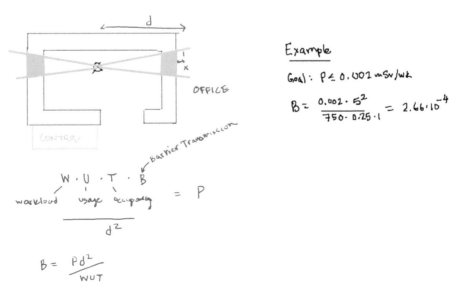

Example

Goal: $P \leq 0.002 \, mSv/wk$

$B = \dfrac{0.002 \cdot 5^2}{750 \cdot 0.25 \cdot 1} = 2.66 \cdot 10^{-4}$

barrier Transmission

$\dfrac{W \cdot U \cdot T \cdot B}{d^2} = P$

workload   usage   occupancy

$B = \dfrac{Pd^2}{WUT}$

FIGURE 25.3.2
Example geometry and shielding calculation.

The formalism for shielding employs the following factors:

- *W*, workload. Gy/week at isocenter. For example, 750 Gy/week.
- *U*, usage. Fraction of time the beam is pointed to that barrier. For example, 0.25 for a wall.
- *T*, occupancy. Fraction of time that the location is occupied. The location here refers to the point for which the shielding calculation is being performed. For example, a point 1 foot from the wall.
- *B*, barrier transmission. The fraction of radiation transmitted through the barrier in question.

The dose rate outside the barrier is represented by the symbol *P*, and the relevant formula is $P = \dfrac{W \cdot U \cdot T \cdot B}{d^2}$. The factor in the denominator is the inverse square factor; *d* is the distance from the isocenter to the point where the shielding calculation is being performed. Typically there is some design goal for *P* based on the allowed rate, and one wants to solve for the required barrier transmission, *B*, so the equation above can be rearranged:

$$B = \frac{Pd^2}{W \cdot U \cdot T} \tag{25.3}$$

25.3.2 Shielding Example: Primary Barrier

To take an example consider the linac vault shown in Figure 25.3.2. The green areas are barriers where the linac beam can point directly and so require a primary shielding calculation. The goal is to calculate the shielding required for the point marked "x" at a distance of 5 m from isocenter (note that this point is 1 foot from the wall, the distance from the barrier recommended by NCRP Report No. 151). The first consideration is what value for *P* is appropriate. NCRP Report No. 151 recommends using one-tenth of the allowed dose limits (Table 25.3) when shielding to conform to the "ALARA" principle of allowed dose, i.e. As Low As Reasonably Achievable. For public areas, this is 0.002 mSv/week (= 1/10×1 mSv/year), and for controlled areas where occupational workers are located and wearing monitoring badges the value is 0.1 mSv/week (= 1/10×50 mSv/year). If the area in Figure 25.3.1 is an office then this would be a public area, and the goal would be $P \leq 0.002$ mSv/week. For the remaining values we take $W = 750$ Gy/week, $U = 0.25$ (i.e. beam points at this particular wall one-quarter of the time) and $T = 1$ (the office is always occupied). This yields $B = 0.002 \cdot 5^2/(750 \cdot 0.025 \cdot 1) = 2.66 \cdot 10^{-4}$.

This value of *B* is the transmission but does not indicate the thickness of the barrier. To calculate thickness, we use the fact that transmission, *B*, is $B = (1/10)^{TVL}$, where "*TVL*" is the number of tenth-value layers. Taking the logarithm base-ten of each side yields

$$TVL = -\log_{10} B \tag{25.4}$$

Continuing the example above, Equation 25.4 yields $TVL = -\log_{10}(2.66 \cdot 10^{-4}) = 3.57$. That is 3.57 TVLs are needed to achieve the required shielding.

To find the required thickness in cm the NCRP Report No. 151 provides tables of TVLs for various energies of beams and shielding materials. If this were an 18 MV beam and the barrier were entirely concrete then the TVL would be 45 cm. The total thickness then would be 3.57×45 cm = 160.7 cm. This, however, is not quite correct because the first TVL is typically larger than the subsequent TVLs in these shielding calculations, due to increased scatter in the beam and a gradual softening of the spectrum. For this example, NCRP151 Table B.2 indicates that the first TVL is 45 cm and subsequent TLVs are 43 cm. The total required thickness, therefore, is 45 + 2.57·43 cm = 155.5 cm = 5.1 feet. A 1.6-meter-thick concrete shield is very thick, so primary barriers are often constructed at least partly of steel. The TVL for an 18 MV beam in steel is 11 cm.

In addition to the requirements for shielding over a long time (e.g. 0.002 mSv/week for public areas), there is also a requirement for dose rate, namely that it should be less than 2 mrem (0.02 mSv) in any one hour. This is often the most stringent specification in public areas. Dose rates through the barrier can be calculated with the formula for transmission,

$$\dot{D}_P = \frac{\dot{D} \cdot U \cdot T \cdot B}{d^2},$$ where $\dot{D}$ is the dose rate (Gy/hr) at isocenter.

25.3.3 Shielding for Leakage and Scatter Radiation

The example above demonstrates the method for calculating shielding for a primary barrier, but there are other sources of radiation as well, namely scatter and leakage (Figure 25.3.1). The formalism for leakage is very similar to primary radiation with two minor changes. First, the regulatory limit for medical linacs is that they must be self-shielded to produce no more than 0.1% of the primary dose rate. Therefore, the W factor in Equation 25.3 is replaced by 0.001·W. Second the usage factor is one. That is, leakage radiation is always directed at the barrier under consideration because it is assumed to emanate isotropically from the linac. Equation 25.3 then becomes

$$B_{\text{leakage}} = \frac{Pd^2}{0.001 W \cdot T} \tag{25.4}$$

One further factor is needed, the "IMRT factor" which accounts for the fact that more MU per Gy of dose at isocenter may be required for IMRT deliveries. More MU results in higher rates of leakage radiation. W is increased by this IMRT factor and a factor of 5 is often used.

The formula for scatter radiation follows similar formalism. Again, all barriers are exposed to scatter. The formulae for scatter are somewhat more complex, and more details can be found in NCRP151. NCRP151 recommends that scatter and leakage both be calculated and the larger of the two be used to define the barrier thickness. If the two are within one TVL of each other then one HVL is added to thickness.

Note that if the door is located such that there is a direct path of scatter or leakage photons (such as in Figure 25.3.2) then the required door can become very thick. In this case, a maze is helpful. This consists of another wall which shields the door from direct radiation.

25.3.4 Shielding for Neutrons

Photon beams with energies greater than approximately 10 MV are capable of producing neutrons through the photonuclear dissociation process. These beams, therefore, require considerations for neutron shielding. The most salient points are: lead or steel do not work well as a shielding material (a material with high-hydrogen content is required for inelastic scattering, see Section 7.3), concrete barriers usually provide a thickness that is sufficient, and doors can be a problem since they are often constructed from metal. Solutions for this latter problem are to use a maze or to make a door from boronated polyethelene (recall that boron has a large cross-section for neutron capture). In this latter case, the steel part of the door should be on the outside to shield for photons created from neutron capture reactions.

25.3.5 Survey Meters

Various survey meters are available in order to measure the exposure at a barrier. The most common are handheld ion chamber devices which can be calibrated for exposure rate (mR/hr) and are relatively independent of energy. Other meters operate in Geiger–Muller mode, that is, with large voltages and large charge amplification, and though these are not calibrated to give the exposure rate they are useful for locating hot spots around shielding or for locating lost seeds in brachytherapy procedures. More details and images are available in the video.

Further Reading

Committee to Assess Health Risks from Exposure to Low Levels of Ionizing Radiation and National Research Council. 2006. *Health Risks from Exposure to low levels of Ionizing Radiation: BEIR VII – Phase 2*. Washington, DC: National Academies Press.

Deye, J., et al. 2005. *NCRP Report No. 151, Structural Shielding Design and Evaluation for Megavoltage X- and Gamma-Ray Radiotherapy Facilities*. Bethesda, MD: National Council on Radiation Protection Report.

Dieterich, S., E. Ford, D. Pavord and J. Zeng. 2016. *Practical Radiation Oncology Physics*. Chapter 10. Philadelphia, PA: Elsevier.

Gregg, E.C., et al. 1989. *NCRP Report No. 102, Medical X-Ray, Electron Beam and Gamma-Ray Protection for Energies Up to 50 MeV*. Bethesda, MD: National Council on Radiation Protection Report.

Kahn, F.M. and J.P. Gibbons. 2014. *Kahn's The Physics of Radiation Therapy*. 5th Edition. Chapter 16. Philadelphia, PA: Wolters Kluwer.

McDermott, P.N. and C.G. Orton. 2010. *The Physics of Radiation Therapy*. Chapters 8 and 17. Madison, WI: Medical Physics Publishing.

Meinhold, C.B. et al. 1993. *NCRP Report No. 116, Limitation of Exposure to Ionizing Radiation*. Bethesda, MD: National Council on Radiation Protection Report.

Metcalfe, P., T. Kron and P. Hoban. 2007. *The Physics of Radiotherapy X-rays and Electrons*. Chapter 13. Madison, WI: Medical Physics Publishing.

United States Nuclear Regulatory Commission. 2019. Consolidated guidance about materials licenses program-specific guidance about medical use licenses: Final report. *NUREG*, 9(3).

Chapter 25 Problem Sets

*Note: * indicates harder problems.*

Useful materials are NCRP Report #151 (Deye et al. 2005) especially Table B.1 for occupancy factors, Table B.2 for TVL values in the primary beam, and Table B.5a or B.7 for TVLs of secondary radiation.

1. What is the dose equivalent of 5 cGy delivered with 9 MeV electrons?
   a. 5 mSv
   b. 50 mSv
   c. 5 cGy
   d. 5.5 cGy

2. What is the effective dose equivalent of the whole lung receiving a uniform dose of 1 cGy with 1 MeV neutrons? Assume values from the NCRP116 Report.
   a. 0.12 mSv
   b. 0.6 mSv
   c. 24 mSv
   d. 48 mSv

3. What is the maximum permissible dose for an occupational worker who is pregnant according to NCRP 116?
   a. 0.5 µSv per month
   b. 0.5 mSv per month
   c. 50 mSv over the pregnancy
   d. 500 mSv over the pregnancy

4. According to the US NRC, what are the minimum requirements for releasing a patient who has received a brachytherapy implant with ^{125}I seeds for prostate cancer with a total activity of 20 mCi if the exposure rate measured at a distance of 1 m from the anterior surface is 1.5 mR/hr? How would this change if this was a ^{103}Pd implant?
   a. Release with no specific instructions
   b. Release with instructions to patient on reducing dose to family and others
   c. Hold in facility for ten half-lives of the isotope
   d. Hold in facility for one half-life of the isotope

For the following problems refer to Figure PS25.1 and assume the following for the linac unless otherwise specified: 6 MV only, use factor of 1/4 for each wall, workload of 750 Gy/week. Distances listed are to the point labeled. Note that all points as listed are 1 foot away from the wall. The arrow indicates the rotation of the linac, i.e. beam can be directed toward point A and the opposite wall and ceiling.

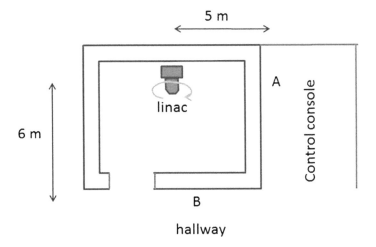

FIGURE PS25.1
Example linac vault.

5. What is the thickness of steel required to keep point A in Figure PS25.1 below 1/10 of allowable limits?
   a. 15.7 cm
   b. 18.8 cm
   c. 24.8 cm
   d. 48.7 cm

6. How do the numbers of TVLs required to shield point A in Figure PS25.1 change if the distance from isocenter to point A is increased to 8 m?
   a. Decreased to 0.77 of original value
   b. Decreased to 0.92 of original value
   c. Increased to 1.09 of original value
   d. Increased to 1.31 of original value

7. In what way might the barrier need to be changed if a CyberKnife unit were placed in this vault?

8. Discuss the shielding requirements for the ceiling if the vault is in a stand-alone facility with no direct access to the roof. How might this vary in a rural vs. urban area?

9. * If the steel barrier in Problem 5 is accidentally installed at half the thickness, what will be the result of the dose rate at point A in Figure PS25.1?
   a. 2 times higher
   b. 4 times higher
   c. 9 times higher
   d. 270 times higher

10. * What is the thickness of concrete required to shield point B shown in Figure PS25.1 to allowable limits?
    a. 14.2 cm
    b. 43.2 cm
    c. 60.7 cm
    d. 74.4 cm

26

Brachytherapy Applications and Radiopharmaceuticals

26.1 Planar Implants

Chapter 4 presents a method for calculating dose from a single brachytherapy source but, clinically, single sources are rarely used. Rather, sources are most often implanted in arrays. In this section we will consider 2D planar implants which illustrate physical principles that apply to many brachytherapy treatments.

26.1.1 Quimby System: Uniform Loading

Figure 26.1.1 shows an example planar implant using brachytherapy line sources. Historically these would be radium needles. There were several systems for determining the appropriate activity in the needles. The Quimby system, developed by Edith Quimby and colleagues at Memorial Hospital in New York in the 1930s, called for a uniform loading of activity in the needles. Uniform activity, however, results in a non-uniform dose which can be understood by considering a point at the center of the implant. This point receives contributions from activity at all four quadrants of the implant, while for a point near the periphery three of the quadrants are farther away and therefore contribute less dose due to inverse square falloff. This effect is large for high-energy sources like ^{226}Ra (mean energy 830 keV) or ^{137}Cs (energy 662 keV), since activity will contribute to dose at distant parts of the implant.

26.1.2 Manchester System: Uniform Dose

An alternative system for planar implants is the Manchester system, or Patterson–Parker system, also developed in the 1930s. This specifies a recipe for spacing between the line sources of 1 cm, appropriate for high-energy sources. **Also, this system calls for peripheral loading, which is a key concept.** Sources near the periphery are higher activity as shown in Figure 26.1.1. The result of this is that a point near the center will receive a dose that is approximately the same as a point near the periphery. The point near the center has a contribution from more nearby sources, but the point near the periphery is nearer to the higher activity sources. Tables are available providing the optimal loading of activity to receive a uniform dose (specified as ±10% in this system). The amount of peripheral loading depends on the size of the implant. The specification is that the dose should be uniform along a line drawn perpendicular to the sources at 0.5 cm from the center of the sources.

A Quimby system

B Manchester system (Patterson-Parker)

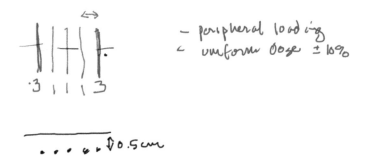

FIGURE 26.1.1
Planar implant systems with brachytherapy line sources. (A) The Quimby system developed results in a non-uniform dose, higher in the middle (green) and lower at the periphery (red). (B) The Manchester system calls for peripheral loading and results in a uniform dose within 10%.

26.1.3 Historical Systems in Perspective

The Quimby and Manchester systems from the 1930s as well as the Paris system from the 1960s (not described here) have many limitations. They ignore filtration, scatter, and attenuation in tissue. Thus, while they work reasonably well for high-energy sources, they are not applicable for lower energy sources like ^{125}I. Also, the dosimetry does not follow modern dosimetry protocols like AAPM TG-43. These systems are no longer used for brachytherapy implants but illustrate the important concepts of uniform loading vs. peripheral loading.

26.2 Prostate Brachytherapy

The prostate is a small organ anterior to the rectum and is accessible anatomically lending itself to a brachytherapy approach (Figure 26.2.1A). The most common form of brachy-therapy treatment is a low-dose rate (LDR) implant that is left in permanently, though it is also possible to treat with HDR (for a description of LDR and HDR see Section 4.1).

26.2.1 TRUS-Guided LDR Prostate Implants

Prostate LDR implants are performed with a trans-rectal ultrasound probe (TRUS) which provides imaging of the gland both prior to the implant for treatment planning and

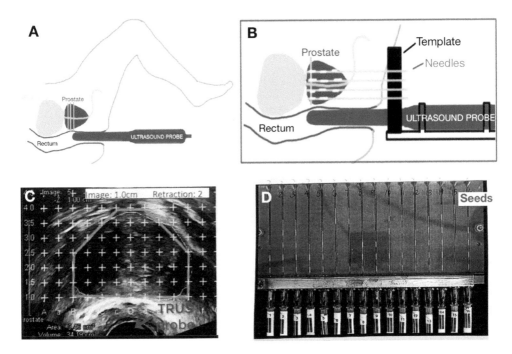

FIGURE 26.2.1
Prostate implant. (A) Transrectal ultrasound probe producing axial image slices (green). (B) Implant template with grid of holes. (C) Pre-implant image with template grid overlaid. (D) Pre-loaded needles with seeds required per plan.

during the implant procedures. Axial images are acquired with the probe (green slices in Figure 26.2.1A and C) stepping from the base of the gland to the apex, that is, from the superior aspect near the bladder to the inferior point. From these images a treatment plan is created which determines the required activity of the seeds and the location of each seed (Figure 26.2.2). The goal of the plan is to achieve uniform coverage of the gland but to avoid hot spots in the center (where the urethra is) and in the posterior aspect near the rectum. To achieve uniform dose peripheral loaded is often employed, that is, more seeds in the peripheral aspects (see Figure 26.2.2). The most common isotopes are ^{125}I and ^{103}Pd, but ^{131}Cs and ^{198}Au are also available though less widely used. Common prescription doses are shown in Table 26.1. Note the lower prescription dose with ^{103}Pd vs. ^{125}I due to the higher biologically effective dose due to the shorter half-life.

For an implant, the probe is connected to a template (Figure 26.2.1B) which consists of a grid with holes. Needles are inserted through the holes in the template as needed. The stepper is used to control the in/out position of the probe and to verify the location of the tip of the needle. Seeds can be loose or stranded. Figure 26.2.1D shows pre-loaded with the spacing between seeds as determined in the treatment plan.

26.2.2 Quality Assurance and Safety for Prostate Brachytherapy

Key resources for quality assurance of prostate brachytherapy implants and programs are AAPM TG-64 on prostate brachytherapy, AAPM TG-128 on QA of ultrasound systems (Pfeffer et al. 2008), AAPM TG-56 on brachytherapy QA generally (Nath et al. 1997), and ACR-ABS Practice Parameter for Prostate Brachytherapy (2015).

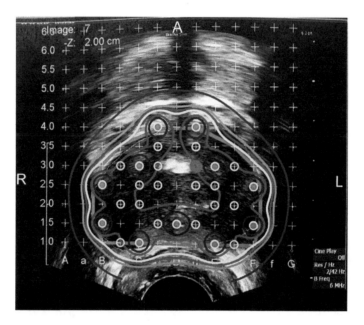

FIGURE 26.2.2
Treatment plan for a prostate implant. Seed locations (green) are determined in order to provide uniform dose coverage.

TABLE 26.1

Common Prescriptions for Prostate Brachytherapy Implants

Isotope	Typical Prescription Dose for Monotherapy	Typical Prescription Dose in Combination with External Beam
^{125}I	145 Gy	110 Gy
^{103}Pd	125 Gy	100 Gy

The activity of the seeds must be independently verified. Ten percent of the seeds are assayed and must be within 3% of the activity specified by the seed manufacturer. These assays are performed with a well chamber where the current is measured and the activity calculated, given a known calibration of activity per current (mCi/nA) for the particular model of seed. If pre-loaded needles are used, the loading is verified via an autoradiograph. For details see the video. Another important QA measure is surveys. The room where the implant will be performed is surveyed prior to the procedure. During the procedure, the waste is surveyed to ensure there are no lost seeds, and after the procedure the patient is surveyed to ensure they meet release criteria (see Sections 25.2.5 and 25.3.5). Finally, a post-implant verification study is performed, typically performed 30 days after the implant when swelling has largely resolved. This needs to include CT imaging in order to verify the location of the seeds.

26.3 HDR Brachytherapy

HDR brachytherapy is used in a number of applications, the most common of which is the treatment of cervical cancer. For a review of HDR physics see Section 4.1. The

TABLE 26.2

Clinical Use of HDR Brachytherapy for Cervical Cancer

Stage	Therapies
IA1 and IA2	Brachytherapy
IB1	EBRT + brachytherapy (+ chemo)
IB2—IVA	Chemo + RT + brachytherapy
IVB	Brachytherapy (+EBRT)—palliative

most commonly used isotope is ^{192}Ir which undergoes β^- decay with a half-life of 73.8 days, producing decay photons from the daughter product with an average energy of 380 keV.

26.3.1 HDR for Cervical Cancer: Clinical Indications

HDR brachytherapy is used as a monotherapy for early-stage cancers, or in later-stage cancers in combination with external beam radiotherapy. In the latest stages brachytherapy is used, sometimes with EBRT, with a palliative intent (Table 26.2).

26.3.2 Cervical Cancer: Applicators

HDR treatment consists of an applicator which is inserted into the patient, the remote afterloader which provides remote control of the high-activity source, and the transfer tube which connects the two (Figure 26.3.1A). The afterloader positions the source at various dwell positions in the applicator in order to achieve the desired dose distribution. These dwell positions are equivalent to seed locations. Various applicators are available. The vaginal cylinder is shown in Figure 26.3.1A which comes in diameters of 20 mm to 40 mm; dose is prescribed to the surface or to 5 mm depth. The tandem-and-ring and tandem-and-ovoid applicators (Figure 26.3.1B) are used to treat disease in the uterus (via the tandem) and around the cervical os (via the ring or ovoids). For more extensive disease interstitial applicators are used, e.g. Syed applicator. These combine needles for brachytherapy sources with the intracavitary applicators. See the video for further details on the geometry as well as a demonstration of the assembly of the applicators.

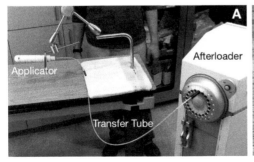 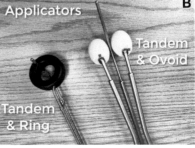

FIGURE 26.3.1

HDR treatment. (A) HDR afterloader connected to the applicator, here a vaginal cylinder. (B) Other applicators for intracavitary HDR, the tandem-and-ring and tandem-and-ovoid.

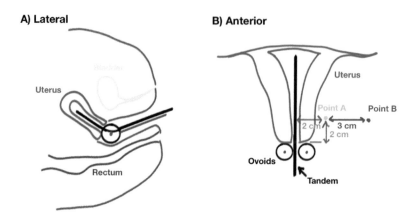

A) Lateral

B) Anterior

FIGURE 26.3.2
Geometry for a tandem-and-ovoid-based HDR treatment and Manchester dose points.

26.3.3 Dose Specification Systems

The Manchester system provides a method for recording and reporting dose in HDR gynecological brachytherapy. This is described in ICRU Report No. 38 (Wyckoff et al. 1985) and updated in ICRU Report No. 89 (Potter et al. 2013) which also discusses the more modern approach of volumetric imaging via CT or MRI. Figure 26.3.2 shows the geometry of the tandem-and-ovoid applicator during treatment and the associated dose points in the Manchester system. These points are localized using AP and lateral radiographs. "Point A" is 2 cm up from the top of the ovoid and cervical os and 2 cm lateral from the tandem. "Point B" is 3 cm lateral from Point A corresponding to the pelvic sidewall. The bladder point is defined by placing a Foley catheter with 7 cc of contrast; the point is on the posterior aspect of this. The rectal point is 0.5 cm posterior to the vaginal wall visualized with radio-opaque gauze. Typical prescriptions to Point A are 5.5 Gy×5 fractions, 6 Gy×5, or 7 Gy×4. Common limits for doses are: Point B, 30–40% of Point A dose; rectum <4.1 Gy/fxn (<70% Rx); bladder <4.6 Gy/fxn (<75% Rx); and mucosa <120 Gy (<140% of Point A dose).

26.3.4 Other Applications of HDR

Although HDR is most commonly employed for gynecological cancers, it is used in other oncologic applications as well including prostate cancer, breast cancer, and intraoperative radiation therapy. Several applicators are available for breast, an example of which is the Savi applicator (Cianna Medical Inc.). A dummy applicator is put in during lumpectomy surgery, and this is later replaced by a six-channel applicator which is expanded. This provides fine control of the dose distribution by controlling dwell times. Further details can be found in the video and the reading materials.

26.4 Radionuclide Therapy

Isotopes used for radionuclide therapy have several key features: short half-life (5–15 hours), short-range particles to deliver dose (typically β or α particles), and photons for imaging. Table 2.1 shows properties of some of the relevant isotopes.

- ^{131}I. Approved by the FDA in the United States in 1974, it is used as a targeted treatment for thyroid cancer and hyperthyroidism because iodine accumulates naturally in the thyroid gland. Activities of 75 to 200 mCi are used depending on the disease.

- ^{90}Y. Used for liver metastases and hepatocellular carcinoma, ^{90}Y is attached to small glass beads (20–30 mm) and injected intraarterially. Targeting is achieved because the tumor blood supply is preferentially from the artery vs. normal liver from the vein. Two formulations are Therasphere®, approved for use in 1999, and SIR-sphere, approved in 2002.

- ^{223}Ra-Chloride. This is used to treat castrate-resistant prostate cancers. Bony metastases are targeted due to the chemical similarity of chloride to calcium. This has some advantages over ^{90}Sr-chloride which was approved for used in 1993.

- ^{177}Lu-dotatate. This compound is a somastatin analog which targets neuroendocrine tumors which overexpress the somastatin receptor.

Radionuclide therapy is a complex topic. Further detailed information on isotopes and MIRD dosimetry for dose calculations can be found in the video.

Further Reading

ACR-ABS Practice Parameter for Transperineal Permanent Brachytherapy of Prostate Cancer. 2015. https://www.acr.org/-/media/ACR/Files/Practice-Parameters/brachy-prostate.pdf.

Dieterich, S., E. Ford, D. Pavord and J. Zeng. 2016. *Practical Radiation Oncology Physics*. Chapter 8. Philadelphia, PA: Elsevier.

Kahn, F.M. and J.P. Gibbons. 2014. *Kahn's The Physics of Radiation Therapy*. 5th Edition. Chapters 15 and 23. Philadelphia, PA: Wolters Kluwer.

Kubo, H.D., et al. 1998. High dose-rate brachytherapy treatment delivery: Report of the AAPM Radiation Therapy Committee Task Group No. 59. *Med Phys* 25(4):375–403.

McDermott, P.N. and C.G. Orton. 2010. *The Physics of Radiation Therapy*. Chapter 16. Madison, WI: Medical Physics Publishing.

Nath, R. et al. 1997. Code of practice for brachytherapy physics: Report of the AAPM Radiation Therapy Committee Task Group No. 56. American Association of Physicists in Medicine. *Med Phys* 24(10):1557–1598.

Pfeiffer, D., et al. 2008. AAPM Task Group 128: Quality assurance tests for prostate brachytherapy ultrasound systems. *Med Phys* 35(12):5471–5489.

Potter, R., et al. 2013. ICRU Report No. 89, Prescribing, Recording and Reporting Brachytherapy for Cancer of the Cervix. *ICRU* 13(1):1–258.

Wyckoff, H.O., et al. 1985. *ICRU Report No. 38, Dose and Volume Specification for Reporting Intracavitary Therapy in Gynecology*. Bethesda, MD: ICRU.

Yu, Y., et al. 1999. Permanent prostate seed implant brachytherapy: Report of the American Association of Physicists in Medicine Task Group No. 64. *Med Phys* 26(10):2054–2076.

Chapter 26 Problem Sets

*Note: * indicates harder problems.*

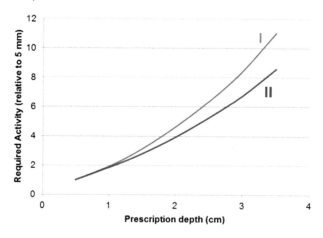

FIGURE PS26.1
Planar implant activity.

1. Match the curves in Figure PS26.1 for a ^{192}Ir planar implant with the size of the implant. Depth is the depth in the implant perpendicular to the sources.
   a. 5×5 cm²
   b. 10×10 cm²

2. Which isotope will have the most gradual dose falloff with depth in a planar implant?
   a. ^{192}Ir
   b. ^{226}Ra
   c. ^{137}Cs
   d. ^{125}I

3. How does the D90% for a prostate implant on the day of implant compare to the D90% at 30 days post-implant?
   a. Higher
   b. Lower
   c. The same

4. What is the dose rate for the vaginal cylinder plan in Figure PS26.2 at Point 1 if the dose rate at the applicator surface is 10 Gy/hr?
   a. 4.4 Gy/hr
   b. 6.7 Gy/hr
   c. 11.0 Gy/hr
   d. 22.5 Gy/hr

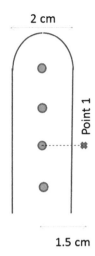

2 cm

Point 1

1.5 cm

FIGURE PS26.2
HDR cylinder dwell positions.

5. In the applicator shown in Figure PS26.2, which dwell position has the longest time if the dose distribution at the surface is to be uniform?

    a. Bottom most (most proximal)
    b. Top most (near tip)
    c. All are the same
    d. Depends on applicator diameter

6. List three methods of reducing rectal dose in a tandem-and-ovoid treatment.

7. What is the treatment time for a vaginal cylinder case with a ^{192}Ir afterloader three months after the source is installed if the initial treatment time is 3.5 min?

    a. 1.0 min
    b. 1.5 min
    c. 8.1 min
    d. 11.8 min

8. List the advantages and disadvantages of a tandem and ring applicator vs. a tandem-and-ovoid applicator for the treatment of intermediate stage cervical cancer.

9. * What is one advantage of a multichannel applicator for ^{192}Ir treatment of the post-surgical cavity in breast cancer?

    a. Shorter treatment time
    b. Smaller treatment volume
    c. Improved skin sparing
    d. Simplicity

10. * List the advantages and disadvantages of HDR for intra-operative radiation therapy (IORT) over radiotherapy with electron beams.

27

Patient Safety and Quality Improvement

The topic of patient safety and quality improvement is vast and many-pronged, but this chapter will focus on two specific techniques for maintaining and improving the quality of care: incident learning (Section 27.1) and risk analysis (Section 27.2). While not the domain of physics only, this is a natural topic for this text since a key job description of medical physicists is quality management oversight.

27.1 Incident Learning and Root-Cause Analysis (RCA)

27.1.1 Example Error and Nomenclature

Figure 27.1.1 shows an error (a wrong prescription not per intended protocol) that propagates through the workflow and eventually reaches the patient before it is caught after four treatment fractions on a routine chart review. This event would be classified as an *incident*, that is, it reached the patient. If the event were identified prior to treatment, say by review of a qualified medical physicist (QMP), then it would be classified as a *near-miss*. Since the dose error is a 25% overdose ($250/200 = 1.25$), this would, in theory be reportable to a regulatory agency since the threshold for reporting is 20%. In the United States, this agency would be the state department of health. However, note that the treatment is in 25 fractions and if the dose were corrected after four fractions the total error would be $(4 \times 250 + 21 \times 200)/5000 = 4\%$ which is not reportable. However, this still would be an incident that reached the patient.

27.1.2 Swiss Cheese Model of Accidents

Related to the above example is the Swiss cheese model of accidents (Figure 27.1.2) due to James Reasons, Professor of Psychology from Manchester University. Here an error happens (left), but before it reaches the patient it may be interrupted by one of the many safety barriers in place, e.g. review by a physicist or therapist or imaging. These barriers are the slices of cheese. However, the barriers do not operate perfectly. There are holes, and the holes may line up in such a way as to let the error through.

27.1.3 Root-Cause Analysis

Error scenarios like that outlined above present an opportunity for learning and improvement. A key tool for this is root-cause analysis (RCA) in which one attempts to understand the underlying causal factors that drive error in order to prevent them. In RCA one investigates the event to identify the care delivery problems ("what" happened) and the contributing factors ("why" it happened). In spite of the name "root cause," there is often no one single cause but many causes conspiring together. In the example above one factor might

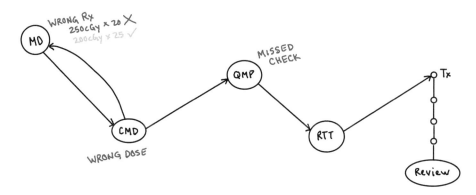

FIGURE 27.1.1
Example error scenario.

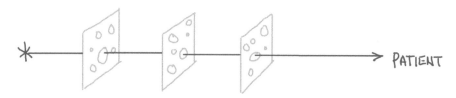

FIGURE 27.1.2
Swiss cheese model of accidents.

be a lack of review by radiation therapists. In RCA one asks why repeatedly, to uncover deeper causal factors. Perhaps there was a new therapist who did not understand that review should be done. That, in turn, may be due to a lack of formal policies around this and so on. In conducting RCA it is helpful to adopt a "systems thinking" approach whereby one views not only the people and their actions but the system they work within. This is related to the concept of Just Culture, whereby one does not take a punitive approach to human error but rather attempts to understand what drives it. **Just Culture, incident learning, and RCA are key concepts**.

27.2 Incident Learning

There are strong data and recommendations to support the use of incident learning in healthcare and radiation oncology. Data show that hospitals with higher incident reporting rates have safer operations due to the associated learning that happens. This is especially true when using near-miss events as a learning tool. Heinrich's triangle from industrial engineering notes that for every one incident of harm there are many near-misses and faulty practices.

Incident learning is recommended by national and international associations including ASTRO in the Safety Is No Accident Report. Of note is the Radiation Oncology Incident Learning System (RO-ILS™), a national system sponsored by ASTRO and AAPM for practices in the United States. This system, launched in 2014, provides a platform for shared learning in a protective and non-punitive environment. This is conducted under the auspices of a Patient Safety Organization (PSO) which is an entity protected under US law.

These and other issues are discussed further in the video, and more information can be found in a recent review paper on incident learning in radiation oncology (Ford & Evans 2018).

27.3 Failure Mode and Effects Analysis (FMEA)

27.3.1 Failure Modes and Risk

The goal of FMEA risk analysis is to understand and identify risk in the complex process of care. A standard reference for this is the AAPM Task Group 100 report (Huq et al. 2016) which provides a definitive guide to FMEA in the radiation therapy context. More detail can be found in the video as well.

A failure mode is essentially anything that can go wrong in the process of care. Examples include a wrong treatment location or a miscommunication about the start of chemotherapy. FMEA consists of identifying these potential failure modes and then assigning each of them a numerical score for risk.

27.3.2 The FMEA Process

There are several key steps in an FMEA exercise. First, decide on a process to study. Typically, a more focused process is better (e.g. SBRT treatment planning, not all of treatment planning). Next, assemble a team which should include all professional groups involved. Third, draw a workflow map of the process under consideration. Fourth, brainstorm on possible failure modes. Finally, score these failure modes for risk.

27.3.3 The FMEA Scoring System

FMEA scoring considers three variables: severity, S (how severe might the failure mode be if it reached the patient), occurrence, O (how often it occurs in practice), and detectability, D (how difficult the failure is to identify). These three scores are typically on a 1–10 scale (see TG-100 for tables of suggested scores). Note that the detectability score of 10 means that the failure is difficult to detect. The S, O, and D scores are multiplied together to arrive at a risk priority number, $RPN = S \times O \times D$. Table 27.1 shows two example failure modes and scoring. Here the risk score for the first example failure mode is higher even though the severity is lower. That is because it is more difficult to detect. The end result of FMEA is a list of failure modes ranked by risk which can then be used for further quality improvement.

TABLE 27.1

Example Failure Modes and Associated FMEA Scores

Failure Mode	Severity	Occurrence	Detectability	$RPN = S \times O \times D$
Miscommunication about a repeat treatment case	7	7	6	294
Wrong treatment location	8	6	2	96

27.3.4 FMEA vs. Incident Learning

FMEA can be challenging, but need not be very time consuming. It serves as a useful adjunct to incident learning especially to identify potential problems that do not often occur in clinical practice but might pose a risk. More information is available in the video.

Further Reading

ASTRO. 2019. Safety is no accident. https://www.astro.org/Patient-Care-and-Research/Patient-Safety/Safety-is-no-Accident.

Dieterich, S., E. Ford, D. Pavord and J. Zeng. 2016. *Practical Radiation Oncology Physics*. Chapter 12. Philadelphia, PA: Elsevier.

Ford, E.C. and S.B. Evans. 2018. Incident learning in radiation oncology: A review. *Med Phys* 45(5):e100–e119. doi:10.1002/mp.12800.

Huq, M.S., et al. 2016. The report of Task Group 100 of the AAPM: Application of risk analysis methods to radiation therapy quality management. *Med Phys* 43(7):4209–4262.

Chapter 27 Problem Sets

*Note: * indicates harder problems.*

1. What characteristic best describes a near-miss event?
   a. Reaches the patient but is not an adverse event with impact
   b. Reaches the patient but harm is avoided by medical intervention
   c. Does not reach the patient but represents a risk
   d. Does not reach the patient but did in the past

2. What is a characteristic of an effective root-cause analysis effort?
   a. Led by senior manager
   b. Undertaken on all near-miss events
   c. Identifies blame in an objective manner
   d. Reveals more than one causal factor

3. Which of the following characterizes "systems thinking" and a "just culture" view of error?
   a. Human error is identified as a key causal factor
   b. Disciplinary action is applied equally to all staff
   c. A "blame-free" environment with respect to error
   d. Personal responsibility is considered along with the systems of care delivery

4. According to Heinrich's triangle which event in radiation therapy is most common?

   a. Overdose of 20%

   b. Treatment with incorrect isotope

   c. Forget policy on ID verification

   d. Film incorrect isocenter

5. What is one advantage of operating an incident learning system under the auspices of a Patient Safety Organization?

   a. Protection of data

   b. Quicker to set up

   c. Can be used outside the US

   d. Access event reports from other clinics

6. Which of the following QI tools could be used to evaluate safety risks in a new SBRT program?

   a. Failure mode and effects analysis (FMEA)

   b. Root-cause analysis

   c. Incident learning

   d. Checklists

7. Rank the following failure modes in terms of overall risk.

Failure Mode	Severity	Occurrence	Detectability
Failure mode 1	9	2	5
Failure mode 2	3	9	2
Failure mode 3	7	5	5

8. Yearly external audits of machine output reduce which component of risk?

   a. Severity

   b. Occurrence

   c. Detectability

   d. Financial

9. Which quality improvement steps would help prevent the treatment of a patient with a wrong MU calculation ("hand calc")? Check all that apply. Label each as preventing the error vs. identifying the error.

   a. Validation of commissioning

   b. Annual QA

   c. Review by a second person

   d. Training of new staff on the procedure

10. * List all of the potential QA steps that might have identified the IMRT misdelivery error reported by the *New York Times* in 2010: www.nytimes.com/2010/01/24/health/24radiation.html.

Index

For Product Safety Concerns and Information please contact our EU representative GPSR@taylorandfrancis.com Taylor & Francis Verlag GmbH, Kaufingerstraße 24, 80331 München, Germany

T - #0266 - 160425 - C374 - 254/178/18 - PB - 9781138591707 - Gloss Lamination